CAMBRIDGE LIBRARY COLLECTION

Books of enduring scholarly value

History of Medicine

It is sobering to realise that as recently as the year in which On the Origin of Species was published, learned opinion was that diseases such as typhus and cholera were spread by a 'miasma', and suggestions that doctors should wash their hands before examining patients were greeted with mockery by the profession. The Cambridge Library Collection reissues milestone publications in the history of Western medicine as well as studies of other medical traditions. Its coverage ranges from Galen on anatomical procedures to Florence Nightingale's common-sense advice to nurses, and includes early research into genetics and mental health, colonial reports on tropical diseases, documents on public health and military medicine, and publications on spa culture and medicinal plants.

The Life of Edward Jenner M.D.

Written by his friend, the physician John Baron (1786–1851), this laudatory biography of the 'father of immunology' did much to enhance the reputation of Edward Jenner (1749–1823) upon its publication in two volumes between 1827 and 1838. The work covers Jenner's personal and professional life both before and after his development of the vaccine for smallpox, as well as touching on the vaccine's reception and use around the world. Thoroughly explaining the history and facts of vaccination, Baron established himself as an authority on the subject. Although criticised by some for its unquestioning praise of Jenner's genius, the work is valuable for its use of primary sources, drawing heavily on correspondence and personal notes, excerpts of which appear throughout the text. Volume 1, published in 1827, focuses on Jenner's early life and the history and science of vaccination.

The Life of Edward Jenner M.D.

With Illustrations of his Doctrines, and Selections from his Correspondence

VOLUME 1

JOHN BARON

CAMBRIDGE
UNIVERSITY PRESS

University Printing House, Cambridge, CB2 8BS, United Kingdom

Published in the United States of America by Cambridge University Press, New York

Cambridge University Press is part of the University of Cambridge.
It furthers the University's mission by disseminating knowledge in the pursuit of education, learning and research at the highest international levels of excellence.

www.cambridge.org
Information on this title: www.cambridge.org/9781108071130

This edition first published 1827
This digitally printed version 2014

ISBN 978-1-108-07113-0 Paperback

EDWARD JENNER, M.D.

L.L.D. F.R.S. &c. &c.

Drawn from the Bust by H. CORBOULD, and on Stone by R. J. LANE.

Printed by C. Hullmandel.

THE

LIFE

OF

EDWARD JENNER, M.D.

LL.D., F.R.S.,

PHYSICIAN EXTRAORDINARY TO THE KING, &c. &c.

WITH

ILLUSTRATIONS OF HIS DOCTRINES,

AND

SELECTIONS FROM HIS CORRESPONDENCE.

BY

JOHN BARON, M.D., F.R.S.

LONDON:
HENRY COLBURN, NEW BURLINGTON STREET.
1827.

LONDON:
PRINTED BY S. AND R. BENTLEY, DORSET-STREET.

TO THE

KING'S MOST EXCELLENT MAJESTY.

SIRE,

THE incalculable advantages that the world in general was likely to derive from the indefatigable perseverance of the author of Vaccination at a very early period attracted your MAJESTY's notice: and the beneficent designs of Jenner called forth, on more occasions than one, the unequivocal and generous expression of your Royal favour.

The discovery which your MAJESTY condescended so to signalize must ever be memorable in the annals of your MAJESTY's dominions, and of mankind.

A work devoted to the elucidation of that discovery, and to the personal history of its author, will, I trust, prove not unacceptable to your MAJESTY. I therefore beg leave humbly to lay it at your MAJESTY's feet.

I have the honour to be,

Sire,

Your MAJESTY's

Most humble servant,

And dutiful subject,

JOHN BARON.

Gloucester, January 6, 1827.

CONTENTS.

CHAPTER VII.

CHAPTER VIII.

CHAPTER IX.

CHAPTER X.

CHAPTER XI.

CHAPTER XII.

CHAPTER XIII.

CHAPTER XIV.

INTRODUCTION.

THIS work has been composed from materials of the most authentic description, the whole of the notes and correspondence of Dr. Jenner having been put into my hands by his executors. My close and unreserved intercourse with him, and my intimate knowledge of his sentiments and habits of thinking on all subjects during the last fifteen years of his life, probably induced them to believe that I was not an unfit person to draw from such sources an accurate delineation of his character and opinions.

Many reasons, with which I need not trouble the public, would have induced me to shrink from the labours of such a work; and nothing certainly could have reconciled me to the attempt had I not been influenced by the most sincere veneration for the name of Jenner, and by the conviction that the confidence with which he honoured me *did* afford me facilities for acquiring an insight into his feelings and motives, by which I have been enabled to speak without hesitation or doubt on all those points that most concerned either his conduct as a man, or the nature of his doctrines.

Notwithstanding these encouragements, I cannot but own that I have entered upon this undertaking with a degree of anxiety in which I can scarcely expect any to sympathize. I trust that I am not deceiving myself when I say that nothing of a personal nature prompts this avowal. It is of moment that a true and faithful portrait should be drawn of so distinguished an individual; that those, who have admired and extolled him as a great benefactor of our race, may know that on many other grounds he was worthy of their highest regard and warmest affection. It cannot be expected that there should be an uniformity of sentiment on momentous questions of a professional or scientific nature, but I shall ever have cause to lament if, through any defects of mine, the kindness, the rectitude, the consistency, and the unextinguishable ardour and devotedness of Jenner in a glorious cause do not shine conspicuous in every act of his long and laborious life.

On a knowledge of these things my pretensions as his biographer chiefly rest. If I have failed in imparting that knowledge I shall have a cause for regret which no deficiency in any other part of my design could occasion. The world at large has felt and acknowledged the blessings of his great discovery; but few are aware how numerous were his claims to admiration. For these reasons I have been anxious in the early part of this work to disjoin, as much as possible, his private character, and his acquirements as a naturalist, from that important subject which has so extensively occupied the public mind and caused his other numerous titles to consideration to be overlooked.

It must, at the same time, be remembered that Doctor Jenner was nearly fifty years of age before he published his first work on the Variolæ Vaccinæ. The whole of the early part of his life having been spent in comparative seclusion, it cannot be expected that it should afford those materials which best suit the purposes of the biographer. His epistolary intercourse with Mr. Hunter has enabled me to fill up a space in his life that could not otherwise have been supplied. Unfortunately, all Jenner's replies to Mr. Hunter have been destroyed; and had it not been for his printed papers, we should have been left in total ignorance of the result of the inquiries to which his letters to Jenner refer. Notwithstanding this circumstance, I have not refrained from publishing many of those letters. Mr. Hunter was too remarkable a character in many respects, and his name is too intimately associated with the progress of natural history and physiology in this country, to permit me to doubt for a moment, that whatever fell from him in his correspondence with such a man as Jenner, will be favourably received by the public.

After thus endeavouring to trace Jenner's history in early life, I have brought together various incidents to illustrate the progress of his mind in effecting the discovery of Vaccination. This detail will, I trust, show alike the force and originality of his genius and the benevolence of his purpose. From the time of his first successful vaccination in 1796 to the last hour of his existence he laboured incessantly to disseminate the practice. In every instance where it was had recourse to; in whatever clime, or under whatsoever circumstances it was performed, his name and reputation were either directly or indirectly associated with it. This peculiarity has so much identified vaccination with Jenner that it is impossible to think of him or to speak of him, as he deserved, except in conjunction with its magnificent and animating course. In this respect he stands pre-eminent; and it cannot but be interesting to those, who have reflected upon what has been brought about by his means, to be assured that his private demeanour well accorded with his public reputation; that he lived with the generosity of a good man, and the simplicity which befits a great one.

While thus delineating his personal qualities I have endeavoured to elucidate his doctrines relating to that subject in which all must feel so deep an interest. My duty on this point has led to a discussion that some may think foreign to the object of a biographical work. I am induced to hope, however, when all things are duly considered, that I shall stand excused not only for bestowing pains in placing Dr. Jenner's opinions in a proper light, but for collecting from different sources such scattered rays of knowledge as may tend to explain and confirm them. He drew his conclusions from the pathological facts which he had an opportunity of observing. These conclusions, it will be found, are verified in a remarkable manner by observations made in different ages, and in different countries.

Had I, therefore, been guided merely by what was due to him, I could not but attempt to explain and vindicate his views. In so doing, I trust it will be found that I am assisting that great cause in which his life was spent; that I am bringing past experience and the unprejudiced testimony of impartial witnesses to confirm the decisions which his own investigations led him to; and that therefore, in rendering justice to him, I am giving greater force to his doctrines, and increased confidence in the practice founded on them.

Influenced by these considerations I have, after detailing the early history of vaccination, brought together much that seemed to bear upon the literary and medical history of cow-pox, as well as of small-pox. The nature of the former cannot be understood without well examining the properties of the latter. They agree in some very essential particulars; but the points in which they differ, so far as the welfare of the community is concerned, it is of more importance to insist upon. At present it may be sufficient to remark that the coincidences, as well as the peculiarities, of each affection are best elucidated by combining the examination of their pathological character with their literary history. Under this last head, I would hope that some information has been collected not uninteresting in itself, but possessing a higher value from its connection with the grand results of Dr. Jenner's investigation.

Great care has been bestowed in tracing this

history; and though many may be surprised at the conclusions to which it leads, I nevertheless trust it will be found that nothing has been lightly advanced. My own mind was quite unprejudiced, and the views which I am inclined to adopt have arisen entirely from a close examination of the evidence, and not from any preconceived opinions. They were first suggested on examining the description of Philo Judæus, as referred to in Dr. Willan's posthumous work on the antiquity of small-pox; and from meditating on the nature and properties of cow-pox. My own attention having been thus excited, the natural and medical history of the eruptive diseases of man, and of the inferior animals, necessarily became an object of inquiry. The number of instances in which it appears that a disease of this class affects different orders of animals, and is communicable to the human species, has given a degree of interest to this inquiry which I could scarcely have anticipated. That interest is infinitely enhanced when we consider how much the safety and happiness of mankind are connected with the great discovery which this discussion is meant to elucidate.

The whole subject is a curious and important one; and it may with truth be affirmed that in no former instance did historical evidence and remarkable pathological phenomena so singularly and beneficially throw light on each other. In this part of my subject, more especially, I have to acknowledge the deepest sense of obligation to two kind friends, Richard Gamble, M.D. Oxon. and the Reverend John Webb, A.M. Oxon. who have aided me by their learning and research.

Throughout this discussion my first object has been to endeavour to throw some light on the nature of cow-pox itself. My next has been to prove the justness and accuracy of Dr. Jenner's main doctrines. As the benefits which might have resulted from his discovery have, manifestly, been circumscribed by erroneous views on both these points, I was solicitous to collect such a body of evidence as might tend to remove these errors. In executing this design I have been compelled to break in upon the chronological order of events, by deducing from the whole experience of vaccination such an accumulation of facts as ought to convince the most sceptical that nothing but the proper extension of the practice is necessary to accomplish all that its benevolent author promised.

Intimately as Jenner's name and character were connected, and always must be, with every instance of vaccination, I wish it to be distinctly understood that I am not to be considered as the *historian* of *that practice.* The events which arose from its first promulgation were unspeakably interesting to him. These, therefore, have been recorded with great care and fidelity. I have felt this to be a delicate part of my duty, because it required me to bring forward many incidents which I would gladly have allowed to fall into oblivion. In placing his conduct in its true light I have studied to be equally just to that of others, as no statement has been delivered that is not corroborated by original and authentic documents.

I have abstained entirely from taking any part in the violent and discreditable controversy which arose out of the vaccine discovery. Although Dr. Jenner was the object of many harsh and unfounded aspersions, he never thought it necessary to weaken that strong position, which truth and knowledge had enabled him to take, by replying to them. The utmost vigilance of those who ignorantly assailed either his conduct or his doctrines, has left no stain upon his name. All, therefore, that is required of me is plainly and distinctly to describe his actions; and to leave them to speak both for his genius and his virtues.

In treating of the progress of vaccination I have confined myself to such incidents. as were either immediately directed by Dr. Jenner himself, or were of a striking nature from their magnitude, or the station of those engaged in them. Had not my design been limited in this way, it would have been a sincere gratification to me to have made mention of many individuals who distinguished themselves by zealously promoting the practice. Their co-operation was gratefully felt by Dr. Jenner; and whenever the history of vaccination shall be fully recounted, their services will not be overlooked.

The private history of Jenner and of his labours could only be fully derived through those channels of information which have been open to me. I have selected such facts as are for the most part new, I believe, to the public. In order to authenticate the narrative, and to impart to it that spirit which original documents alone can give, I have embodied such as appeared to me most interesting, in the text. I have preferred this method either to that of printing them as an appendix, or in the form of notes. One of the chief reasons for this decision arose from the nature of the transactions that I was called on to record. Many of them regarded the conduct of individuals; and I saw no method of escaping charges of partiality or unfairness except by bringing forward proofs that cannot be denied, and which will show that I have dealt honestly by all. In a question which has in a peculiar degree excited strong feelings, both with the public and in the profession, I can scarcely hope to have written, on all occasions, so as not to have called up recollections of an unpleasant kind in the minds of some. However this may be, I trust it will be apparent that truth and moderation have guided me throughout the whole work.

Many of Dr. Jenner's own letters are published from copies transcribed by himself into his note-books. He appears sometimes to have omitted the introductory and concluding sentences; and very frequently the precise dates. These circumstances will account for an abrupt termination of some; and likewise for any deviation that may be noticed between the manner of expression in the copy and the original.

For almost all his early letters I am indebted to his friend the late Edward Gardner, of Frampton-upon-Severn, who bequeathed them to me on his death-bed. These, together with those addressed to myself, form a series which touches on almost every subject of interest, whether of a public or domestic nature, during the last forty years of his life.

I have also had letters and extracts of letters transmitted to me by Colonel Berkeley, Thomas Paytherus, Esq., Henry Hicks, Esq., James Carrick Moore, Esq., Charles Murray, Esq., and Henry Jones Shrapnell, Esq., to all of whom I beg to return my thanks.

Animated as I have been by the most ardent and devoted attachment to the memory of Jenner, it cannot be expected that I should either repress my feelings, or employ cold and measured language to mark my sense of the value of his labours, and the importance of their results. I can scarcely expect that my reader will go along with me on all such occasions, but I *do* indulge the hope that he will see reason to forgive that warmth in which he may not be able to participate. Jenner's nature was mild, unobtrusive, unambitious; and many who have done justice to his discovery have still to learn how beautifully the singleness of his heart and his genuine modesty graced and adorned that splendid reputation which the wonderful consequences of his labours had acquired for him. In every private affair, in every public transaction one principle guided him. The purity of his motives and the disinterestedness of his actions have, by no means, yet been duly acknowledged: Had those who opposed him and Vaccination known how little of selfishness, of vanity, or of pride entered into his character, they would, I am persuaded, deeply lament the wounds which they inflicted; and in the place of bitterness and reproach would have found cause for unmixed esteem and approbation.

Before I conclude these prefatory remarks I must offer some explanation of the delay that has taken place in the publication of this work. Obstacles were at the very outset thrown in my way, which I need not here specify. The papers too were extremely voluminous, and in the greatest disorder. To bring them into a state capable of affording me any assistance in constructing the narrative with fidelity, required a degree of labour much greater than I could have anticipated. These difficulties not a little increased the really arduous duty that I had to perform. So seriously did I at one time feel this that I anxiously wished and, indeed, had determined, to relinquish my task altogether: in addition therefore to the exertion demanded by the subject itself I may be permitted to state that my professional avocations necessarily precluded me from giving that unbroken and undivided attention indispensable to the rapid progress of a work of this nature.

The publication of the first part, without waiting for the completion of the second, seemed under such circumstances to be expedient, both to the executors of Dr. Jenner and to myself. Other reasons concurred to give strength to this decision. The recent prevalence of small-pox in different parts of Europe, and the corresponding diminution of confidence in the virtues of the Variolæ Vaccinæ, rendered it an object of no inconsiderable importance to endeavour to restore and increase that confidence, by showing that Dr. Jenner clearly foresaw the deviations which have been observed; that his doctrines, if properly understood, satisfactorily account for them; and that nothing, in fact, has occurred which does not strengthen and confirm his original opinions both with regard to the Variola and the Variolæ Vaccinæ. I would hope that something may have been done in these respects, that shall tend to promote the universal adoption of a practice capable of effecting so much good.

Nothing, I am persuaded, can ever accomplish this object except a real knowledge of the nature of that affection which might be made to take place of small-pox. A very sincere wish to accelerate this event has led me to the discussions contained in the present volume, the publication of which, at this time, I would humbly hope may not be without its use.

As there is great reason to fear that Dr. Jenner's views are not sufficiently understood, so in like manner it is to be apprehended that his disinterested efforts, and the formidable difficulties which he overcame, are still very imperfectly appreciated. Due pains have been bestowed to represent both in their true colours. The picture cannot be finished until the subsequent events of his life are recorded: but as the principles which guided him in his early days retained their influence to the last, and as they are set forth in what is already written, the reader will be enabled to form a just estimate of his moral, as well as of his intellectual character.

I own that I have been chiefly solicitous that the true and genuine lineaments of his mind should stand forth in all their fair and just proportions. Had he merely coveted a wide-spread reputation, the voice of the world proclaiming a great epoch in the physical history of man, produced through his instrumentality, might have satisfied an ambition much more greedy of fame. Gratifying as such an acknowledgment must have been, Jenner's heart was too pure not to seek its enjoyment from a constant devotion to higher and better things than those which centre in mere human authority or approbation. Such, in truth, he was; as such it is my earnest desire that he should be remembered. "Ut vultus hominum, ita simulacra vultûs imbecilla ac mortalia sunt; FORMA MENTIS ÆTERNA."

LIFE

OF

DR. JENNER.

CHAPTER I.

HISTORY OF HIS EARLY LIFE.

EDWARD JENNER was born in the vicarage at Berkeley, in Gloucestershire, on the 17th of May, 1749. He was the third son of the Reverend Stephen Jenner, A. M. of the University of Oxford, Rector of Rockhampton, and Vicar of Berkeley. His mother was the daughter of the Rev. Henry Head, of an ancient and respectable family in Berkshire. This clergyman once held the living of Berkeley, and had, at the same time, a prebendal stall in the Cathedral of Bristol.

Besides his church-preferments, the father of Jenner possessed considerable landed property, the family being of great antiquity in Gloucestershire and the neighbouring county of Worcester. It has produced several eminent men, among whom may be mentioned Dr. Thomas Jenner, President of Magdalen College, Oxford, the immediate predecessor of the pious and learned Dr. George Horne. Jenner's father had been tutor to a former Earl of Berkeley; and the late earl, his brother the admiral, and, indeed, the whole of that noble house always evinced a very strong regard to him and to his family. This excellent and devout man was cut off not long after the birth of his son Edward, at the age of 52, in the year 1754. This heavy loss was as much as possible alleviated by the affectionate care and judicious guidance of his eldest brother, the Rev. Stephen Jenner,* who brought him up with paternal tenderness. He had another brother, the Rev. Henry Jenner, M. A., Oxon, Rector of Rockhampton, Gloucestershire, Vicar of Little Bedwin, Wiltshire; and domestic chaplain to the Earl of Aylesbury. From this gentleman are sprung the Rev. George C. Jenner, and Mr. Henry Jenner, who, as will hereafter be seen, assisted their uncle in his interesting pursuits and inquiries.

Dr. Jenner had three sisters, Mary; Sarah; and Ann, who was married to the Rev. Wm. Davies, Rector of Eastington, in the County of Gloucester.

* B. D. Fellow of Magdalen College, Oxford; and Rector of Fittleton, Wiltshire. For many years he was Rector of Rockhampton, and perpetual Curate of Stone, both in Gloucestershire.

He left three sons, the Rev. William Davies, D.D. Rector of Rockhampton; Robert Stephens Davies, Esq., of Stonehouse; and Edward Davies, Esq., of Ebley House, in the same county.

When about the age of eight years, Jenner was put to school at Wotton-under-Edge, under the Rev. Mr. Clissold. He was next placed under the tuition of the Rev. Dr. Washbourn, at Cirencester, where he made a respectable proficiency in the classics, and laid the foundation of some of those friendships which continued throughout life. His taste for natural history began to show itself at a very early period. Before he was nine years of age, he had made a collection of the nests of the dormouse; and when at Cirencester, he spent the hours devoted by the other boys to play or recreation, in searching for fossils, which abound in the oölitic formation in that neighbourhood. His scholastic education being finished, he was removed to Sodbury near Bristol, in order to be instructed in the elements of surgery and pharmacy by Mr. Ludlow, an eminent surgeon there. On the expiration of his term with this gentleman, he went to London to prosecute his professional studies under the direction and instruction of the celebrated John Hunter, in whose family he resided for two years, a favourite pupil.

The energy and originality of Mr. Hunter's character, had already commanded the respect of his professional brethren, and secured to him a large share of public confidence and attention. He unquestionably belonged to that family of genius, whose works, whatever may be their nature, have not merely a temporary and local interest, but an abiding and universal one; because they are founded upon principles which regulate the progress of truth in all branches of knowledge, and they would not have failed to have rendered him a distinguished man in any situation in life. He was not less vigilant in his observation, than he was scrupulous and accurate in his examination, of the objects of his studies. He became thereby a penetrating and original thinker, and being at the same time gifted with much enterprise and perseverance, he mastered difficulties which for ever would have obstructed the progress of inferior minds.

When Jenner went to London, he was in the twenty-first year of his age, Mr. Hunter in the forty-second. He was not at that time a public lecturer, but he had been about two years Surgeon to St. George's Hospital, and for a considerably longer period he had established his menagerie at Brompton, where he so successfully and perseveringly carried on his inquiries respecting the habits and structure of animals.

The boldness and independence of Mr. Hunter's character produced deep and permanent effects on the minds of all who witnessed them. Jenner, in particular, felt their power; he saw a master-spirit advancing steadily in that walk of knowledge to which he himself was led by all the predilections of his taste, and all the influence of his early habits. He saw a kind, free, and manly nature devoted to the acquisition of science, and putting away from him entirely the selfish and personal considerations, which are too apt to encumber the researches, and to circumscribe the objects, of less enlightened minds. The heart of Jenner was peculiarly alive to virtues of this kind, and he had moreover an intellect fully capable of appreciating and admiring the other qualities of his master: it was a singular felicity which brought such men together. The pupil not only respected the teacher, but he loved the man; there was in both, a directness and plainness of conduct, an unquenchable desire of knowledge, and a congenial love of truth. An unfeigned and unchangeable regard to this life-giving principle, adds a peculiar dignity to all human researches, and its influence subsists long after our common occupations, and all the objects of worldly ambition, have passed away.

These remarks will receive full and pleasing confirmation, in the personal history of the eminent individual whose life we are considering. After completing his professional studies in London, he retired from his preceptor's house; but he did not retire from his good-will and affection, nor from his anxious guidance and direction in his scientific pursuits. An uninterrupted epistolary correspondence was kept up between them, till within a short period of Mr. Hunter's death. A very considerable number of his letters have been preserved. The reader can scarcely fail to be interested in those which I mean to present to him, truly characteristic as they are of the writer's mind, as well as illustrative of the nature and progress of the inquiries of Jenner.

Dr. Jenner set a great value upon these letters. They were carefully preserved in a cover, which was inscribed in his own hand-writing "Letters from Mr. Hunter to E. Jenner;" an honour which he was not always in the habit of conferring on more dignified communications.

During the time of his residence with Mr. Hunter, in 1771, Captain Cook returned from his first voyage of discovery. The valuable specimens of Natural History which had been collected by Sir Joseph Banks, were in a great measure arranged and prepared * by Jenner, who was recommended by Mr. Hunter for that purpose. He evinced so much

* The knowledge which he thus acquired he always retained. In the dissection of tender and delicate organs, and in minute injections, he was almost unrivalled, and displayed the parts intended to be shown with the greatest accuracy and elegance. He had the kindness to bequeath to me a preparation which combines all these qualities. It represents the progress of the ovum in our common domestic fowl, from its first developement to its full and complete growth, when it is about to be dropped from the oviduct. The dissection is beautiful, and the vascularity of the membrane which invests the ova, as well as the internal state of the oviduct, where the shell is formed, are all exhibited with masterly skill.

dexterity and knowledge in executing this duty, that he was offered the appointment of Naturalist in the next expedition, which sailed in 1772. But neither this, nor other prospects of a more enticing nature, could draw him from his purpose of fixing his abode in the place of his birth. In this determination he was partly guided by the deep and grateful affection he felt for his eldest brother, who had been his guide and director when deprived of parental care; and partly by an attachment to the rural scenes and habits of his early youth. Possibly in this decision we may now be permitted to trace the agency of a higher power, which induced a young man frequently to reject most flattering prospects of wealth and distinction that he might be enabled to follow up the leading object of his mind in the seclusion of a country village. It was in this situation that the great purpose of his life was to be fulfilled. It was in such a combination of circumstances as was here presented, and in none other, that the discovery of vaccination could have been effected. In this respect it differed from most other investigations. The facts which have led to the knowledge of the principles of the different sciences, are scattered widely over the works of the creation, and may be found out by all who, with competent faculties, set themselves assiduously and patiently to read the volume that is spread out before them. It must be confessed that it is a rare gift among men, to be able to decipher, with profit to themselves and advantage to others, this great book, which every where teems with wondrous instruction. It nevertheless is open to all, and all may peruse it freely; but the page upon which the virtues of vaccination were inscribed, could never have been seen by many individuals. The existence of such an affection as cow-pox was known only in a few districts; it therefore could not become a subject of common observation, nor challenge the keen scrutiny of inquiring intellects to its elucidation. Its reported prophylactic powers, it is true, had not altogether escaped popular notice; but no one had arisen to ascertain the correctness of this rumour, or to investigate the source and accuracy of the tradition, till Jenner was led to the pursuit; and to an almost unlooked-for, and unparalleled extent, rendered it available to the subjugation of the greatest scourge of mankind. It is manifest, therefore, that in the very essence of the inquiry itself, and in the character of the genius of him by whom it was conducted, there was a suitableness and an accommodation, without which it neither could have been begun nor accomplished. This peculiarity will be rendered still more apparent when we come to trace the progress of his mind in maturing the discovery. He mentioned the subject to Mr. Hunter while he was his pupil; and often attempted to arouse the attention of his professional brethren in the country to it, but without success. The merit of persevering in his labours, and the honour of his triumph, rest therefore, in an exclusive manner, with himself.

In attempting to unfold character, it is not less instructive than it is interesting, to find in the private history of a distinguished individual, the successive links in the chain of events, by which it pleaseth Providence to conduct him to that eminence where shines the splendour of his genius and his intellect. This progress in the case of Jenner can luckily be delineated with much accuracy. While yet a youth, and just entering on his elementary studies, that impression was made upon his mind which laid the foundation of all his future researches respecting vaccination; and, with the constancy of a character fitted and fashioned for great achievements, it was never permitted to escape from his consideration till it terminated in that wonderful discovery, the effects of which all nations have enjoyed. It is probable, therefore, that the seed which was sown before his intercourse with Mr. Hunter commenced, would in some future time have germinated, even though he had never witnessed the animating and encouraging example afforded by his prolific and indefatigable genius.

While thus ascribing its natural influence to a fact which will be more fully elucidated in the course of this narrative, we must not underrate the effects of the culture of such a mind as Jenner's, when conducted by a spirit so inquisitive, so searching, and so skilful, as that of Mr. Hunter.

It was a truly interesting thing to hear Dr. Jenner, in the evening of his days, descanting, with all the fervour of youthful friendship and attachment, on the commanding and engaging peculiarities of Mr. Hunter's mind. He generally called him the "dear man," and when he described the honesty and warmth of his heart, and his never-ceasing energy in the pursuit of knowledge, it was impossible not to be animated by the recital, and to perceive that something more than esteem for high intellectual attainments, was required to form that bond of union which, to the last hour of his life, joined the affectionate recollections of the pupil with the memory of the master.

Immediately after his return from London, Jenner commenced the active duties of his profession. Those who know the painful and laborious exertions of a country surgeon, will see with interest and satisfaction that the love of knowledge can overcome all obstacles; that the daily demands of a toilsome and anxious calling, may be duly and vigilantly fulfilled, and buoyancy enough of character left, to enable a young man in a secluded situation, with little aid from books or society, to rise above every discouragement, to keep his mind constantly alive to every new source of information, to commence and carry on original investigations in many branches of physiology and natural history, and ultimately, by patience and humility of mind, to bring forward for the use and unspeakable advantage of his fellow creatures, one of the most valuable discoveries that ever rewarded the exertions of man.

When Jenner returned to Berkeley, he took up his residence with his brother Stephen. His talents, distinguished as they had been by the favour and approbation of the best judges in London, soon gained him confidence and esteem in the country. His practice rapidly increased, and he acquired a degree of reputation, which, at so early an age, seldom attends the character of medical men. Addicted as he was to the study of natural history from his earliest years, it was not likely that his fondness for that pursuit should have been checked by what he had witnessed in London. On the contrary, he applied himself to the prosecution of it with redoubled ardour, and contrived, in a short time, to accumulate a series of specimens illustrative of comparative anatomy and natural history, which, had they been more ostentatiously displayed, would have formed a museum of no inconsiderable magnitude. Had not his mind in latter years, been necessarily drawn to objects of deeper and more universal interest, his researches in this most interesting field of knowledge could not have failed to have placed him in a very distinguished situation among those who have most successfully cultivated it in modern times.

About this period an incident occurred, which might have dissevered his connexion with his native country. He was dining with a large party at

Bath, when a question arose whether the temperature was highest in the centre of the flame of a candle, or at some small distance from its apex. Various opinions were delivered, but Jenner, with his usual ingenuity and readiness, soon settled the dispute. He placed the candle before him, and inserting his finger into the middle of the flame, he retained it in this situation for a short time. He then placed it a little above the flame, but was compelled immediately to withdraw it. "There, gentlemen," he observed, "the question is settled."

His manner on this occasion indicated so much talent and good sense, that a gentleman of considerable political influence, who happened to be of the party, was particularly struck with him. The next day he sought for Jenner, and offered to procure for him an appointment of emolument and distinction in the East Indies. This, however, like a former offer, was declined, as well as another of a still more enticing nature, which was shortly after made by his friend and preceptor, Mr. Hunter.

After what has been said, the reader will easily believe that Dr. Jenner, at this early period of his life, had given indications of genius, which all good judges of character did not fail to recognize as the harbingers of much future reputation. His knowledge and dexterity as a surgeon, his manners as a gentleman, and his general information as a man of science, rendered his company always acceptable in the families most distinguished by rank or talent in the district where he lived. But there were other qualities, of a personal nature, which more peculiarly endeared him to his intimate associates. He not only commanded confidence by his skill, and respect and esteem by his acquirements, but also secured to himself good-will and affection, by the tenderness, and kindness, and benevolence of his nature, and the meekness with which he carried all his faculties in the sight of his fellow men. To much depth and accuracy of observation and uncommon delicacy of feeling, which at times cast a shade of pensiveness and sorrow over his mind, there was added a liveliness of disposition, which rendered him a friend capable alike of entering into the deepest and saddest emotions of the soul, or participating in all the joys of its gayest and happiest moments.

In following the calls of his profession through the "alleys brown" and shady lanes of the beautiful vale where he resided, he kept a constant eye to the varying scenes which were passing before him; he had the keenest relish for picturesque beauty, and in his excursions alike gratified his taste in this respect, and increased his knowledge by pursuing the details of natural history. He thus contrived to combine the labours of his profession with the truest pleasure and instruction. On such occasions he encouraged his friends to join him in his rides. I have known, and do now know, those who have been favoured with such happiness, who have accompanied him for twenty or thirty miles in a morning, and listened with the highest interest at one time to the overflowing of his mind, while with a vivid and imaginative fervour he shadowed forth his own feelings, or with a painter's eye and poet's tongue delineated the beauties around them. He would then descend to less impassioned themes, and explain, with the most captivating simplicity and ingenuity, the economy of vegetables and animals, or the various productions that came within observation.

His manners in every respect corresponded with a mind so given to such objects. He never met with any one without endeavouring to gain or to impart instruction. In natural history, in particular, he wished from his earliest years to show how much information and amusement lie scattered around us, how bountifully the sublimest sources of gratification are supplied, and how desirable it is that all should be taught to taste them.

With some of his particular friends he often, at this period of his life, spent days in their houses, especially in cases of sickness of a serious nature. In this way he made their home for a season his head-quarters, and from thence went to visit his patients in the surrounding district. In a situation like the Vale of Gloucester, this temporary sojourning at a few miles distance from his own abode, was not attended with much inconvenience in a professional point of view, and it was often a source of recreation and amusement to himself,—and I may add of unmingled satisfaction to his friends.

The cottage is still standing, in the neighbourhood of his friend Henry Hicks's house at Eastington, where he, on one of these occasions, put together his remarks on the cuckoo, and prepared that paper for publication. But it was chiefly at Clapton, a farm belonging to his aunt Hooper,* near Berkeley, that his observations on that subject were made.

His appearance and manner during that portion of his life which we have just been surveying, has been described to me by one of his earliest friends. A delineation so characteristic it is not right to omit:—

"His height was rather under the middle size, his person was robust, but active, and well-formed. In his dress he was peculiarly neat, and every thing about him showed the man intent and serious, and well prepared to meet the duties of his calling.

"When I first saw him it was on Frampton Green. I was somewhat his junior in years, and had heard so much of Mr. Jenner of Berkeley, that I had no small curiosity to see him. He was dressed in a blue coat and yellow buttons, buckskins, well-polished jockey boots, with handsome silver spurs, and he carried a smart whip with a silver handle. His hair, after the fashion of the times,

* This lady was Deborah Jenner, youngest sister of Jenner's father, and married to Thomas Hooper, Gent. of Clapton. She was particularly attached to her nephew Edward: and he spent many of his boyish days at her house. She died at the age of 80, in the year 1784.

was done up in a club, and he wore a broad-brimmed hat.

"We were introduced on that occasion, and I was delighted and astonished. I was prepared to find an accomplished man, and all the country spoke of him as a skilful surgeon and a great naturalist; but I did not expect to find him so much at home on other matters. I, who had been spending my time in cultivating my judgment by abstract study, and smit from my boyhood with the love of song, had sought my amusement in the rosy fields of imagination, was not less surprised than gratified to find that the ancient affinity between Apollo and Esculapius was so well maintained in his person."

In words such as these did poor Edward Gardner, but a short time before his death, describe to me his first interview with Jenner. The acquaintance thus begun, soon ripened into a cordial friendship, which existed for more than forty years. During the whole of that time he enjoyed his confidence, and, as will hereafter appear, was intrusted with all his most secret and private affairs.

Such was the attachment of Jenner's friends to him at this time, so much did they covet and prize his society, and so highly did they value his amusing and interesting conversation, that when he, either as a visitor or in his professional capacity, had been at their houses, they would accompany him on horseback many miles on his way home, and this, too, often at midnight, that the pleasure derived from his company might be prolonged. This arose from the singular and happy union of scientific and original observation, with the playfulness, and mirth, and wit, of familiar intercourse. Profound and inquisitive minds discovered in his society wherewithal to be pleased and instructed—the most superficial and gay wherewith to be amused. At one time he would be dealing out abstract propositions with a clearness and distinctness peculiarly his own, and with the precision of a Franklin, but with more imagination, he would render these propositions applicable to the common concerns of life. At another time the truest and most illustrative delineations of manners and character would flash from his mind. His practical humour, too, was often most enlivening and descriptive. It was the more engaging as it was alike free from all manner of impurity and malevolence. In these respects he was, as honest Izaak Walton says—neither beholden to the devil nor his own corruptions, but kept clear of both. Whether he was mirthful or grave, he could blend both his serious admonitions and his jocose remarks, so as to produce a most harmonious combination.

His recreations from his more severe studies consisted, at this time, in the cultivation of polite literature; and occasionally he sought an acquaintance with the Muses. His imagination, indeed, was always singularly vivid; and he had a peculiar facility, even in common conversation, of clothing his remarks in the gay and lively colours of poetry. His knowledge of the economy of plants and animals, and his vigilant attention to all the varied forms and properties of surrounding objects, supplied him with an inexhaustible fund of analogies and imagery, which alike animated and adorned every subject that he touched upon.

His conceptions of this sort were frequently embodied in little fugitive pieces, which were sometimes read at convivial meetings, or passed between himself and his friends in the ordinary interchange of their correspondence. Gardner, who was no mean judge of matters of this kind, who had been the school-fellow of Chatterton, and who had with no inconsiderable success devoted himself to the study of poetry, often declared that Jenner in becoming a distinguished physician, had lost the opportunity of acquiring renown as a poet.

Without, by any means, wishing the reader to take this opinion in its literal sense, the present is not an unsuitable time to introduce a few of Jenner's poetical *jeux d'esprit*, merely to illustrate the character of his mind and the faithfulness of the preceding delineation. His own estimate of his powers and acquirements in this field may be gathered from what he says to his friend Gardner :—

Berkeley, Thursday Night.

Dear Gardner,

Enclosed is the medley and the song I sang to the gallant Bowmen; miserably scrawled; indeed you will hardly make it out. If your brain is not too much in a whirl, let me remind you of the Ranunculuses for Mr. Nelmes. Did not you promise me some, and some rose-trees? These are flowery subjects, and I hope, in harmony with your mind.

I have thrown in a few more stanzas to "Ladbroke's Entire:" it is my best song. I wish I could give up, or, at least, suspend the little acquaintance I have made with the Muses. Every time I begin a bagatelle, I almost swear it shall be the last; and hardly steer clear of perjury, you see. But when I see that the resolves of the greatest philosophers can be set aside by the most gentle means in the world, I, who am among the lowest of the order, should not repine at my lot. As I once told you, we are certainly puppets danced about by wires that reach the skies. For my own part, I rejoice at your thinking it wrong to dance without a partner, and shall be among the first to congratulate you on the great Master of the Ceremonies indulging you with the hand of a fair one.

Sincerely yours,

E. Jenner.

Instead of presenting either of the songs spoken of in this letter, I think the reader will be more gratified by perusing two addresses to a robin. Both the style and sentiments are in strong contrast, and show the versatility of his powers.

ADDRESS TO A ROBIN.

Come, sweetest of the feather'd throng!
And soothe me with thy plaintive song:
Come to my cot, devoid of fear,
No danger shall await thee here:
No prowling cat, with whisker'd face,
Approaches this sequester'd place:
No schoolboy with his willow-bow
Shall aim at thee a murd'rous blow:
No wily lim'd twig ere molest
Thy olive wing or crimson breast:
Thy cup, sweet bird! I'll daily fill
At yonder cressy, bubbling rill;
Thy board shall plenteously be spread
With crumblets of the nicest bread;
And when rude winter comes and shows
His icicles and shivering snows,
Hop o'er my cheering hearth and be
One of my peaceful family:
Then soothe me with thy plaintive song,
Thou sweetest of the feather'd throng!

ADDRESS TO A ROBIN.

IN ANSWER TO ONE BY CAPTAIN SNELL.

Begone this instant from my door!
Nor plague me with thy canting more.
Hop off! I say, nor in this place
Dare show thy hypocritic face.
Pray do'st thou think, ungrateful fellow,
Because thy voice is somewhat mellow,

Or that thou hither com'st assuming
A kind of modesty in pluming;
Wilt thou allure me, whining beggar?
Or my true notions of thee stagger?
Have I not seen thee, sturdy ruffian,
With impious claw thy father cuffing? *
Seen thee, thou vile impostor, blackguard,
With many a blow thy mother smack hard?
Strip from her back the downy feather,
Spite of inclemency of weather;
Nay, threaten her with instant killing,
If thy full platter she put bill in:
Why then how dar'st thou thus from me
To ask for hospitality?
  Disdainful wretch! when smiling spring
Bids every bird tune up and sing,
Though the sweet orchestra should want ye
To take a part, a soft andante,
The lark, who leads the band, in vain
Solicits thy assisting strain,
For slily thou leav'st all their chanting,
Deep in the woods to go gallanting.
  Long have I known thy ready knack 'tis
A thousand wily tricks to practise.
Did'st thou not use deception vile
A bard † to cozen and beguile,
Draw by a kind of *hocus pocus*
His rays poetic to a focus,
Then craftily divert the flame
To blaze upon thy worthless name?

* Unum arbustum non alit duos Erythacos."

† Captain Snell.

Think'st thou I know not, rogue ungrateful,
Of mischief thou hast got a pateful?
Do qualms of conscience ne'er molest thee?
No retrospective thoughts infest thee?
Hast thou not entered farmer's houses,
Annoying oft their lawful spouses?
Deform'd their butter, peck'd their cheese,
And robb'd them of their market fees?
Though ne'er did they deny thy asking,
(Villain, a hypocritic mask in!)
But ever ready were to pour
Around thy head the crumby show'r.
And pray another thing—but 'sdeath!
Why do I thus consume my breath?"—
Once more I say, Hop off!—hoh! hoh!
'Tis well thou thought'st it time to go:
And this I tell thee, little blade,
If ever on my palisade
Again I catch thee—by the law
Thy grave shall be Grimalkin's maw!

The "Signs of Rain" exhibits much of the minute painting of Cowper, or of Crabbe; and shows, in pleasing combination, the accuracy of the naturalist and the fancy of the poet.

SIGNS OF RAIN.

AN EXCUSE FOR NOT ACCEPTING THE INVITATION OF A FRIEND TO MAKE A COUNTRY EXCURSION.

The hollow winds begin to blow,
The clouds look black, the glass is low,

The soot falls down, the spaniels sleep,
And spiders from their cobwebs creep.
Last night the sun went pale to bed,
The moon in halos hid her head.
The boding shepherd heaves a sigh,
For see! a rainbow spans the sky.
The walls are damp, the ditches smell,
Clos'd is the pink-ey'd pimpernel.
Hark! how the chairs and tables crack;
Old Betty's joints are on the rack.
Loud quack the ducks, the peacocks cry,
The distant hills are looking nigh.
How restless are the snorting swine—
The busy flies disturb the kine.
Low o'er the grass the swallow wings;
The cricket too, how loud it sings.
Puss on the hearth, with velvet paws,
Sits smoothing o'er her whiskered jaws.
Thro' the clear stream the fishes rise,
And nimbly catch the incautious flies.
The sheep were seen at early light
Cropping the meads with eager bite.
Tho' June, the air is cold and chill;
The mellow black-bird's voice is still.
The glow-worms, numerous and bright,
Illumed the dewy dell last night.
At dusk the squalid toad was seen
Hopping, crawling, o'er the green.
The frog has lost his yellow vest,
And in a dingy suit is dress'd.
The leech, disturb'd, is newly risen
Quite to the summit of his prison.

The whirling winds the dust obeys,
And in the rapid eddy plays.
My dog, so altered is his taste,
Quits mutton bones on grass to feast;
And see yon rooks, how odd their flight,
They imitate the gliding kite,
Or seem precipitate to fall,
As if they felt the piercing ball.
'Twill surely rain—I see with sorrow,
Our jaunt must be put off to-morrow.

He frequently indulged his fondness for Epigram. His trifles in this way possess both point and humour.

DEATH AND MR. PEACH.

A SHORT DIALOGUE.—N. B. MR. P. DIED IN APRIL.

P.—Awhile forbear thy horrid gripe,
Do pray, dread Sir! remember
Peaches are never fairly ripe
'Till August or September.

D.—To gratify my longing taste,
And make thy flavour fine,
I had thee in a hot-house placed,
And moistened well with wine.

Mr. Peach shortened his life by the too free use of the bottle.

ON THE DEATH OF A SHERIFF'S BAILIFF.

Arrested by Death! cries John; I'll give bail;
No! no! replies Death, I must take you to gaol;
To gaol friend? quoth John, why sure there's a flaw,
Look in statute the —— Stuff! nonsense! pshaw, pshaw!
Rejoins the grim tyrant, you've done with your tricks:
Contention give o'er, and go headlong to Styx.

ON THE DEATH OF AN OLD WOMAN NAMED HEYWOOD,

NOT REMARKABLE FOR HAVING LED AN EXEMPLARY LIFE.

Tho' some may exclaim 'Twas strange, or 'twas cruel,
Yet 'tis said to be true; Old Nick wanting fuel,
Gave an order for faggots well-season'd and good;
So Death took his hatchet and cut down Hey-wood.

ON THE DEATH OF A MISER.

Tom at last has laid by his old niggardly forms,
And now gives good dinners; to whom pray?—the worms.

ON LORD BERKELEY'S HUNTSMAN, WHO DIED IN THE CHACE.

Determined much higher to hoist up his name,
Than Nimrod the hunter, in annals of fame,
Hark forward! cried Charles, and gallantly whirled
His high-mettled steed o'er the gates of the world.

Knowing that good company and good discourse were the sinews of virtue, he promoted them on all occasions. He was especially fond of music, and was a member of a catch club that met at Cam. He could also play on the violin and flute; and he was in the habit of forming select musical parties, where he occasionally was a performer.

I have seen him in his latter years, after his renown had filled the world, and after the many cares attendant upon vaccination had often weighed heavy upon him, shake them entirely off, he would then take up a humorous strain, and sing one of his own ballads with all the mirth and gaiety of his youthful days.

He had a particular dislike to cards, both because they interfered with a much more instructive employment of time, and often led to evils of a much more serious nature.

CHAPTER II.

1773—1783.

INCLUDING LETTERS FROM JOHN HUNTER.

I SHALL now proceed to illustrate the foregoing sketch, by a reference to those facts and incidents which a knowledge of existing documents enables me to present. A great deal of the interest of his early professional life is naturally derived from his epistolary intercourse with Mr. Hunter. Unfortunately many of the letters are without date, and I have been under the necessity, therefore, of endeavouring to ascertain the periods at which they were sent, from collateral evidence. The first appears to have been written in 1773; it refers to some of Jenner's observations on the cuckoo. I may also mention that the picture alluded to became his property, and remained in his possession till his death. He bequeathed it to his friend Henry Hicks, in whose house at the Leaze it now remains.

Mr. Hunter to E. Jenner.

Dear Jenner,

I received yours, and was extremely happy to hear of your success in business: I hope it will continue. I am obliged to you for thinking of me, especially in my Natural History. I shall be glad of your observations on the cuckoo, and upon the breeding of toads: be as particular as you possibly can. If you can pick me up any thing that is curious, and prepare it for me, do it, either in the flesh or fish way. Pictures have been very cheap, but the season is now over. There will be but one sale, viz. Fordyce's; but I believe all his pictures are exquisite, and will go beyond you or me. Since you wrote to me, I purchased up a small landscape of Barrett's, of cattle and herd; I gave five pounds seven shillings and sixpence: it is one of his eight guinea pictures. You shall have it or not as you please. I have one of the same size that I bought of him some time ago.

I saw the young lady, your patient. I do not know well what can be done. If it was possible to pass a bougie from the nose up the duct to the sack, it might be of service; but nothing but a solid can be of any use as a local application. Her general habit should be attended to, such as sea bathing, or cold bath; using a good deal of gentle exercise, such as getting up early in the morning, riding, &c.; she might take gentle mercurials with the bark and the cicuta. Let me hear from you soon.

Ever yours,

John Hunter.

The three subsequent letters appear to have been sent not long after the preceding. Mr. Hunter had a short time before commenced lecturing to the pupils at St. George's Hospital. He afterwards extended his plan, and threw his class open to the public, and began steadfastly to execute that great scheme, for elucidating the structure and functions of organised bodies, which subsequently became the main object of his life. The ardent, energetic, and original character of the writer, is seen in every sentence. Although all Jenner's replies have been destroyed, the results of some of the experiments, to which the following letters refer, were published in 1778 in the "Philosophical Transactions," in Mr. Hunter's paper on the heat of animals and vegetables; and others will be found in "The Observations on some parts of the Animal Economy,"* which were published in 1792.

Among Dr. Jenner's papers I find a manuscript, detailing many of the experiments which he made at this time, at the instigation of Mr. Hunter, on hedge-hogs; but I deem it expedient to delay its publication till it is found convenient to collect and print all his medical and philosophical papers.

Mr. Hunter to E. Jenner.

Dear Jenner,

I received yours, as also the cuckoo's stomach. I should like to have a few more of them, for I find they do not all show the same thing. If possible, I wish you could

* See pp. 112, 156, 195, 233.

remove the cuckoo's egg into another bird's nest, and tame the young one to see what note it has. There is employment for you, young man! If you collect eggs, you should also collect the nests; and I do not care how many you send. I wanted a crow's nest, as also a magpie's, in the branches of the tree where they are built; but I am afraid it is now too late.

This evening, looking into my book of patients, to scratch out the name of one who had just paid me, and whose name began with an M, I saw a Mr. Mathews of Berkeley, recommended by you. He did not pay me. I forget whether he was recommended by you as a friend, to serve him, or me: if it was to serve him, I scratch him out of my book. Do you keep an account of the observations of the cuckoo; or must I refer to your letters?—I want a nest with the egg in it; also a nest with a young cuckoo, and also an old cuckoo.

I hear you saying, there is no end to your wants.

Ever yours,

John Hunter.

Mr. Hunter to E. Jenner.

Dear Jenner,

I received your salmon and very fresh, and just examined enough to want another, but will wait till another season. If I was to have another, it should be one that had just spawned; I will take a cock salmon when you please.

If you catch any bats let me have some of them; and those you try yourself, open a hole in the belly, just size enough to admit the ball; put the ball down towards the pelvis, and observe the heat there; then up towards the diaphragm, and observe the heat there; observe the fluidity of the blood; do all this in a cold place. Extraneous fossils are all vegetable and animal productions, found in a fossil state. See if you can catch the number of pulsations and the frequency of breathing in the bat, without torture. If the frost is hard, see what vegetables freeze; bore holes in large trees, and see whether the sap runs out, which will show it is not frozen. I am afraid you have not a proper thermometer. I will send you one.

Your very much obliged servant,

J. HUNTER.

I have not seen Dr. H. but I dare say he will be glad to have the cases.

MR. HUNTER TO E. JENNER.

DEAR JENNER,

I do not know any one I would sooner write to than you: I do not know any body I am so much obliged to. I thank you for a fish, but I should thank you more if you had let me know who it comes from.

I beg for the future you will always write when you send me any thing. Somebody sent me a cheese with a fish upon it; it was perhaps you: you know I hate to be puzzled. Also let me know what things you have sent me lately. I have not received the cuckoo's nest yet. Now for your patient. I believe the best thing you can do, is to do little. I would not touch the fungus with an escharotic, for fear the brain should be near; I would also use but a very slight compression, as the fungus will be a bandage to the brain; and as to the fungus itself, you have nothing to fear; for whenever the parts underneath are sound, the fungus will subside of itself. Keep your patient rather low and quiet. Let me know how he goes on, and any thing else you can.

Ever yours,

JOHN HUNTER.

The next letter, I presume, was written in 1775. It refers to his design of establishing a school of natural history, on a scale till that time unknown in this country. He wished Jenner to join him in the undertaking.

MR. HUNTER TO E. JENNER.

DEAR JENNER,

I have received many things from you, and will thank you in the lump; but while I thank you, let me know what I owe you. I have a great scheme to communicate to you, and want you to take part in it; but remember it is as yet a most profound secret. My scheme is to teach natural history, in which will be included anatomy, both human and comparative. The labour of it is too much for one man, therefore I must have some person to assist; but who that person shall be is the difficulty. When running over a variety of people, you have come into my mind among the rest. Now if it is a scheme you would like, and a possibility of your leaving the country, at the same time able and willing to lay down one thousand guineas, I will send you the whole proposals: but if you cannot leave the country upon any terms, then it is unnecessary to go any farther: and all I have to beg is to keep it a secret. I would not have you mention it to Ludlow ——— &c. I proposed it to L——— before he left London; but his father objected, I believe, to the money. I know the scheme itself will be to your taste. Before you consult with any of your friends, just consult with yourself, and ask, can I go to London, and can I give one thousand guineas for any chance that can be worth it? Let me hear from you soon.

Yours,

London, May 24th. J. HUNTER.

Much as Jenner was attached to Mr. Hunter, and interested as he was in the pursuits to which the preceding letter refers, and flattered as he could not but be by taking a conspicuous station in the metropolis, under circumstances most promising to a young man, he nevertheless declined the proposal. I do not know the cause that was assigned for this determination. Mr. Hunter, from his knowledge of Jenner's character, seems to have anticipated it.

MR. HUNTER TO E. JENNER.

DEAR JENNER,

I received yours in answer to mine, which I should have answered. I own I suspected it would not do; yet as I did intend such a scheme, I was inclinable to give you the offer.

I thank you for your experiment on the hedge-hog; but why do you ask me a question by the way of solving it? I think your solution is just; but why think,—why not try the experiment? Repeat all the experiments upon a hedge-hog as soon as you receive this, and they will give you the solution. Try the heat. Cut off a leg at the same place: cut off the head, and expose the heart, and let me know the result of the whole.

I am, dear Jenner,

Ever yours,

August 2nd. JOHN HUNTER.

There are a considerable number of letters written, I believe, between the last-mentioned period and that which I am now about to specify. They refer to professional subjects, experiments on the heat of animals and vegetables, and the collection of fossils and other subjects of natural history for the Museum. Mr. Hunter kindly acknowledged Jenner's labours in his behalf, as may be seen by what he says on sending him a picture.

MR. HUNTER TO E. JENNER.

DEAR JENNER,

I received yours by Dr. Hicks with the hedge-hog alive; I put it into my garden, but I want more. I will send you the picture, but by what conveyance or by what place? I have a picture of Barret and Stubbs. The landscape, by Barret; a horse frightened at the first seeing of a lion, by Stubbs. I got it for five guineas: will you have it? I have a dearer one, and no use for two of the same masters; but do not have it excepting you would like it, for I can get my money for it. I am glad you have got black-birds' nests. Let me know the expenses you are at, for I do not mean that the picture is to go for any of it, only for your trouble.

Ever yours,

JOHN HUNTER.

N. B. I should suppose hedge-hogs would come in a box, full of holes all round, filled with hay, and some fresh meat put into it.

MR. HUNTER TO E. JENNER.

DEAR JENNER,

I received the box, also your letter. I am very much obliged to you for your kind attention to me, and how to reward you I do not know. Let that be as it will, I must still give you commissions. If you can get me easily salmon spawn I should like to have it, and out of different places, as it will be of different ages. It should be put into bottles immediately with spirits, each parcel separate from the other. The spirits should be proof, and there should be rather more spirit than spawn.

I will also take any specimens of fossils you may send me, or indeed any thing else. Did I send you any of my publications in the Philosophical Transactions? If I have not, let me know. I want to put you upon some experiments this winter. What do you think of examining eels? Their sexes have not yet been found out, nor their mode of propagation; it is a thing of consequence in Natural History. I began it, but I could not get eels immediately from the river, and to get them from fishmongers, who buy them in custom, does not do. My intention was to examine several pretty large eels on the first and fifteenth of every month in the year. If eels are in plenty with you, and if you like the proposal, let me know, and I will give you full instructions how to proceed; also next spring I would have you make experiments upon the growth of vegetables; and, if you have no objection, I will set you upon a set of experiments upon the heat of vegetables in the winter. If in any of these pursuits you discover any principle worthy the public, I will give it into the Royal Society for you. I must pick you up a picture this winter. I saw Mrs. Black * at Mr. Drummond's the other day. I suspect Mr. Black is dead, but I durst not inquire. Cannot you get me a large porpoise for either love or money? What is the bird you sent me, also the two young animals, which I imagine to be Guinea pigs?

Ever yours,

John Hunter.

* Jenner's eldest sister Mary, who was married to the Rev. Geo. Charles Black, LL.B. of Norwood, Middlesex.

On the eleventh of May, of this year, he gave Jenner the first intimation of his illness. He mentions it, as will be perceived, very casually, although it was of such a nature as to excite very serious apprehensions in the breasts of all his friends.

MR. HUNTER TO E. JENNER.

DEAR JENNER,

I have before me now two letters of yours, which I should have answered much sooner. Your friend Dr. Hicks I have not seen. I was not at home when he called, and I have had no time to wait upon him as he lived entirely out of my walk. I should have been glad to have seen him, but I suppose he stood upon ceremony. I received the fossils, and should be glad of any that you can get. If any bones of animals are found, be sure to get them for me. I should be glad to have some of the salmon fry. I had the pleasure of seeing your brother, but only for a short time. I received the bird; I am not acquainted with it: send me some more, if you can get them readily. I sent with Mr. Jenner the thermometer; if you do not understand it, let me know.

Not two hours after I saw your brother I was taken very ill with a swimming in my head, and could not raise it off the pillow for ten days: it is still not perfectly recovered. Have you begun the eels? No porpoises. No salmon spawn before it has hatched. You see I am very greedy. Be sure to keep an account of all the out-goings.

My compliments to Mrs. Black and to your brother, and let me hear from you.

Ever yours,

London, May 11th, 1777. J. HUNTER.

MR. HUNTER TO E. JENNER.

DEAR JENNER,

Excuse me for not answering your letters so soon as I could wish: send me all the fossils you find. What I meant by bones, was all the bones that are found any depth under the surface of the earth: many are found in stones, &c. I suppose those skeletons are not complete, but send me some of them; and if any history can be given, send it also. The thermometer is a very useful one when understood. You will observe the scratch; upon the glass stalk; perhaps about two inches from the globe, which is the freezing point; put 0, or nought, which is upon the ivory scale, two degrees below the scratch, then 0 becomes the thirtieth degree, and the scratch being two degrees above it, stands at the freezing point; then from that count upwards; or if the cold is below 30, then put No. 1 or No. 2, or 3 at the scratch, and count down: every No. is 10 degrees. What the d—l becomes of your eels in the winter? but try them in the summer, and see what you can make of them.

I do not remember Dr. Fordyce's ever supposing a polypus vascular. I rather should believe that he supposed the contrary; you know it comes near my idea, viz. that the blood is alive, and is the bond of union everywhere. But I should very much suspect that a polypus formed after death is not of that kind. I am pretty certain that I have injected them in arteries after amputation. I have a preparation which shows it, and which supports my theory.

Yours,

JOHN HUNTER.

London, *July 6th*, 1777.

Mr. Hunter felt very much disposed to continue to treat his indisposition lightly. He found it necessary, however, to seek relief by retiring from his labours for a season. Bath was fixed upon for this purpose, and he arrived there in the end of August. Jenner visited him soon afterwards. He was most sensibly affected by his appearance, and concluded at this early period, that that very disease existed, which, in 1793, suddenly extinguished Mr. Hunter's most valuable life. This impression so much distressed Jenner, that he abstained from giving to the world his observations on the affection under which he believed Mr. Hunter to labour, lest his attention should be drawn to it, and his fears excited by its truly formidable nature. He could not, however, refrain from communicating his alarm to some of Mr. Hunter's friends. He wrote the following interesting letter to the late Dr. Heberden on the subject. From some cause or other it was not forwarded; but it finds a place here with much propriety, because it shows the delicacy of his feeling, and moreover establishes his accuracy as a pathologist; and contains the first account of a morbid affection, which had escaped the observation of former inquirers. In a subsequent page the reader will find a letter from Sir Everard Home, which fully confirms the justness and accuracy of Jenner's diagnosis. It will be seen, also, that the letter to Dr. Heberden contains the substance of the communication which he made to his friend Dr. Parry of Bath, and which was published in his work in 1799.

E. JENNER, TO DR. HEBERDEN.—1778.

SIR,

When you are acquainted with my motives, I presume you will pardon the liberty I take in addressing you. I am prompted to it from a knowledge of the mutual regard that subsists between you and my worthy friend Mr. Hunter. When I had the pleasure of seeing him at Bath last Autumn, I thought he was affected with many symptoms of the Angina Pectoris. The dissections (as far as I have seen) of those who have died of it, throw but little light upon the subject. Though in the course of my practice I have seen many fall victims to this dreadful disease, yet I have only had two opportunities of an examination after death. In the first of these I found no material disease of the heart, except that the coronary artery appeared thickened.

As no notice had been taken of such a circumstance by any body who had written on the subject, I concluded that we must still seek for other causes as productive of the disease: but about three weeks ago, Mr. Paytherus, a surgeon at Ross, in Herefordshire, desired me to examine with him the heart of a person who had died of the Angina Pectoris a few days before. Here we found the same appearance of the coronary arteries as in the former case. But what I had taken to be an ossification of the vessel itself, Mr. P. discovered to be a kind of firm fleshy tube, formed within the vessel, with a considerable quantity of ossific matter dispersed irregularly through it. This tube did not appear to have any vascular connection with the coats of the artery, but seemed to lie merely in simple contact with it.

As the heart, I believe, in every subject that has died of the Angina Pectoris, has been found extremely loaded with fat; and as these vessels lie quite concealed in that substance, is it possible this appearance may have been overlooked? The importance of the coronary arteries, and how much the heart must suffer from their not being able duly to perform their functions, (we cannot be surprised at the painful spasms) is a subject I need not enlarge upon, therefore shall only just remark that it is possible that all the symptoms may arise from this one circumstance.

As I frequently write to Mr. H. I have been some time in hesitation respecting the propriety of communicating the matter to him, and should be exceedingly thankful to you, Sir, for your advice upon the subject. Should it be admitted that this is the cause of the disease, I fear the medical world may seek in vain for a remedy, and I am fearful, (if Mr. H. should admit this to be the cause of the disease) that it may deprive him of the hopes of a recovery.

Mr. Hunter remained in Bath till the middle of November. He continued to have frequent intercourse with Jenner. Some of his letters written at this time are of an unusually playful character, and show that his mind was not at all depressed by his illness, and that the spirit of inquiry was kept as much awake as when he was in perfect health.

MR. HUNTER TO E. JENNER.

DEAR JENNER,

Till yesterday we did not know from whom the hare came, but the cook found it out; we thank you; it was a very fine one. By your not taking any notice of my letter, I do suppose you did not receive it. Near three weeks ago, I wrote to you to meet us at the Hot Wells, Bristol. Some days after the date of the letter we went to the place appointed, by 10 o'clock in the morning, but no Jenner there ; we breakfasted, we dined, we went to Kings-Weston, we drank tea, we supped, we staid all night, and set out for Bath next day. We would have come on to Berkeley, but we were afraid that you might not be there. I am afraid it will not be in my power to come to see you, although I wish it much. I shall be obliged to take Southampton in my way home. Are hedge-hogs so saucy as to refuse coming, without coming for them? See if you can coax them. We are all alive here. The Downs look like a bee-hive. Let me hear from you. Mrs. H. gives her compliments to you.

Yours,

J. HUNTER.

Bath, 18*th*.

My letter was sent to your friend in Bristol by the coach, but perhaps the coachman did not deliver it.

It has been already mentioned that Mr. Hunter published his paper on the heat of animals and vegetables in the "Transactions of the Royal Society" in 1778. Two letters written, one in March, another in November of that year, have a reference to that paper, as well as to other pursuits in which he was engaged. They, like all the others, afford very striking displays of the writer's character and mode of thinking and expression. The latter not always strictly grammatical, but nevertheless full of energy and meaning. His ideas, though distinguished by their novelty, and obviously the result of his own unaided reflection, are often given in an abrupt and laconic manner; but there is a clearness of conception, which enabled him to compress into a small space what would have required from others tedious and minute detail.

The kindness also, with which he acknowledges Jenner's assistance, and the little pledges of respect which he from time to time offered to his friend, form incidents not unworthy of notice in the history of two such men.

The picture by Bassan, which is mentioned below, this instant hangs in the dining-room of the Chauntry (Dr. Jenner's house) at Berkeley. It has been often pointed out to me by Dr. Jenner as the gift of Mr. Hunter. It has been preserved with great care, and seems a genuine specimen of the painter whose name it bears.

Mr. Hunter to E. Jenner.

Dear Jenner,

Your letter of December has lain before me ever since I received it, to put me in mind that it was not answered. I am glad you liked the candlesticks; I thought them pretty. The fossils were none of the best; but I know you did not make them, therefore not your fault. The particular one you put the Q? upon is only a cast of a bivalve. I wish I had seen E——'s collection. I am matching my fossils as far as I can with the recent. Have you made any experiments with hedge-hogs? and can you send me some this spring? for all those sent me died, so that I am hedge-hogless.

Mr. Luders sent me the bone; it is a very curious one; whether he will let me keep it or not I do not know.

I received yours by Mr. Jones, with the bird; I thank you for thinking of me. Frogs live an amazing while after they are dead, as also all animals of that tribe. The directions I gave you about black-birds were, when you have a black-bird's nest, viz. with four young ones, take one just hatched, and put it bodily into spirits by the head, extending the wings and legs. Observe when the feathers are beginning to sprout; then take another, and serve it the same way: then a third and a fourth, so as to get a series of the growth of the feather, but the last or fourth must not be so old as the feathers to cover the other parts where feathers do not grow. This you will better understand when you come to make the trial. I have a picture of Bassan's, that I lent a poor d——l three guineas upon: he died, and never redeemed the picture. I intend sending it to you; it is a good deal damaged, but some of the figures are very good. Get a frame for it, and hang it in a strong light. There are some experiments of mine publishing in the Philosophical Transactions, which I will also send you with the picture; accept of them as a remembrance of the trouble I put you to. Let me hear from you when convenient. Mrs. Hunter desires her compliments to you.

I am, dear Jenner, your

Most obedient and most

Humble Servant

JOHN HUNTER.

London, March 29th, 1778.

MR. HUNTER TO E. JENNER.

DEAR JENNER,

I thank you for all the trouble you have taken. I have interspersed your experiments among my works, and I wish I had more, but I do not know well what to set you about. If you could make some experiments upon the increased heat of inflammation, I should be obliged to you. I have made some, but I am so much hurried that they are but imperfect. To give you an idea of such experiments, I first introduced the thermometer into the anus of an ass; then I injected a solution of corrosive sublimate, above a pint, which it threw out very soon. Some hours after I threw in another, and about twelve hours after I again introduced the thermometer. The same experiment might be made upon a dog. I opened the thorax of a dog, between two ribs, and introduced the thermometer. Then I put some lint into the wound to keep it from healing by the first intention, that the thorax might inflame; but before I had time to try it again (from the hurry of business) my dog died, which was on the fourth day. A deep wound might be made into the thick of a dog's thigh; then introduce the thermometer and some extraneous matter; put in the whole depth, large enough to allow the thermometer to enter when the plug was taken out, which should be taken out in the time of inflammation, and see the difference. If these experiments will amuse you, I should be glad they were made; but take care you do not break your thermometer in the dog's chest, &c.

I am, dear Jenner,

Your much obliged

and most obedient Servant,

London, Nov. 21st. JOHN HUNTER.

About the period we are now considering, Dr. Jenner was instrumental in forming a society which had for its object the improvement of medical science. It was also intended to promote conviviality and good fellowship. The following were among the members: Dr. Parry, of Bath; Dr. Hicks, of Bristol; Dr. Ludlow, of Corsham; Dr. Mathews, of Hereford; Mr. Paytherus, and Jenner. The meetings were chiefly held at the Fleece Inn, at Rodborough, but it was customary to appoint them in other parts of the country, as it suited the convenience of the members.

These meetings were often truly interesting. After the more serious business was finished, the members dined together. They occasionally permitted visitors, who were not medical men, to join them, at this part of their entertainment. No one more frequently enjoyed this indulgence than Jenner's faithful friend Henry Hicks. This gentleman's house lay in the direct road from Berkeley to the place of meeting, and it was often Jenner's custom to call as he passed and carry him with him to Rodborough.

Jenner added largely to the mirth of the party, but he never appeared without contributing his full share to the intellectual banquet also. His paper on the Angina Pectoris, which forms the ground-work of his friend Dr. Parry's book on that subject, was read at the meeting. He also communicated others on many points of physiology and pathology. These papers fell into the hands of some of the members, and he could never recover them: I have often heard him lament the loss of one of them in particular. It contained observations respecting a disease of the heart, which frequently comes on during attacks of acute rheumatism, and leads to enlargement and disorganization of the part. This formidable disorder had very much escaped the notice of medical men. Jenner's observations were original, and had they been published at the time they were first communicated to the society, his claims to priority could not have been set aside as they have been since that time by other writers.

He also wrote a treatise on opthalmia, which was read to the society, and was intended for publication. It is drawn up in a systematic form, and was meant to point out a method of treatment which he had found successful in certain stages of the disease. This branch of surgery has been, since the period referred to, cultivated with so much scientific precision as to deprive less perfect observations of much of their value. The essay, however, of Jenner, bears unquestionable marks of close and accurate discrimination both of the symptoms and varieties of the disease; and with his other writings on professional subjects, could not have failed to have secured to him a sound reputation as a medical philosopher, had not the splendour of his other discoveries cast all inferior objects into obscurity. This letter I presume refers to the society I have just been describing.

MR. HUNTER TO E. JENNER.

DEAR JENNER,

I received your account of your experiments on the hedge-hog, also the dog-fish, for which I thank you. I have now received your account of the aneurismal vein with the cast, and showed it to my pupils this evening with the description.

I hope you will be able to procure the arm when the man dies. If you would choose to have it published, I would either give it to the Medical Society here, or send it to Edinburgh to be published in their commentaries. Let me know your inclination, and I will add whatever I may think wanting, and give it your name. I am very happy to hear that some of you have wished to communicate your ideas to one another. If I can give you any assistance, command me: I shall always be glad to hear from you as an individual, or as from the society. Mrs. H. desires her compliments to you. Have you left off fossilizing?

I am, dear Jenner,

Your much obliged and humble servant,

London, April 28th. JOHN HUNTER.

Jenner was also a member of another society which assembled chiefly at the Ship, at Alveston, a village about ten miles from Bristol. This society he denominated Convivio-Medical, in opposition to the one that met at Rodborough, which he called Medico-Convivial. Among the members of the former were the Ludlows, (E. and D.) of Sodbury, Shute of Winterbourne, Bradford of Frenchay, Fewster of Thornbury, Pountney of Hensbury, Davies of Bristol, and Richardson and Taylor of Wotton-under-edge. I am indebted to the last-named gentleman for the preceding list, and he, with Dr. Bradford, are now, I believe, the only survivors of their former associates.

Dr. Jenner has frequently told me that at the meetings of this society he was accustomed to bring forward the reported prophylactic virtues of cow-pox, and earnestly to recommend his medical friends to prosecute the inquiry. All his efforts were, however, ineffectual; his brethren were acquainted with the rumour, but they looked upon it as one of those vague notions from which no accurate or valuable information could be gathered, especially as most of them had met with cases in which those who were supposed to have had cow-pox, had subsequently been affected with small-pox. These discouragements, as will be seen more at large hereafter, did not suppress the ardour of Jenner's mind. He often recurred to the subject in these meetings; at length it became so distasteful to his companions, that I have many times heard him declare that they threatened to expel him if he continued to harass them with so unprofitable a subject. This society was in existence so late as 1789.

As it has often been asserted that the late Mr. Fewster of Thornbury, has a claim to be considered as the inventor of vaccination, I take this opportunity of saying a few words on that subject.

That gentleman in his early days was associated with Sutton in the practice of small-pox inoculation, and had frequent opportunities of hearing the popular notion that those who had had cow-pox could not be infected with small-pox. He, as has just been said, was one of the members of the medical society which met at Alveston. The subject was often brought before the meetings, but neither this circumstance, nor his own previous acquaintance with the reports of the country, and his experience that they were not altogether without foundation, could induce him to prosecute the investigation further, or to countenance Jenner's efforts. On the contrary, he certainly undervalued them, and continued to do so even after the Inquiry was published. In a letter to Mr. Rolph, surgeon, in Peckham, dated at Thornbury on the 11th of October 1798, and published in Dr. Pearson's inquiry concerning the history of the cow-pox, he has the following words:—"I think it (*i. e.* the cow-pox in the natural way) is a much more severe disease in general than the inoculated small-pox. I do not see any great advantage from inoculation for the cow-pox: inoculation for the small-pox seems so well understood, that there is very little need of a substitute. It is curious, however, and may lead to improvements."

I am the more induced to record the above statement, because I have recently witnessed attempts to assert this gentleman's claim to merit as an investigator and discoverer in the field where Jenner acquired that reputation to which Mr. Fewster unquestionably had no just claims. His own opinion, as given in the preceding extract, is quite conclusive upon this point, and must for ever silence all further doubts respecting it. If other evidence were wanting, it may be mentioned that there are letters from Mr. Fewster to Dr. Jenner, now in existence, which clearly assign all the merit to Dr. Jenner, without in the most remote degree alluding to any pretensions of his own. I may also state, that after Dr. Jenner's death, I wrote to Mr. Fewster for some information concerning the Alveston meeting and the early history of vaccination, and that I received a very civil letter, in his name, from his son, dated Thornbury, Sept. 3d, 1823, ; but not the most distant insinuation of any claims, such as I have been considering, was advanced.

I cannot refer to these meetings, but especially to the one at Rodborough, without more particularly mentioning one of its members, the late Dr. Parry, of Bath. His acquaintance with Jenner commenced when they were at school together at Cirencester. Early intimacy and corresponding studies and tastes, laid the basis of a friendship which endured through every vicissitude of life. By means of letters, and as frequent personal intercourse as the nature of their occupations would permit, they stimulated each other in the pursuit of knowledge whether it respected professional or other subjects ; and it was in furtherance of this design that the little society at Rodborough was instituted.

Not long after this period Jenner appears to have experienced some disappointment in his affections, which for a considerable time materially impaired his happiness. Mr. Hunter alludes to this occurrence in one of his letters in a manner rather more playful and jocular, than was quite in unison with the deep and tender emotion, which for a time filled the heart of his correspondent.

Mr. Hunter to E. Jenner.

Dear Jenner,

I own I was at a loss to account for your silence, and I was sorry at the cause. I can easily conceive how you must feel, for you have two passions to cope with, viz. that of being disappointed in love, and that of being defeated; but both will wear out, perhaps the first soonest. I own I was glad, when I heard that you was married to a woman of fortune; but let her go, never mind her. I shall employ you with hedge-hogs, for I do not know how far I may trust mine. I want you to get a hedge-hog in the beginning of winter and weigh him; put him in your garden, and let him have some leaves, hay, or straw, to cover himself with, which he will do: then weigh him in the spring, and see what he has lost. Secondly, I want you to kill one at the beginning of winter to see how fat he is, and another in the spring, to see what he has lost of his fat. Thirdly, when the weather is very cold, and about the month of January, I could wish you would make a hole in one of their bellies, and put the thermometer down into the pelvis, and see the height of the mercury; then turn it up towards the diaphragm and observe the heat there. So much at present for hedge-hogs. I beg pardon, examine the stomach and ntestines. If Hewson's things go cheap, I will purchase some that I think proper for you; those you mention I am afraid will be every body's money, and go dear.

Ever yours,

JOHN HUNTER.

London, Sept. 25th, 1778.

I am afraid that the hedge-hogs in this case had not quite the power ascribed to them by Mr. Hunter. For several years afterwards Jenner certainly continued to feel the mortification which he had experienced. Two of his letters to his friend Gardner, written in 1783, seem to refer to this subject, and paint very strongly the intensity of his feelings. In one he says, "I am jaded almost to death, my dear Gardner, by constant fatigue: that of the body I must endure; but how long I shall be able to bear that of the mind, I know not. Still the same dead weight hangs upon my heart. Would to God it would drag it from its unhappy mansion! then with what pleasure could I see an end of this silly dream of life."

Again, on the 8th of April of this year, he writes thus to the same friend: "As for myself, the same stream of unhappiness is still flowing in upon me; its source seems inexhaustible; but there is a soothing consolation in it; all little disquietudes are sunk or washed away. I feel their influence no more."

Though he thus occasionally indulged his melancholy feelings, and poured them into the bosom of a very kind and attached friend, it is a satisfaction to know that they did not materially interfere with his usefulness as a professional man, or retard his scientific pursuits. He alludes, in one of the preceding extracts, to the fatigue occasioned by his labours. An incident which occurred about this time, affords a pleasing proof of the extent of his reputation as a surgeon, and will easily account for the avidity with which his assistance was sought for in the surrounding district. A patient in the Infirmary at Gloucester was afflicted with a disorder which required immediate relief, by an operation of a delicate nature. Both the surgeons of the institution were so circumstanced as to be unable to lend their assistance. In this perplexity Mr. Jenner, of Berkeley, was sent for. He performed the operation, and saved the man's life.

To judge of an occurrence of this kind properly, it must be remembered that Berkeley is sixteen miles distant from Gloucester, and that surgeons were then less skilled than they are now; and that though at the present time there are few, or none who are not prepared to undertake the most hazardous operations, such knowledge, when Jenner was called upon from his native village, was chiefly confined to hospital surgeons.

His intercourse with Mr. Hunter was kept up as usual; and I introduce a few more letters written about this period, which, though they do not differ much from others that have been presented to the reader, will nevertheless, I trust, be found to possess sufficient interest to justify their insertion.

MR. HUNTER TO E. JENNER.

DEAR JENNER,

I received yours with the eel. The spawn of the salmon was lost. I shall send you back the eel again, with the liver, stomach, and gut removed; and nothing left but a fringe, which passes down the sides of the back-bone, which I took and still take to be the spawn; but I never saw any difference in it at any time of the year, and this one you have sent is similar to all I have yet seen. I think your stopping the eels is a good scheme, if you can; but I should suspect that they would be more slippery than hedge-hogs. I do not know if hedge-hogs burrow. About a month hence examine another, and compare him with your notes, and memory also. Examine his heat in the pelvis, diaphragm, &c.; a month after that, another, &c. I like your experiment on the toad and snake, but bury them rather deeper, and let the ground be kept moist about them, especially in the summer. I shall keep your letters, but I expect in the end all your notes. I like your friend Ludlow; he is a lively sensible fellow. I have got a few preparations for you; I am getting them put into a little order for you before I send them. Are there no bats in the old castle of Berkeley? I should like similar experiments to be made upon them to those of the hedge-hog. Mrs. H. desires her compliments to you. Believe me to be most sincerely,

Yours,

JOHN HUNTER.

London, Nov. 9th, 1778.

MR. HUNTER TO E. JENNER.

DEAR JENNER,

What are you doing? how do hedge-hogs go on? do they fall off in their weight? how cold are they in winter? &c. &c. Let me hear from you. I have not yet sent the preparations for you: I have added an eye to them of my own making.

Yours,

JOHN HUNTER.

London, Jan. 16*th*, 1779.

MR. HUNTER TO E. JENNER.

DEAR JENNER,

I thank you for the trouble you have taken. I do not see another experiment to be made with hedge-hogs but one: get a piece of meat into the stomach of one in the very cold weather, and kill him twenty-four hours after, to see if it is digested, which I have done with lizards. This may be difficult; but suppose he was made lively in a warm room, and then fed and put out into the cold immediately, with a little hay over him. If this does, two or three may be served in the same way, and kill them at different distances respecting time; observe their breathing when in the cold; if possible, the quickness of the pulse, and the fluidity of the blood. If you should chance to get more than you can use, I would take a few to put into my garden to walk about in the evenings.

Is there no chance to see you in London, this winter?—do come and see us. I shall send you a paper of mine upon the free martin; also one to Ludlow. I wrote to him in answer to his letter. I hope he received it. If a good deal of that air of the hog's guts could be collected, see if a candle will burn in it as large as in common air. I had a letter from Mr. Cheston of an ossified thoracic duct. I wish he would let me have it; you see how greedy I am. You will hear from me soon.

Ever yours,

JOHN HUNTER.

MR. HUNTER TO E. JENNER.

DEAR JENNER,

I have not troubled you with any letter this long time, nor have I heard from you. This moment I do not know if I sent you the butter-jugs; if they are not sent, they shall this week. I want you to pursue the experiments upon the heat of the hedge-hog this winter, and if you could send me a colony of them, I should be glad, as I have expended all I had, excepting two: one an eagle eat, and a ferret caught the other. Mrs. Hunter and I were at Bath the other day, and came home by the way of Gloucester; wished much we could have stayed a day to have waited upon you. Let me hear from you soon.

I am, dear Jenner,

Yours,

JOHN HUNTER.

London, Nov. 8th, 1779.

Many of the preceding letters refer to organic remains. The situation in which Jenner lived in the vale of Gloucester was particularly favourable to the study of these bodies: he was within reach of the principal oölitic series, and, as will presently appear, he was contiguous to another formation, which is closely allied to the former, and is not less rich in fossils.

Geologists have arranged the oölitic formations into three grand divisions, called upper, middle, and lower. Each is characterized by certain properties, and by peculiar organic remains. These remains consist of many extinct genera of oviparous quadrupeds, and various vertebral fishes—testacea, corolloid zoophytes, &c. &c. The oölitic formations divide the island from north-east, to south-west. They bound the vale of Gloucester, and constitute the Cotswold range of hills.

The Lias formation, on which reposes the whole series of oölitic hills, accompanies the latter across the island from the Humber to St. George's Channel in Dorsetshire.

Its range from its northern extremity to a few miles south of Gloucester is very regular, presenting an average breadth of about six miles. It is bounded on the south-west by the oölites, and on the north-west by the red marl. Its eastern extremity, after passing Gloucester, continues to accompany the oölitic ranges through Somersetshire, to the coast in Dorsetshire. Its western limit is more irregular. It feathers in and out among the coal fields which occur towards the estuary of the Severn, and the upper part of the Bristol Channel, in the counties of Gloucester, Somerset, Monmouth, and Glamorgan. The coal formation alluded to occupies three great basins. The edges of these basins consist of nearly vertical strata, composed of mountain lime-stone and old red sand-stone, and form bold and precipitous anges of hills. Among the valleys are deposited the horizontal strata of lias, with subjacent beds of red marl, and magnesian lime-stone. The manner in which the lias is cut through by the strata of mountain lime and old red sand-stone, above-mentioned, gives its outline in this district a very irregular character.

It is not necessary here to dwell with much greater minuteness on the geological peculiarities of this part of England. In order, however, to connect this subject with the pursuits of Dr. Jenner, it is proper to specify a little more minutely some of the formations in his neighbourhood. With this intention I refer to one of the coal basins already noticed. It occupies part of the south of Gloucester, and of the north of Somerset. It is bounded on the south by the Mendip hills, on the west by the range which forms the defile of the Avon, and on the north-east by a continuation of the same chain. All these chains exhibit inclined strata of mountain lime, and old red sand-stone. The lias and subjacent horizontal beds fill up the interior of this basin in the neighbourhood of Bath and Bristol, and throughout the Somersetshire collieries. On the north-west of the ridge which forms the edge of the coal basin, they are to be seen at Pyrton, at Aust, at the Hock-crib, and on other points along the banks of the Severn. Without pursuing further the various ramifications of the lias, I will merely mention that at the places last enumerated, Dr. Jenner spent a great deal of time in investigating the mineral contents, as well as the organic remains, with which this formation so remarkably abounds. When he began his labours in this part of natural history, geology was in its infancy, and the study of organic remains had scarcely been freed from the absurd conjectures which in the preceding century were adopted by the best naturalists. It is true, that it was not believed at that time that fossil shells were *lapides sui generis,* and that they were formed by a *vis formativa,* peculiarly inherent in certain parts of the earth. But all the important information which has been derived from the study of this subject, whether as it regards the geological relations of rocks that contain organic remains, their connection with the genera and species of animals and vegetables which now inhabit our globe, or the proofs of the existence at some remote period of organic bodies, no traces of which are now to be found in a recent state, was utterly unknown.

The remains of this description, which exist in the lias, are peculiarly interesting, and belong to a greater number of the higher order of animals than are to be found in almost any other formation. It contains two very remarkable extinct genera of oviparous quadrupeds. Besides other remains of this class, large quantities of the testaceous molluscæ, &c. are found. For further information on these subjects, the reader may consult Conybeare and Phillips' Outlines of Geology.

The trap rock in Mickle-wood, and the great variety in its structure, together with the changes produced in the organic remains found in the transition lime-stone, where it comes in contact with the trap, also particularly occupied the attention of Dr. Jenner. Specimens, illustrating these changes, were the ornaments of his study and of his garden.

A large collection of specimens from the oölite and the lias was also made by Jenner. He sent many of them to Mr. Hunter, and to his friend Dr. Parry at Bath, who sometimes came to Pyrton to ransack the treasures which are found in the lias at that place. There was another haunt on the opposite side of the river at Westbury, where Jenner was wont to carry on his pursuit. A section of this cliff is given by Mr. Conybeare to illustrate the lower beds of the lias formation. The following is the order of stratification. White lias, blue shale, black shale, green siliceous grit, containing *black bones*, and known here, and at Aust, by the name of the *bone bed;* black shale, green grit, greenish marl-stone, red marl-stone, of the new red sand-stone formation. A letter from Dr. Parry points to these favourite pursuits.

DR. PARRY TO E. JENNER.

Bath, October 20, 1781.

MY GOOD FRIEND,

I should have written to you before, had I not waited to announce to you the sending off of a small cargo of fossils and insects. But I have been altogether unable to procure any thing, or any number of things, from the neighbourhood, of the former kind, which could at all deserve your notice. Of the latter, I shall send you about a hundred specimens, as soon as I can get them properly disposed in a box, so as to bear the carriage. Have you made any additions to any of your collections? I wish it was it my power to visit you and Pyrton at a season when I could take advantage of the dashing tides. However, I must learn to be content with such things as I have.

* * * * *

CHAPTER III.

LIFE FROM 1783, TO THE PUBLICATION OF THE "INQUIRY."

MANY of the early letters of Mr. Hunter, as well as some of those that are to follow, regard the singular phenomena connected with the torpidity of animals. When they pass into this state, a material change is observable in all their most important functions. The temperature is diminished: the circulation of the blood becomes slower; while respiration, digestion, and the irritability and sensibility of the muscular and nervous system, are either suspended or much diminished. Mr. Hunter's experiments, as well as those which were performed by Dr. Jenner, throw considerable light on these points of physiology. It was found, for instance, that the temperature of the hedge-hog at the diaphragm, in summer, was 97° Fahrenheit, while the thermometer in the shade was 78°. When the temperature of the air fell to 44°, the animal became torpid and its temperature was reduced to 48° 5″. That of the air having fallen to 26°, the heat of the animal was found to be 30°. Facts of this kind satisfactorily prove that there is a great reduction of temperature in the animal during the state of torpidity. The reduction in the velocity of the pulse is, perhaps, still more remarkable; animals whose heart beats more than 100 strokes in the minute, having them lowered to 14 or 15; and in some the circulation is not at all to be discovered. It is, nevertheless, a remarkable fact that the blood, though apparently devoid of motion, never coagulates, nor is it prone to do so, even when the animal is exposed to a degree of cold lower than zero. Spallanzani proved this with regard to the marmot. He likewise proved that the blood drawn from the same animal was very readily frozen when exposed to a temperature higher than that of the lungs. It thus appears that there is even in this living death a principal of vitality which resists the ordinary laws of matter.

The function of respiration in some animals seems to cease altogether. They can exist in the exhausted receiver of an air-pump, or in a medium, which in a very short time would extinguish life if they were not in a state of torpidity. To ascertain whether digestion was carried on under such circumstances, both Mr. Hunter and Dr. Jenner introduced food into the stomach of lizards and hedge-hogs, as they were about to become torpid, and on subsequent examination it was found unchanged. As the ordinary internal stimuli do not affect torpid animals, so they are alike insensible to most of the external agencies. They may be wounded, thrown about with violence, and even be subjected to the penetrating influence of the electric fluid, and yet remain insensible; but what these powerful agents could not effect, may be accomplished by the vivifying energy of heat. It gradually unlocks the secret springs of life, and when the animal is, at the same time, exposed to light and air, all the suspended functions are speedily restored.

Besides carrying on experiments illustrative of these points, in conjunction with Mr. Hunter, he let no opportunity slip of prosecuting other branches of Natural History, and of procuring subjects for Mr. Hunter's Museum. Mr. Hunter was particularly desirous of obtaining a bustard; this is the largest of the British birds: it is likewise, the rarest. Two letters refer to subjects of this kind, as well as to Jenner's inquiries respecting the natural history of the cuckoo.

Mr. Hunter to E. Jenner.

Dear Jenner,

I am very much obliged to you for your attention to me. I will very readily give three guineas for the Bustard, therefore give such orders as you think fit. I request the whole history of cuckoos this summer from you. I have bought a house in Leicester Fields, and shall move this summer, when I shall be able to pick out some things for you. Give my compliments to Clinch, and I hope to see him before he sets out for Newfoundland; if I do not, let him think of the white hares, to tame a buck and doe, and send them to me. Let me know in your next what you are doing. I hope to see you in London about two years hence, when I shall be able to show you something.

I am, dear Jenner,

Ever yours,

JOHN HUNTER.

When the bustard arrives I will write to you.

MR. HUNTER TO E. JENNER.

DEAR JENNER,

I have received the bustard safe, as also the bones. Your friend Mr. Hazeland has been very kind, for which I have wrote to him and thanked him; but when you see him, or write to him, express the same to him, as an indirect thank is better than a thousand direct ones. Are hedge-hogs in great plenty? I should like to have a few. You must pursue the cuckoo this summer.

I am employed building, moving, &c. I wish this summer was well over. When I am fitted up I hope you will come and see me.

Ever yours,

April 22d. JOHN HUNTER.

Their correspondence sometimes took a different turn, and questions of an abstract nature were discussed between them. Whoever has read any of Mr. Hunter's works must have discerned, amid the plain and straightforward information which they contain, a considerable tendency to abstruse, or even metaphysical reasoning. This is the usual bias of minds gifted with powers like his; and had his early education enabled him to draw from collateral sciences such proofs and illustrations of his conceptions as they might have afforded, his genius, great as it was, would have acquired a solidity and an extension which would have raised him still higher as a medical philosopher.

I do not know on what occasion the subsequent letter was written. From the context I should infer that it was suggested by one of those interesting occurrences which are now and then met with, where an individual, who was born blind, has by some means been restored to sight, and is just beginning to acquire ideas of colours; or of an individual who laboured under some extraordinary peculiarity of vision; but, however the case may be, the letter is written with great power and energy.

MR. HUNTER TO E. JENNER.

DEAR JENNER,

I thank you for your last letter. I want you to pursue the inquiry considerably further; and to give you an idea of what I mean, I will first premise, that there are in nature but three colours, viz. *red*, *blue*, and *yellow*, all the others being combinations of these three.

First, present him with these three colours singly, and see what he calls them; then altogether, (not mixed) and see how far they correspond with his first ideas of them: when that is ascertained, then begin to mix them; for instance, a blue and yellow, (which makes a green,) see what he calls that; then a yellow and red, (which makes a scarlet,) next a blue and red, (which make a purple.)

Now to explain the intention of these experiments. Suppose he has a perfect idea only of one colour, and although you mix that colour ever so much, yet he sees none of the other, but only that colour in the mixture. Suppose all the three colours when seen singly, or unmixed, (with him) are blue; mix blue with red, (making a purple,) he will only see the blue, the red not being visible to him: and so on of the others, according as he sees them. Suppose that a simple colour makes no impression; but a compound one does, viz. green, (which is composed of blue and yellow:) then mix blue and red in all proportions, to see what the colour is. Then mix yellow and blue in all proportions, and see what colour these are: if he sees no green in any of them, then mix all the three colours in various proportions, and see what colours those make. When all the colours are mixed in various proportions, and the whole is a green, perhaps, of different shades, according to the quantity of blue and yellow, then you may fairly conclude, that it is the mixture of the blue and the yellow which produces it; the red never making any impression. If there is any other simple compound that he sees, as scarlet, which is yellow and the Modena red; or a purple, which is blue and red; see if when those two are predominent in the mixture, (although there are all the three colours in the mixture) that the compound becomes the visible colour.

The drawing of the scull has been made ever since you desired it; but I had forgot it. I have a cast for you of the aneurismal varix described by Dr. Hunter. How shall I send both? Let me know.

Ever your much obliged

J. HUNTER.

While Jenner was obeying the injunctions of Mr. Hunter, by performing the experiments which he pointed out to him, he continued to busy himself with ardour in forwarding useful knowledge on all professional subjects. We have already seen how wide a range his mind had taken, but it extended also to objects which are with difficulty followed in a situation such as his. Among these may be mentioned pharmaceutical chemistry. Some of the processes recommended for the preparation of several of the most powerful medicines, he found to be less perfect than could have been desired. This was particularly the case with the emetic tartar then commonly used. He had been often foiled and disappointed in practice, by the uncertainty of the action of this very useful medicine, arising from the imperfection of its preparation. He therefore instituted some experiments for the purpose of obtaining one more regular in its strength, and consequently more uniform in its operation. The process which he ultimately adopted for this purpose is contained in a letter addressed to Mr. Hunter, which was read before the Society for the Improvement of Medical Knowledge, and is published in the first volume of their Transactions. Much intercourse had, however, previously taken place between him and Mr. Hunter on the subject. The *facetiæ* of the following induce me to print it.

MR. HUNTER TO E. JENNER.

DEAR JENNER,

I am puffing off your tartar, as the tartar of all tartars, and have given it to several physicians to make trial, but have had no account yet of the success. Had you not better let a bookseller have it to sell, as Glass of Oxford did his magnesia? Let it be called Jenner's Tartar Emetic, or any body's else you please. If that mode would do, I will speak to some, viz. Newbery, &c. You are very sly, although you think I cannot see it: you very modestly ask for a thermometer; I will send one, but take care that those d——d clumsy fingers do not break it also. I should be glad to have a true and particular account of the cuckoo, and as far as possible under your own eye. To put all matters out of dispute, if the cuckoo's eggs were taken out of the hedge-sparrow's nest in which they were laid, and put into another by human hands, there could be no supposition that the parent cuckoos would feed, or take any care of them. I also want some young ones. I had a series from you, but a moth got in among them, and plucked them.

Let me hear from you when you can.

Yours,

J. HUNTER.

In the year 1783, Montgolfier attracted the attention of scientific men, by announcing the result of his first experiment in aerostation. The mind of Jenner was excited by the event, and he resolved to exhibit to his gazing neighbours in the vale an illustration of this new species of aerial navigation, as it was then called. He accordingly constructed a balloon: it was filled with hydrogen gas, according to the plan of M. Charles, in the hall of the castle at Berkeley. It rose majestically, and winged its flight over the hilly barrier which bounds the vale, and descended near a place called Symond's Hall.

It was filled again at Kingscote, and a second time let loose to take its flight "through fields of ether."—Affixed to it were the following lines by his friend Gardner.

LINES AFFIXED TO AN AIR-BALLOON.

Stranger, whoe'er thou art, whose gazing eye
Is fixed with wonder on this novel scene,
Ignoble on the ground behold me lie,
And kiss, indignant kiss, the level green.

From Chloe's hand launched forth in fields of air,
Swift as the bolt of Heav'n I took my flight;
Child of the wind, I flutter'd here and there,
Till clouds obscur'd me from the gazer's sight.

Long while held on my daring rapid course,
I travers'd worlds where eagles never flew,
With strengthen'd wing, and undiminish'd force,
Far from the keenest ken of mortal view.

But fate, alas! to check my tow'ring pride,
At length has laid me at thy feet thus low;
Let not thy pity be to me denied,
But on my fate one tender sigh bestow.

Art thou to mad ambition now a slave;
Or dost thou hope in higher walks to shine?
Tutor'd by me, thy dear contentment save,
Or prophesy thy future fate by mine.

If yet a youth, the moral lesson hear;
For, oh! believe thou canst not know too soon
A truth (which added years will make more clear)
"That vain ambition is—an Air-Balloon!"

Hurt not my form; 'twere sacrilege to wound
A form by Chloe's hand so sacred made;
Let not that cruel wretch on earth be found,
That dares, that impious dares, my sides invade.

My flight I took from Kingscote's happy plain,
A daring wand'rer through the ethereal sky;
Then, gentle friend, pray take me back again,
Perhaps, once more, another course to try.

The history of this invention affords one among the many proofs of the near approaches that may be made by different individuals to the truth, without reaching it. The possibility of floating heavy bodies in the atmosphere had been very well ascertained, but no one till the time of Montgolfier had hit upon the right method of doing it. Mr. Cavallo, of London, seems to have fully comprehended the principle, and nothing but the absence of one little precaution in conducting his experiments, prevented him from acquiring the credit of the discovery. He attempted to fill bags of paper with inflammable air, but he forgot that this substance was permeable to the gas. Had he employed varnished silk of sufficient dimensions, or any composition of a similar nature, he could not have failed of success.

In the commencement of the year 1786, Jenner's life was put to extreme hazard by exposure to a great degree of cold. I shall give the recital in his own words.

"January 3d, 1786.—I was under the necessity of going from hence (Berkeley) to Kingscote. The air felt more intensely cold than I ever remember to have experienced it. The ground was deeply covered with snow, and it blew quite a hurricane, accompanied with continual snow. Being well clothed, I did not find the cold make much impression upon me till I ascended the hills, and then I began to feel myself benumbed. There was no possibility of keeping the snow from driving under my hat, so that half my face and my neck was, for a long time, wrapt in ice. There was no retreating, and I had still two miles to go, the greatest part of the way over the highest downs in the country. As the sense of external cold increased, the heat about the stomach seemed to increase. I had the same sensation as if I had drank a considerable quantity of wine or brandy; and my spirits rose in proportion to this sensation.

"I felt, as if it were, like one intoxicated, and could not forbear singing, &c. My hands at last grew extremely painful, and this distressed my spirits in some degree. When I came to the house I was unable to dismount without assistance. I was almost senseless; but I had just recollection and power enough left to prevent the servants from bringing me to a fire. I was carried to the stable first, and from thence was gradually introduced to a warmer atmosphere. I could bear no greater heat than that of the stable for some time. Rubbing my hands in snow took off the pain very quickly. The parts which had been most benumbed, felt for some time afterwards as if they had been slightly burnt. My horse lost part of the cuticle and hair at the upper part of the neck, and also from his ears. I had not the least inclination to take wine, or any kind of refreshment."

"One man perished a few miles from Kingscote, at the same time, and from the same cause."

In the spring of 1787, he was in London. A letter to Sir Joseph Banks relates to a promise which he made at that time, and contains also an account of some experiments on the effect of animal manure on vegetation.

E. Jenner to Sir Joseph Banks.

Sir, *Berkeley*, 1787.

When I had the honour of waiting on you in London, in the spring, I promised to send you an account of the dog and fox; but the gentleman from whom I received it not sending it so soon as I expected, occasioned this long delay. His account is as follows:

"I could not before this day get such intelligence as could be relied on, respecting the dog-fox and terrier bitch, which I have taken the first opportunity of transmitting to you. The bitch did not seem very desirous of receiving the fox at his first approaching her; but after a little amorous dalliance she soon came to. They copulated three times in the course of the day, and each time continued together between ten minutes and a quarter of an hour. This happened sometime in the month of July. It did not appear in consequence of this union that the bitch showed any signs of pregnancy."

Notwithstanding the above, almost every sportsman asserts that foxes and dogs will produce an offspring. But I shall use every endeavour to set the matters clear, by experiments with these animals.

I recollect that I promised to send you an account of some experiments made on vegetables with animal manure. I wish they were more worthy your observation. A person engaged in business cannot conduct these matters as he would wish; his pursuits are too often interrupted. But though they do not go far enough to determine whether animal manure will produce lasting good effects on vegetables, they prove that a superabundance of this substance is destructive to vegetable life. I shall copy the notes as they stand on my journal.

February 10th, 1780.—A small quantity of the serum of human blood was poured over about a square foot of grass on a grass-plat. Three sprinklings were given at the distance of a fortnight each, and the whole of the quantity applied was the serum contained in forty ounces of blood.

April 1st.—The effects it has produced on the vegetation of the grass is astonishing. It is beautifully green and thick, and has sprung up several inches, while the surrounding grass has but just begun to shoot, and looks of a yellowish green.

May, 1781.—Some mustard seed was strewed over thin layers of wool, in three different tea saucers.

The wool in No. 1 was moistened with water.

No. 2, with the serum of blood.

No. 3, with the coagulated part of the blood mixed with the serum.

The seeds in No. 1 sprang up soon.

In No. 2 and 3, they swelled a little, but did not push out their radicle, grew mouldy and died.

This experiment was repeated with equal parts of serum and water. A few of the seeds just showed the radicle and then died.

It was again repeated with one part serum and two parts water; these shot and flourished very well.

A considerable quantity of blood mixed with a little wood ashes and powdered chalk, was applied round the roots of some polyanthus plants. The plants soon assumed a different appearance from their neighbours. The leaves were more luxuriant and green. But about the time when the flower-stems (which were uncommonly vigorous) were pushed up to half their usual height, they suddenly withered away and died.

April 21st, 1782.—Two young peach trees were manured with animal substances. About eight pounds, or perhaps more, as it was not weighed, were applied to the roots of each tree. They were in a sickly state, and had a very small number of blossoms. Two trees adjoining were very similar in appearance. The manured trees are distinguished thus, No. 1—2. The unmanured, No. 3—4.

May 30th.—From the unusual inclemency of the weather during the whole spring, the peach trees in general are greatly injured, and many of them appear to be destroyed.

Yet the manure above-mentioned has produced a wonderful effect.

No. 1 and 2 appear vigorous, and seem to have regained their health, while No. 3 and 4 look sickly, and have pushed out only weak and tender shoots.

With a view of ascertaining in some measure what quantity of animal substance might be applied to a plant with woody roots to its advantage or disadvantage, the following experiments were made.

April 20th, 1782.—I took four young currant trees of the same age, and nearly as possible of an equal growth and appearance. They were planted in large garden pots of an equal size, in the following different substances.

No. 1. was planted in the coagulated part of fresh blood, the surface of the pot only being covered with garden mould.

No. 2. was planted in equal parts of blood and common garden mould mixed together, and the surface being covered with plain mould.

No. 3. was planted in common garden mould, without a mixture of any other substance; but this plant will, from time to time, be moistened with the serum of blood, marking the quantity made use of each time.

No. 4. was planted in garden mould, and is to remain as a standard without the addition of any animal substance, to point out the difference of vegetation.

The four pots were placed under an east wall in the open air.

April 26.—A pint of serum was poured on No. 3.

May 3.—The serum in the same quantity applied again.

June 6.—No. 1. dead.
No. 2. nearly so.
No. 3. sickly—though it vegetates in a small degree.
No. 4. healthy.

July 20.—No. 3. recovered—shoots, and looks healthy.

By a letter from Mr. Blagden, I am informed my letter to Mr. Hunter on the Natural History of the cuckoo is ordered for publication in the Philosophical Transactions. I shall pursue the subject during the summer, and hope to have the honour of presenting you with another paper in the autumn, and also on the exciting cause of emigration in birds.

I am, Sir, &c.

E. Jenner.

The reply of Sir Joseph proves that at this time the paper on the cuckoo had been before the Royal Society, but it was returned to Jenner in order to enable him to record some additional facts which he had ascertained.

Sir Joseph Banks to E. Jenner.

Soho Square, July 7th, 1787.

Sir,

I beg you to accept my thanks for your obliging favour, respecting the fox and bitch, and the experiments on the effect of manure upon plants, which interest me a good deal.

In consequence of your having discovered that the young cuckoo, and not the parent bird, removes the eggs and young from the nest in which it is deposited, the council thought it best to give you a full scope for altering it as you shall choose. Another year we shall be glad to receive it again, and print it. Your other papers I hope you will proceed with when your leisure allows you opportunity.

Mr. Hunter has given us a most excellent paper on the genus of whales. I have no doubt that he will send you a copy, when it is printed, which will soon happen. That genus so difficult to naturalists, because so seldom to be met with, remained to Hunter as a rich field, which he was almost the first to cultivate, and indeed it has produced him a valuable harvest, and I have no doubt but that the migrations of birds will attend the same to you, as no author has yet treated it to advantage.

I am, Sir,
Your most obedient and
Very humble Servant,
JOSEPH BANKS.

The paper on the cuckoo was finally read on the 13th of March, 1788, and printed in the *Transactions* of that year.

MR. HUNTER TO E. JENNER, 1788.

DEAR JENNER,

I have been going to write to you some time past, but business and a very severe indisposition for these three weeks past has prevented me.

Your paper has been read, passed the council and is in print, for I had a proof sheet this day, and I have ordered 50 copies, 25 for you and 25 for myself, to give to friends. I spoke to both Sir Joseph Banks and Dr. Blagden about your wish: Sir Joseph has not the least objection, and will give us all his assistance; but he thinks the paper had better be first printed and delivered, and let the people rest a little upon it,—for he says there are many who can hardly believe it wholly: this will put off the certificate till the beginning of the next winter, when we shall hang you up. I have received a box with a Wapping landlady and two lizards. Mrs. Hunter's and my own compliments to Mrs. Jenner.

I am dear Sir,

Your most obedient servant,

JOHN HUNTER.

The result of the investigations respecting the cuckoo, begun at the request of Mr. Hunter, was presented through him to the Royal Society. As the *Transactions* are not in every one's hands, and the paper itself appears not to be very generally known, I think that some account of it may not be unacceptable in this place. A great deal of care was bestowed both in collecting the facts, and in reporting them with fidelity and accuracy. From one copy of the paper, which I have found in MS. it appears that a scrupulousness was observed in both these respects, which does the greatest honour to the observer.

The peculiar economy of the cuckoo in committing her offspring to the care of a foster-parent, had long excited the attention and wonder of ornithologists, and many reasons had, at different times, been assigned for the signal departure from the common impulses which prompt the feathered tribe to watch over the developement of their offspring and provide for their nourishment and support. The cuckoo in these points is peculiarly unnatural. She does not prepare a place where her young may be brought forth, nor does she regard them when they come into being. There is a double evil involved in her neglect; she is not only an improvident mother herself, but she renders others so likewise. When she furtively deposits her egg in the nest of another bird, it is done not that her offspring may be a sharer of the care of the foster-parent, but that it may engross it entirely to the total destruction of its own natural offspring. A perversion of all the maternal instincts is a most remarkable result of this vicarious incubation.

The hedge-sparrow, or other birds whose nests have been visited by the cuckoo, actually sometimes eject their own eggs to make room for the new guest: but it occasionally happens that this is not done; the eggs are not disturbed, and the process of hatching is allowed to go on regularly, and the young sparrows and the cuckoo emerge from the shell about the same time. This event, when it is permitted to happen, does not at all improve the condition of the former; on the contrary, it only exposes them to greater sufferings. The size of the egg of the cuckoo does not vary much from that of the bird in whose nest it is deposited When the young sparrow, therefore, and the intruder first come to life, they are pretty much on an equality; but unhappily for the foster-brethren, this equality does not last long: the cuckoo's growth rapidly outstrips that of his companions, and he immediately exercises his new powers with abundant selfishness and cruelty. By a singular configuration of his own body he contrives to lodge his companions, one by one, upon his back, and then scrambling up the sides of the nest, he suddenly throws them from their seat, and completely ejects them from their own home to become food for worms. There is reason to believe that the unnatural parent is often an unmoved witness of this atrocity. Her whole care and affection are absorbed by the intruder, and her own flesh and blood literally turned out to perish. It sometimes, though very rarely happens, that two cuckoos' eggs are deposited in the same nest. When this occurs, and they are both hatched together, a bitter feud soon arises. These animals, it seems, show no more regard to their own kindred than they do to the stranger. A battle of this kind is here described.

"June 27, 1787.—Two cuckoos and a hedge-sparrow were hatched in the same nest this morning; one hedge-sparrow's egg remained unhatched. In a few hours afterwards a contest began between the cuckoos for the possession of the nest, which continued undetermined till the next afternoon; when one of them, which was somewhat superior in size, turned out the other, together with the young hedge-sparrow, and the unhatched egg. This contest was very remarkable. The combatants alternately appeared to have the advantage, as each carried the other several times nearly to the top of the nest, and then sunk down again oppressed by the weight of its burden, till at length, after various efforts, the strongest prevailed, and was afterwards brought up by the hedge-sparrows."

The reasons of these strange doings next arrested Jenner's attention. How happened it, he inquired, that the cuckoo neither builds a nest, incubates, nor rears its young?

The Honourable Daines Barrington and Dr. Darwin have both mentioned an instance in which the cuckoo did perform these functions. The first was witnessed by the Rev. W. Stafford, in Blossopdale, Derbyshire; the second by Mr. Wilmot, at Morley, in the same county.

These examples seem to be well authenticated, but they certainly are extremely rare and still leave the questions, enumerated above, fit subjects of curiosity.

There is nothing in the structure of the cuckoo which incapacitates it either from building a nest, or incubating, or procuring food for its young. All naturalists previous to Jenner were inclined to ascribe the peculiarity in the economy of the cuckoo to causes of this kind. It was imagined that the large stomach, which is only protected by a thin covering, rendered the pressure attendant upon this office incompatible with its health. This theory was adopted without due examination. There are many birds that incubate, whose stomachs are quite as capacious and as little protected as that of the cuckoo; and moreover, the stomach of all nestlings, which is always much distended with food, is the point on which the weight of the body is mainly supported while it continues in the nest. It sometimes happens too that the young cuckoo, from an accidental circumstance, performs the office of a sitting bird. Jenner saw an example of this kind in 1786. He found a hedge-sparrow's nest containing a cuckoo, apparently about a fortnight old; on lifting up the bird, he observed under it two unhatched eggs. On breaking one it was found to contain a living fœtus. These eggs must have been laid some time after the cuckoo was hatched. It completely filled the nest, and by reason of this peculiar incident, it was, in fact, sitting on the eggs of the hedge-sparrow.

Facts, such as these just enumerated, satisfied Jenner that the causes usually assigned were not adequate to explain the peculiarities in the habits of the cuckoo. He therefore, with his characteristic modesty, proposes this solution. May they not, he observes, be owing to the following circumstances? namely, the short residence this bird is allowed to make in this country, where it is destined to propagate its species, and the call that nature has upon it, during that short residence, to produce a numerous progeny. The cuckoo's first appearance here is about the middle of April. Its egg is not ready for incubation till some weeks after its arrival. A fortnight is taken up by the sitting bird in hatching the egg. The young bird generally continues three weeks in the nest before it flies, and the foster-parents feed it more than five weeks after this period; so that if a cuckoo should be ready with an egg much sooner than the time pointed out, not a single nestling, even one of the oldest, would be fit to provide for itself, before its parent would be instinctively directed to seek a new residence, and be thus compelled to abandon its young; for the old cuckoos take their final leave of this country the first week in July.

Had they remained as long as other migrating birds, (such as the swift and the nightingale,) which produce, and rear a single set of young ones, foster-parents evidently would have been unnecessary. It is not known how many eggs the cuckoo generally lays, but that they do lay a great number is very probable, as appears by dissection of the ovaries.

There is no precise time for the disappearance of the young cuckoos. They seem to go off in succession, as soon as they are capable of taking care of themselves. But this is by no means so early as their growth and the fulness of their plumage would seem to indicate. Dr. Jenner has frequently seen a young one so large that the little fostering hedge-sparrow has found it necessary to perch itself upon its back on half extended wing, in order to obtain elevation enough to put the food into its mouth.

The same impulse which prompts the cuckoo to deposit her egg in the nest of another bird, directs the young one also to secure the entire possession of it to herself. There would be neither room nor food for any other inhabitant, and she is quite as rigorous in enforcing this law in regard to her own kindred, as she is in respect to those of a different species.

This very elegant and satisfactory paper, of which a brief account is now presented to the reader, has ever been looked upon as a specimen of accurate and successful investigation. It explained the habits of the cuckoo with much clearness, and has afforded to every subsequent naturalist a plain, convincing, and instructive account of a subject which, till that time, had been involved in the greatest obscurity.

His nephew Henry became his apprentice in 1783. Jenner was then in extensive practice, and the former tells me his situation was no sinecure. Besides attending to the numberless duties of a strictly professional nature he found in the other pursuits of his uncle many calls upon his time which he was obliged to answer. One of his occupations was to pay a daily morning visit to the nests which contained the young cuckoos. This generally required a ramble of four or five miles in the neighbourhood, and, although Henry had a most inquisitive mind, and was himself particularly given to the study of natural history, he often found his task somewhat tiresome, and would have been perfectly satisfied now and then to permit the hedge-sparrow and the cuckoos to adjust their affairs without any interference on his part. He had at last, however, the happiness to see the object of his journeys successfully accomplished. The reports which he daily brought to his uncle were duly examined, and their accuracy ascertained. The points of chief moment having been elucidated, the time came for arranging the observations and preparing the paper for the Royal Society, for which it was intended.

"Come, Henry, our work is coming to a close," he said, "I have now drawn up the paper, read it, and find all the faults you can." In such kind and familiar language did he encourage his pupil in his studies, and showed him that his humble efforts in promoting the object of the inquiry were justly felt, and duly acknowledged.

The cuckoos' nests were chiefly at Clapton, a small estate belonging to Mrs. Hooper, Jenner's aunt, where, as has been said, he spent much of his time when he was a boy. But he laid all his friends under contribution to him while carrying on his pursuits. Some he employed in adding articles to his museum; others in watching the habits of plants and animals when he himself could not attend.

It was chiefly at Sannighar, a property belonging to Lord Berkeley, but in the occupation of Mr. Pearce, that he carried on his inquiries respecting the migration of birds. There he marked the swallows and the swifts, and Mr. Pearce watched for their re-appearance and communicated with Jenner.

On the sixth of March, 1788, an event of great moment to him took place. He was on that day united in marriage to Miss Catharine Kingscote, a lady on whom his affections had been long fixed, and in whose counsel and sympathy he found his surest solace in many of the most trying scenes of his future life. She was elegant in her manners, accomplished in her mind, and possessed an understanding of great vigour. She had been an invalid for a considerable time before her marriage, and she never at any time, after her early years, enjoyed robust health. The family of the Kingscotes is one of the most ancient and respectable in this county; and it has received additional distinction from their personal merits as well as from their alliances with eminent individuals. Anthony Kingscote, Esq. was a kinsman of the great Sir Matthew Hale, and became his guardian after the death of his parents; and the affinity contracted by the union with Jenner will not reflect less lustre on their name.

His eldest son Edward was born on the 24th of January, 1789.* This is alluded to in a letter which I subjoin as a specimen of his style of correspondence with one of his earliest and best beloved friends.

It is long; but it contains so much interesting domestic intelligence, and exhibits such a picture of the kindness and benevolence of the writer's mind, that I will not withhold from others what has afforded me so much gratification in the perusal.

* Mr. John Hunter was this child's godfather.

E. Jenner to the Rev. John Clinch, Newfoundland.

Berkeley, 7th February, 1789.

My Dearest Friend,

Your rebuke on account of my not answering the long and affectionate letter I received from you in the autumn is far milder than I deserve; you whip me with a feather when the severest lash might with all justice be applied as a punishment due to me. Not a single excuse have I to offer, and I must beg you to accept my best thanks for your extreme lenity. It is a maxim universally admitted that of all the ill habits a man may fall into, indolence is the most difficult to get rid of. I for one am a sad example of the truth of this position, and this very sin, (for I will not call it by a milder name) has got me into more scrapes than all the rest put together. You, my dear Clinch, may bless your stars that you have not an atom of it in your composition. It was this spiteful demon that bid me defer, and defer it till every ship was sailed that could give one a chance of conveying a letter: a thousand and a thousand times my heart smote me for it, and I seem at this time as if I never should do the like again; yet conscious that I am still under the dominion of indolence, it would be imprudent in me to make a rash vow. I should not think of troubling you with a preface of such a length, but it is my intention to let the letter bear an equal proportion. You will find it disjointed and *unconcatenated*, containing a variety of subjects; and you must not be surprised to find me pass hastily from one to another, as I shall at my leisure take up the paper and set down matters as they occur to my recollection. As it is uppermost in my thoughts I must in the first place tell you that I have a son. My dear Catharine has lain in about a fortnight. The child, though small, appears to be remarkably healthy; and I ardently hope there may as much affection subsist between the young Edward Jenners as between their fathers: there will be no room for more. Do not you think one of these days to send him to England for his education? I should be happy in taking him entirely under my own inspection, and would do by him exactly the same as if he were my own. I hear by the captain, you and Mrs. Clinch keep him under no kind of restraint, but indulge him with a full gratification of all his wants and wishes. Let me intreat you both to take care what you are about:—remember the path of life is full of *thorns*, and if you keep him upon *velvet* till the day arrives when he must begin to feel their points, think how much more poignant must be his feelings. On this subject one might enlarge, but what I have said I trust will be sufficient to awaken every proper sentiment. I shall only just add that severity to children I utterly abhor, and my observation leads me to consider it as being more injurious to them than the contrary extreme; but either one or the other is so hurtful to the mind that we feel its baneful effects to the very latest period of our lives.

I sincerely lament your having been again beset by that dreadful fever the typhus, and as sincerely congratulate you on your recovery. There is no laying down, I believe, any general mode of treatment in this disease. A man must be guided by his own genius; indeed, without a good portion of this, a physician must ever cut a poor figure; and if he should be a man of fine feelings, he must often be subject to unpleasant sensations within himself. Something new is for ever presenting itself—neither books, lectures, nor the longest experience are sufficient to store his mind with the undescribable something a man of our profession should possess. It not unfrequently happens that at the onset of typhus, the patient becomes affected with extreme nausea, and oftentimes vomiting; this was my own case in the dreadful fever I suffered in the year —86. Taking up the common idea that nature was wishing to throw something off the stomach I took emetic tartar, and increased by this means the malady; for I did not cease vomiting ten minutes for several days. Last April, after taking cold, I felt exactly the same symptoms coming on; the same confusion in my head, and the same distressful prostration of strength. Being pre-determined, if ever the like should happen, to treat it in a very different way, instead of increasing the nausea, I took something immediately to give a check to it. First, I swallowed a glass of wine, and at once felt all the symptoms give way. I then got on horse-back, but before I had gone two miles the effects of the wine were at an end, and I was forced to support myself by leaning on my horse's neck. I then got a glass of brandy, and in five minutes was quite well; my head, stomach, and muscles, all acting or ready for acting very properly. I had five miles to ride home, and by the time I could reach it, the debility was as great as ever. I constantly observed that my headache, sickness, and weakened muscular action, all came on and went off (after the use of the cordials) at the same time. I went at once to bed, and took a pretty large dose of laudanum with some *spiritus vitrioli dulcis:* this, like the brandy, gave immediate ease. I went to sleep, and awoke quite well. I am particular in reciting this case to you, knowing that your constitution subjects you to fevers of this kind; but I am well aware that the attack is so different in different habits that the mode of treatment here mentioned cannot be *generally* adopted. I wish the term *putrid* applied to this fever could be discarded, for I really think it leads the practitioner into errors. I do not recollect ever seeing a case that arose from the vapour of putrid animal substances, nor do I think that putrefaction, in the worst of cases, ever takes place before death, either in the fluids or solids.—If mortifications come on in consequence of too languid a circulation, or any thing else, these may be the effect and not the cause. The giving antiseptics, as they are called, with a view of preventing one's carcase from becoming putrid while alive, is too ridiculous. The worst of it is, that these fevers, at first, often commence with symptoms evidently inflammatory, and for a long time before extreme debility comes on, the disease appears to be of a mixed nature. It is at this period that I find *spiritus vitrioli dulcis* so useful; that is, when on the one hand the antimonial plan cannot be admitted, and, on the other hand, the bark plan. And I may add, that I have seen good effects arise from the use of antimony and *spiritus vitrioli dulcis* given together at a time when inflammatory symptoms seem to preclude the use of the latter given by itself.

George has at length left us, to take leave of his friends elsewhere before he departs for your snowy shores. Your offer was in every respect so liberal that it would have been unjust in me to have said any thing to have damped his ardour for catching at so good an opportunity of improving his fortune. It was only in my power to improve his education; and the progress he made during the time he staid here was extremely rapid. You will find him a youth of extraordinary talents: his penetration and discernment not limited to his profession only, but general. As a medical character we shall one day or another see him shine;—as a natural historian he already ranks extremely high; and, as he is extremely industrious and active, he will doubtless bring to light many things that have lain hid in darkness about your island from the beginning of time. During his absence, Henry and myself must fag on as well as we can. I am sorry to say Henry's health is not in so good a state as I could wish it; nor do I know what his complaint arises from. But before I say any thing of Henry let me just tell you that what I have said of George is according to my judgment, the just outline of his character, and not an eulogy made up of falsehood. Henry is much as you left him, the same simple, inoffensive lad, and indeed, I think as boyish almost as ever; and though his mind is stored with ideas that do him the greatest credit, yet his general appearance and manner is so very *fifteenish*, that a poor mortal on the bed of sickness will hardly look up to him with that eye of confidence and hope that might safely be placed in him. For it is by appearances, my dear friend, not from a real knowledge of things, that the world (at least the major part of them) form a judgment. A look of significance, a peculiar habit, and a very scanty acquaintance with the human machine, will make a man pass current for a great physician. This you and I know to be an unfortunate fact; while, without these auxiliaries, a man with the knowledge of the *hall* and *college* concentrated will be looked upon as a mere pretender. 'Tis almost midnight; and from the confused manner in which I write, it is certainly time to give my thoughts a respite till "Nature's great restorer sleep" shall give them fresh vigour.

To return, How is your head? I fear but so so. I am still of opinion that the complaint you have so long experienced arises from the blood-vessels, carrying blood to the head, overacting their part; you should certainly try every means to lessen this morbid determination. Your feet and legs should be kept as warm as conveniently may be, for when the vessels of the head are too full, the vessels of other parts are generally too empty; and this occasions coldness. I don't see how heat can arise from any other cause than increased flow of blood upon a part. For example, the degree of heat in the highest state of inflammation is never more than the heat of the blood about the heart. This I have repeatedly ascertained by experiment. I fear you are too free with the use of tobacco; can't you have resolution enough to give it up? Spirits, even diluted as you drink them, I think, are bad; and so are liquors containing much fixed air, and consequently spruce beer. 'Tis hard to say what you can substitute if you relinquish those things; but you should make a thousand *innocent* experiments on your constitution, and see if this distressing complaint is owing to any error in your mode of living. How far the state of your stomach is connected with it should be studiously made out. You know it has long been my creed that *stomach* is the governor of the whole machine, the mind as well as the body. The seat of action is certainly in the brain; but the stomach gives the word of command, and tells it how it shall act. I have taken up a notion, that every one of the large arteries may, in consequence of the stomach being affected in a peculiar way, take upon itself an action quite independent of the heart, except in its number of pulsations. This subject is too wide for the confined plan of a letter; and I shall therefore touch it very briefly. The blood-vessels are, you know, all connected with nerves; and consequently, irritable. You will think it odd in me to assert such a thing; but I really believe that most of the diseases human nature is liable to arise from these nerves being affected, and consequently subjecting the blood-vessels to a diseased action. The skin may be affected by certain applications, and these may produce the same effects as if they were applied to the stomach: cold, for example. I mention this because I have called the stomach the ruling principle: but the skin and stomach have a peculiar connection one with the other. If you will turn your thoughts to the subject a multitude of illustrations will arise, to make you sensible of the truth of the observation respecting diseases arising from improper action in the blood-vessels; or, to simplify the idea, from too large a quantity being thrown upon one part, while another part is deprived of its due quantity. Don't you find that when one part is too hot (remember I have said that heat depends entirely upon an increased quantity of blood in a part) that another part is too cold? When the face is flushed with heat, the feet and hands are generally freezing with cold; and what does this flushing arise from but from the arteries of the face feeling a stimulus? and this quite independent of the heart itself being acted upon; for we often find that while this appearance exists, the heart is beating slowly. I have selected this appearance because it is one of the most familiar.

But in a word, what disease is there, either acute or chronic, that does not arise from these irregular *dashes* (if I may use the expression) of blood upon different parts of the body? This question I shall leave you to answer; and at the same time quit the subject, which I fear is here spoken of with too little perspicuity for you to comprehend.

One little hint arising from it I must not omit; and that is, if your feet are in the least disposed to be cold, that you use a tepid bath (about 100 degrees) for ten minutes every night. Plunge in both legs as well as your feet. This reminds me of the thermometer, which I hope you have received long ago; for though I did not write to you I was not so bad as not to write to London about it: this I did immediately on receiving your letter.

Mrs. Jenner begs Mrs. Clinch to accept her thanks for her civilities, to which I beg leave to add mine. The cranberries are, as usual, excellent and I paint my mouth with them frequently. But, alas! the Parson's fish! George will tell you all about it. I have been week after week waiting to go to town, where I thought to have picked up something worth sending to you; but I now begin to fear that the ship will sail ere I shall be able to go. But you know you are too often in my thoughts to be forgotten. Pray send me the catalogue of your books that I may know what medical works may at any time be useful to you. But must this monstrous piece of water for ever stand betwixt us? are we to have no intercourse but with pens and ink? Among your numerous callings is there not one that would enable you to find ample succour for your family, breathing at the same time the balmy air of your native Isle?

With an earnest request that you would seriously ponder over these questions I shall conclude this long epistle: I don't urge your speedily relinquishing the situation you are placed in; but only to have it in contemplation to be, one day or another, my companion as well as friend.

Till this hour arrives, and to the latest period of your existence, may every earthly comfort be your portion. No one more ardently wishes happiness to you and yours, than your ever

Faithful and affectionate friend,

EDWARD JENNER.

To the Rev. John Clinch,
Newfoundland.

The gentleman to whom this was addressed had been Jenner's schoolfellow: they afterwards studied together under Mr. Hunter's eye, and contracted for each other the warmest and most sincere friendship. Clinch, being destined for a distant colony, took orders as a clergyman, and when he settled at Trinity in Newfoundland, he combined the practice of medicine with the exercise of his sacred functions.

A pleasing account of Jenner's domestic happiness, and of his contentment in his sequestered situation, is found in a communication to his friend Gardner. To explain the latter part of it, I may remark that Gardner, during the early portion of his life was a wine and spirit merchant; and Jenner's cellar was usually supplied from his store.

E. Jenner to E. Gardner.

Dear Gardner,

When I consider what miserable weather we have had for some time past, I can readily forgive your not coming to see me. You know I sat half an hour with your father to-day in his closet. It was (if I mistake not) the third time of my calling at your house without the pleasure of finding you at home. I thought your father seemed to be quite forsaken by his usual vivacity, and that the infirmities of our nature were beginning to squeeze him too hard.

My place of residence, though unfinished, is extremely comfortable; and I can with truth assure you the last year of my life, dating it from the month of March, has been the happiest beyond all comparison I ever experienced: and

I will take upon me to aver (nay I would swear it) that if you could be lucky enough to connect yourself with a woman of such a disposition as kind fortune has, at last, given to me, you will find a vast addition to your stock of happiness.

You have rallied me lately on a change of sentiment with regard to a rural life. There were no fair grounds for it, for I feel my mind as *cottagish* as ever; and when you see my habitation finished, you will allow me to be quite as retired and countryfied as yourself. Whether a man sits on a sofa or a bench, dines off a plate or a trencher, it does not, according to my ideas, make much difference. Still I think his mind may delight in the very opposite to that which mankind in general run after.

My chain of philosophy is hastily broken by the intrusion of a busy thought; and indeed it is rather lucky, as I took up my pen for that purpose. It was to tell you that my brandy is out, and my cellar getting dry. I must still trespass upon your patience in sending me such small quantities of malt, but as yet I have no materials for brewing put up. Send the brandy according to quality. If it is *apt* (as we say) send a large hamper; if not, send some for present use only. Where is my bill? If you don't send it soon, it may be worse for you. I expect to go through your turnpike to-morrow, where I shall probably drop this scrawl; for I fear my time will be so taken up that I can't see you.

Mrs. J. jogs my elbow to say she wishes to hear a duet; and to give her compliments to you.

Adieu! Yours ever very sincerely,

E. JENNER.

The correspondence with Mr. Hunter was maintained, but the declining health of that gentleman rendered his communications less frequent. One of them regards a curious pathological fact, which Jenner had observed concerning the dislodgement of hydatids from the kidneys. Dr. Jenner continued, notwithstanding Mr. Hunter's scepticism, to believe that the oil of turpentine was the agent of their removal.

MR. HUNTER TO E. JENNER.

DEAR JENNER,

I have just received the favour of yours. I have just now forgot the case of the hydatids; but if there was any thing that struck me, I dare say it was laid by. They are frequently in the kidneys, but I should doubt your oil of turpentine having any merit in bringing them away. My reason for supposing them animals is because they move after they have been extracted. I have taken them out of the head or brain of a sheep, and they have contracted in different parts of them when put into warm water. I should be glad to employ you if I knew in what; but if any thing comes across my imagination, I will think of you. The measly pork are hydatids.

I am afraid of your friend Mrs. L. There is a hard tumour that almost fills the pelvis, most probably the uterus. How does Mrs. Jenner do? do you bring her to London? What family have you got? My compliments to Mrs. Jenner, and believe me to be, dear Sir,

Your most obedient, and
Most humble servant,

December 8th, 1790. JOHN HUNTER.

This letter is interesting on two accounts. It alludes to the care with which all documents of any moment were 'laid by,' and preserved by Mr. Hunter. It was his habit to have such portions of the communications of his correspondents transcribed as contained any new information. His collections of this kind must, therefore, have been highly interesting and very voluminous, but I have some reason to fear that they have all perished. The other point to which I refer, concerns Dr. Jenner's inquiries on the subject of hydatids. The natural history of these bodies, and their influence in producing disorganizations of the most formidable kind, began about this period to attract much of his attention. It led him likewise to investigate with great care the healthy functions and diseased conditions of the lymphatic system. To gain information on these points, he had recourse to comparative anatomy and pathology; and derived from the inferior animals most satisfactory evidence of the accuracy of the opinions he had been led to adopt.

The important illustrations of the origin and progress of diseases to be drawn from these sources continued to occupy his mind at different intervals throughout all the remaining years of his life. Some of his early opinions were communicated to Dr. Beddoes, in order to explain the formation of tubercles in the lungs: and were published by that physician in his work on factitious airs. At a subsequent period they became the subject of much intercourse and correspondence between Dr. Jenner and myself. This is not the place to dwell at large on the doctrines connected with it His opinions will be found in two works which I published in the years 1819 and 1822. Some of the views disclosed in these works were taken by the best physiologists and pathologists of former days, but in consequence of the overwhelming interest which was attached to other parts of the animal economy, they were permitted to fall into neglect. They will doubtless, however, revive, and Dr. Jenner's claims as a pathologist will be admitted, for though he was not the *first*, he certainly was an *original* inquirer in this field ; he having been quite unacquainted with the writings of which I speak. Like his great master, Hunter, he was more given to trust to observation than to literary research. Besides, the publications were in some measure of a fugitive nature, consisting chiefly of short dissertations: and had not Haller deemed them worthy of a place in his collections, they would probably have sunk into oblivion.

The successful experiments of Ruysch on the sanguiferous system led physiologists to pay almost exclusive attention to that part of the animal economy. Thus the excitement which had arisen from the discovery of the lymphatics was allowed to subside, and all the functions of the body, whether in a healthy or diseased state, were entirely ascribed to some modification of sanguiferous action, to the almost total exclusion of other agents. At a subsequent period Mr. Hunter's own writings contributed not a little to produce a similar result. It is, nevertheless, certain that he attempted to guard against an error of this kind, though his followers did not always perceive it, and have employed the authority of his name to support opinions which he has explicitly disowned.

The origin of the many changes of structure that affect organized beings, has ever been a subject of interest, because it, in truth, includes many of the most serious "ills that the flesh is heir to." Tuberculous disorganizations, in one shape or other, destroy an immense proportion of the inhabitants of these islands: and the number of kindred changes of structure that appear in the shape of tumours of various denominations, producing either deformity or death, invest this branch of pathology with a degree of importance second to none. This was the opinion of Jenner, and he laboured with unceasing perseverance to throw light upon the subject, earnestly hoping that accurate knowledge on the origin and progress of this class of diseases would lead to more successful means of treating them.

It has been the custom to ascribe them all to inflammatory action. There is great reason to fear that this general term has been too indiscriminately used, and that the ideas attached to it are very far from being precise. The opinion seems to have arisen from examining changes of structure in an advanced period, when the original distinction of parts is lost, instead of watching them throughout their progressive stages, when the real condition of the structures can be traced. Errors of this kind can scarcely be avoided if we confine our researches to human pathology. There we only see the disease in its last stage, and, it may be, invert the order of the phenomena, putting down for a cause what possibly was a consequence.

This conviction induced Dr. Jenner to seek information elsewhere. He examined very carefully the diseases of the inferior animals, and was enabled to prove that the usual course assigned to them was by no means correct. He found, especially in that class of diseases of which we are speaking, that the primary state did not denote any thing akin to inflammation; that the changes were in the first instance obviously connected not so much with the blood-vessels, as with another part of the system.

The Hydatids, of which naturalists now enumerate several species, afforded to him the means of explaining many morbid phenomena. These bodies are generated in almost every texture, and are themselves liable to changes which totally alter their original character, and give rise to diseases most formidable in appearance and fatal in their nature. By far the greater number of encysted tumours, whether found within the abdomen or elsewhere, have had an origin of this kind. The demonstration of this fact can only be obtained by tracing the progress of these bodies in different animals through all their stages. The information derived from examinations of this kind casts a bright and steady light over some of the most obscure parts of human pathology; and has enabled us to explain the formation and growth of morbid masses, which could not be accounted for, rationally, in any other way.

The kindly intercourse which had so long subsisted between Jenner and Mr. Hunter was now soon to cease for ever; and I believe not more than three or four letters were interchanged subsequently to this time. One of them playfully upbraids Jenner for not coming to town. The reason of it is detected in another.—

MR. HUNTER TO E. JENNER.

DEAR JENNER,

What are your about? I have not heard of you, nor from you for this long time. You must certainly be about some mischief that keeps you so quiet. Let me know what you are doing, or else I will blow you, and have you brought to town as a criminal.

Yours, &c.

JOHN HUNTER.

MR. HUNTER TO E. JENNER.

DEAR JENNER,

I was in hopes of having seen you in London long before now; but I have been informed that Mrs. Jenner has been extremely ill. When you come I hope you will bring her with you: we will take all the care we can of her.

Now that the hedge-hogs are gone to sleep, I could wish you would get some of them for me; and put them in a box with very loose and coarse straw, so that they might have air, and not tumble about on each other. I want to open a few. When you catch them, do not put them into a warm place to make them lively: the more stupid the better they will carry and live.

Ever yours,

JOHN HUNTER.

December 10*th*, 1791.

Mr. Hunter's last letter to Jenner was dated on the 12th of August, 1793, a little more than two months before he suddenly expired in St. George's hospital. As it refers entirely to a private consultation, I forbear to print it; and shall only observe, that it is written with his usual power and vigour.

It will easily be believed that Jenner was deeply affected by the loss of such a friend, although his knowledge of the disease under which he believed Mr. Hunter to have long laboured had too well prepared him for such an event. An account of the appearances found on examining the body was transmitted by Sir Everard Home to Dr. Jenner.

SIR EVERARD HOME TO DR. JENNER.

Leicester Square, February 18*th*, 1794.

MY DEAR SIR,

I have sent you by the Major the numbers due to you from the Royal Society. I am well assured that you were sincerely afflicted at the death of your old and most valuable friend, whose death, although we all looked for it, was more sudden than could have been imagined. It is singular that the circumstance you mentioned to me, and was always afraid to touch upon with Mr. Hunter, should have been a particular part of his own complaints, as the coronary arteries of the heart were considerably ossified.

As I am about to publish a life of Mr. Hunter, which will contain both the symptoms of the disease, and the dissection, I shall not say more about it at present; it will be prefixed to the work on inflammation, and we hope to have it printed by the end of next month.

I cannot say that I have met with the ossification of these arteries so frequently as other alterations of structure in the heart, but this case is very much in favour of your theory.

* * * * * * *

The fatigues of general practice having become irksome to Jenner he resolved to abandon one branch of it, and to confine himself to medicine. With that view he obtained in 1792 a degree of Doctor of Physic from St. Andrew's. I cordially wish that all the honours of that ancient University had been as well bestowed.

Towards the conclusion of 1794, he was attacked with typhus fever, which nearly proved fatal. He was visited by Dr. Parry from Bath, and Dr. Hicks from Bristol; but his constant attendant was his nephew George C. Jenner.

He has himself left a very striking picture of his own sufferings, as well as those of his family, at this time.

Doctor Jenner to W. F. Shrapnell, Esq.

Dear Shrapnell,

Having no further use for the small remnant of brain the grim-visaged typhus has left me than to employ it in this way among my friends, it would be highly improper in me to withhold an answer to your obliging favour, for a single post.

Indeed, my good friend, I have been a terrible sufferer since you were at Berkeley, at which time I never experienced such health, my stomach and skin being in such a state from late autumnal bathing, that I scarcely felt the cold you were all complaining of. But who in the front of the battle shall say his breast shall not be pierced by the sword of the enemy? I little thought there was so much malignancy in the disease that cut off the poor dear boy; as it was clearly generated in his constitution, and not communicated.

You shall hear the history of our calamities. First fell Henry's wife and sister. From the early use of bark, they both appeared to recover; but the former, after going about her ordinary business for some days, had a dreadful relapse, which nearly destroyed her. It was during my attendance on this case that the venomed arrow wounded me: I felt it in the form of a nausea, as if I had eaten some indigestible substance, for several days, without the system being in the least affected; and had it not been for a dreary, wearisome ride over mountains of ice, without being able to come at succour, no mischief might have ensued; for I have long been assured that infection may be received and again expelled from the constitution, if no debilitating power should arise to call it into action. Like

Mrs. Jenner's fever, at an early period there was a clear intermission for four days; but, doctor-like, not a single grain of bark, or any thing else, did I take during this period. On the 8th day after the first seizure it again set in, in good earnest, and continued one-and-twenty days! Great were the efforts of those who kindly and humanely attended me. Dr. Parry was with me from Bath five times, Dr. Hicks and Dr. Ludlow as many, and my friend George was never absent from my bed-side. Thank Heaven! these efforts were crowned with success, and I am now enjoying that happy state, an uninterrupted convalescence; yet still so feeble that I think a great matter has been achieved when I quit my bed for an hour or two, and hobble across my chamber. My medicine is the Cascarilla bark; all other barks disordering me much. I sleep well, and my appetite is good, too good! for what is human reason, when opposed to the fury of an empty stomach? But to return to that mansion of melancholy, Henry's. His infant girl has now the fever; a servant maid in the house is dying with it; and, to complete this tragical narrative, about five days ago fell poor Henry himself. His symptoms, at present, are such as one might expect; violent pain in the head, vertigo, debility, transient shiverings, &c.

George is the only person among us that keeps his legs. His report of his brother is rather pleasant this evening; his pulse being sunk from 125 to 100. The stench from the poor girl is so great as to fill the house with putrid vapour; and I shall remove him this morning by means of a sedan-chair to a cottage near my own house. Happier hours are, I hope, on the wing for us; and that you will, ere long, receive from me more agreeable communications. Poor Cooks! I sincerely lament his fate. You have lost such a friend as will not be easily found again. Do you recollect a pithy line in Young on the subject of making friends?

"But friends how mortal! dangerous the desire!"

I heartily wish you well through the ensuing campaign. If your swords are ever to be unsheathed, it will be now. H—— is, and ever was, a vain hypocritical——. I have lived, Sir, in this wicked world, man and boy, these five-and-forty years, and no act of mine during that space ever pleased me more than that of driving the monster from this place; but alas! the immoralities he inculcated have taken too deep a root for even the arm of an Hercules ever to pluck up. To dissipate sadness, and fill up my leisure hours, I have been collecting my scattered thoughts on the subject of the migration of birds, having long promised a paper to the Royal Society. Can you furnish me with any facts respecting this matter that I do not possess? Make my respects to Mrs. Shrapnell, Colonel Wall, and any officer of the battalion I have the honour of being acquainted with, and

Believe me yours, most sincerely,

E. JENNER.

The effects of this illness continued for a considerable time. The feelings of the neighbourhood may be gathered from some expressions of one of his old associates, the Rev. Nathaniel Thornbury, of Avening, a village about fifteen miles distant from Berkeley. He was a learned and somewhat eccentric man, fond of natural history, and a shrewd observer. He makes mention of a fact which Jenner has employed in elucidating his doctrine of the cause of the migration of birds.

THE REV. NATHANIEL THORNBURY, TO DR. JENNER.

Avening, March 22, 1795.

DEAR SIR,

I sincerely congratulate you on your recovery, and let me add, the surrounding neighbourhood also. When such an individual as you migrates to a better planet, (for surely this, as friend Yorick says, is a very scurvy and disastrous one,) the languishing invalid, whose existence you have lengthened, or whose pangs you have alleviated, drops a mournful tear.

Nonentities, like myself, steal from the world, and not a stone tells where we lie. Horace's "in seipso totus teres atque rotundus," is a brief compendium of my present system, which I hereby reduce to the small compass of a ring posy. As I find it improves my aversion to the race to see the yahoos now and then in large groups, I am at this moment setting out for the great Hyrcanian forest, namely Town, to speculate and growl for a few days, and then return to my own den.

I write all this in pure simplicity of heart, well knowing your discretion, as being also one of us, a cynick *in cuore* with the *il volto sciolto.*

The thieves were Dutch ones. I have in your last week's journal been pushing quart and tierce with one of your English ones, for "he that robs me," &c.—not but that the yahoo had literally robbed me by conveying away at different times certain ratios of hay, in which he was detected. The anecdote you inquire after is briefly as follows. Scheveling is a fishing village, on the coast of Holland, about two miles from the Hague. An individual of the place, on whose veracity I think I can rely, informed me that he observed his pigeons, in a morning when the weather was fine, stretch across the ocean in a direction due west, and return in the evening. In the craws or crops of several, on killing them, vetches were found. As to their picking up vetches in Holland, the thing is wholly impossible. Quia, because there are none to pick up. The opposite coasts indeed of Norfolk and Suffolk afford a plentiful supply. You will make the necessary inferences, and conclude accordingly; but, no conclusion you can possibly make, were you to range the whole field of Scibile and Ens, can be more undoubted and certain, than the conclusion of the present, that

I am, dear Sir,

Very truly, very sincerely, yours,

NATHANIEL THORNBURY.

I am not aware of any incident worth recording that befel him or his family, till towards the end of the year 1795, when a fleet of transports under Admiral Christian, destined for the West Indies, was wrecked on the coast near Weymouth. Dr. Jenner's nephew, Stephen, was on board the Catherine. This young officer was on the staff of General Abercromby. A most lively, yet appalling account of this disaster is given in two letters by Mr. Shrapnell, who on the melancholy occasion exerted himself in collecting the stiffened corpses of many of his friends and companions, that lay scattered in terrific confusion along the beach; but he shall speak for himself.

W. F. Shrapnell, Esq. to Dr. Jenner.

Weymouth, November 22, 1795.

My Dear Friend,

Although heartily exhausted with fatigue, I cannot avoid telling you that I have every reason to believe my friend Stephen was unfortunately lost in the Catherine transport. I volunteered the command of a party of forty men of our regiment, to bury the dead, in hopes of finding his body. I have been three days officiating in the melancholy ceremony. I could not distinguish his features, but from size and some resemblance I have lodged in coffins two bodies which I thought resembled him. I will faithfully see them interred, with the bodies of fourteen other officers, with all military honours. Our officers have exerted themselves very much. The labour I and my party have gone through, I look back upon with astonishment, but almost think it miraculous. The cause supported us. We had every day six miles to walk on the bank of pebbles, one mile an hour before we could reach the bodies, and then they lay scattered on the same bank for two miles further. I believe we have buried about 230, but cannot immediately say, as I am much wearied, and have a good way to send to my sergeants for the account. Forgive me if I do not say any thing particular this time. To-morrow I shall again go to the spot with carriages to bring the officers here in coffins to be buried, and on Tuesday the melancholy ceremony will be performed with all military honours.

Mrs. Burns is very well, and in better spirits than we

could have imagined. I have now in my pocket some of her husband's hair for her. I will save some off the body I have taken for Stephen's for you, but it is cropped short.

Adieu,

Your ever sincere

W. F. SHRAPNELL.

W. F. SHRAPNELL, ESQ. TO DR. JENNER.

Weymouth, November 26, 1795.

MY DEAR FRIEND,

I this evening received your answer to my first letter, by which I find my first suggestion was too true, and that my friend Stephen is in reality no more. I persevered at the head of forty men for four days, searching for his body, and interring the soldiers and seamen cast on shore. It would fill your susceptible mind with too much anxiety, was I to attempt a description of a small view of the horrid scene. I selected the bodies of seventeen officers and nine women, who were all interred yesterday in the churchyard of Wyke Regis, with military honours; a solemn and awful sight to survivors.

I selected a body much like my friend in size, height, and, as I thought, features, and preserved it with the bodies of two others, which I thought resembled him also. They were brought to the Wyke yesterday as above. I gave poor Mrs. Burns the satisfaction that I had identified the body of her husband. His marks were so particular that nobody could be deceived.

I also found the body of Captain Creighton, Lieutenant Sutherland, and Lieutenant Kerr. These we are all sure of. I have now been sitting an hour with Mrs. Burns, and making what inquiry I could to satisfy your request. She says that none were sensible of their danger until within five minutes of the ship's striking; that lieutenant Jenner was that morning dressed earlier than usual, and provided (in his turn) breakfast for the mess of officers: that he did not' seem to be alarmed, but went on deck to see which point the wind lay, and immediately came down into the cabin and said in a joking manner, it is now all over with us, for we have not an inch of canvass left. His cot and sleeping place were so much disturbed by the rocking of the ship, that he retired to put his things to rights, or to lie down. Soon after this the mate of the ship looked down into the cabin, and cried out, "gentlemen, save yourselves if you can." Mr. and Mrs. Burns were not dressed, but on the alarm, went on deck with Lieutenant Jenner and Staines, the captain, &c. The ship rocked a good deal, and an immense wave immediately separated Mrs. B. from her husband, and knocked her and the ship's captain down again into the cabin. The captain recovered himself immediately, and went on deck. In about a minute she attempted to follow him, and with difficulty got up and looked on the deck, where no person remained. The ship then struck, and she was hurried a second time into the cabin, surrounded with cracking timbers, and completely enclosed. She imagined her husband and Mr. Jenner had leaped on shore, and attempted a third time in vain to save herself.

She then resigned herself to the will of Heaven, expecting no other than immediate death, and on her knees, half covered with water, and the ship momentarily breaking to pieces, awaited the fatal event. How she was preserved she cannot account. She supposes the wreck must have broken the cabin open, and left a breach wide enough to let her through. A wave threw her some way from the ship, and being perceived by a man at a distance, he saved her at the hazard of his life. She could not be made sensible but her husband must have been saved before her. The fact is that the first time the ship struck every person on deck was washed overboard, and no more seen alive. At this time my unfortunate friend shared the fate of his companions.

I have no doubt but a better chance would have been presented if the unfortunate sufferers had been, able to remain with the ship, as she was driven very far on the land. I saw the next day the whole stern windows of the Catherine, but in a few hours. they were broken up for the iron, by the inhuman plunderers in the neighbourhood. The other parts of the ship were dashed to atoms.

I am,

Your very sincere friend and humble servant,

W. F. Shrapnell.

An account of this calamity was published by Mrs. Charlotte Smith, a lady well known in the literary world. Dr. Jenner, in allusion to her then intended work, observes in a letter to Mr. Shrapnell, " Her pathetic pen will do every thing one could wish with respect to the narrative. Mr. Gardner has been so much engaged with a pamphlet of his own that I did not intend to interrupt him till that was finished; so that no one can feel hurt at Mrs Smith's taking the matter up; and for my own part I am much consoled at the idea."

It will be seen by another part of this work how much Dr. Jenner's mind was, in the course of this year, occupied with that great subject which engaged so much of his time and attention during the rest of his life. He had in the spring performed his first inoculation of the *variolæ vaccinæ* and was preparing to bring the astonishing result of his investigations before the public.

It was in this year, likewise, that he commenced his occasional visits to Cheltenham. A letter from his friend Clinch notices this circumstance, as well as his discovery of vaccination.

THE REV. J. CLINCH TO DR. JENNER.

Trinity, *Dec.* 1, 1796.

MY DEAR FRIEND,

Your kind, friendly, and much-esteemed favours of the 5th and 11th of August last, acknowledging the receipt of my draft on the Society, I have the pleasure to inform you, came safe to hand. I rejoice to hear that your good lady, my worthy and reverend friend, and Mr. Davies were in a better state of health than they had for some time past enjoyed, and I hope your next will announce a perfect re-establishment of that blessing to those much-esteemed friends. As you said nothing of the Rev. Wm. Jenner I flatter myself he was likewise on the mending hand. I was happy to hear that your situation at Cheltenham was both agreeable and salutary to you and your good lady in point of health, and I sincerely trust that your other inducements for making that spot your summer resort are in some degree answerable to your expectations.

I remark what you say of my dear, dear boy; and although I cannot at present form any notions, or give implicit credit to all that philosophers may say, (probably owing to my not having a single spark of philosophy in my composition) respecting the early education of youth, I shall nevertheless, from a full conviction that my Berkeley friends will do every thing for him for the best, always agree with them in the measures most proper to be pursued for his present and future advantage.—You never have, and I trust, my dearest friend, you never will, experience a similar separation. Whenever (and the idea often recurs to my mind) I take a retrospect of the length of time that has already elapsed, and the unavoidable occurrences which have for so long a period as nearly ten years prevented me from visiting those whom I love and esteem on the other side the wide Atlantic—you will readily suppose that my sensations are, at those times, the most poignant and distressing. Could I behold even the smallest glimmer of hope, that a change for the better in the present convulsed state of Europe was likely soon to take place, I should be particularly happy; but as I see no grounds for cherishing so forlorn a hope, the only alternative I have remaining is, that of relinquishing my wished-for intentions of visiting my native shores during the present posture of public affairs.

I am obliged to you for what you say respecting your late discovery. Why not send me a sketch of your idea in print? I wish you could think of something that would repair the organs of my sensorium, as they still continue in a very impaired state.

I have not had an opportunity of seeing any new publication in the medical way since I left England.

As I conceive you have long ere this received the particulars of our late sudden and unexpected alarm, occasioned by the arrival of Citizen Richery, and the powerful squadron under his command having visited this country in the month of September last; and as I would not wish to hurt your feelings on that occasion I shall decline giving you a particular account of our distressed situation during the time they continued to lord it over us. I was twice obliged to remove my family and part of my effects to places of greater safety, that the whole, had they been disposed to pay us a visit, might not have fallen a sacrifice to the fury of the enemy; but I am happy to say we came off much better than could be expected, and have now the pleasure to inform you that we are once more in the possession of peace and tranquillity; but it will be some time before order and regularity is again restored at our deserted cottage. I cannot divest myself of the disagreeable apprehension that as the spring approaches we shall have to experience a renewal of our former troubles. However, we must hope for the best.

Just before the French made their appearance on this coast, my friend George and self had agreed to meet at St. John's; but our plan was soon disconcerted: and indeed the whole of my family have been greatly disappointed in the pleasure of his company at Trinity, by those unwelcome visitors. I heard from him some time ago, at which time I have the pleasure to inform you he was in good health. Your several requests I have sent to him, and from him you will receive a circumstantial account of his situation.

* * * * * * * * *

In the course of the year 1797 Jenner had nearly arranged every thing for the publication of his Inquiry. An incident connected with that event is recorded in the following extract from a letter from his venerable friend and preceptor the late Dr. Hardwicke, of Sodbury. The sentiments expressed are alike honourable to the master and the pupil. "Retired as I have long since been from the arduous task of my profession, and as it were from the world, and full of years, I cannot be ambitious of being brought forward to the public eye; at the same time I confess that I shall feel myself flattered to be noted by a man of your professional character, who was formerly (I am proud to say) my pupil.

That every undertaking of your life may be attended with the success to which you are justly entitled by your indefatigable labours to serve mankind is my ardent wish."

It may be well, before entering on the next great event to which this extract refers, and which so deeply concerned Jenner himself and his fellow-creatures, to take a brief retrospect of the period which has passed in review before us.

We have seen him in circumstances not the most congenial to intellectual efforts, not only actively fulfilling his duty as a professional man, but with great success and perseverance cultivating collateral branches of knowledge, and pushing his inquiries into regions which had formerly been but little explored.

In conjunction with Mr. Hunter he carried on experiments illustrative of the structure and functions of animals. With much industry and ingenuity he explained one of the most unaccountable problems in ornithology; he ascertained the laws which regulate the migration of birds; he made considerable advances in geology, and in the knowledge of organic remains; he amended several pharmaceutical processes; he was an accurate anatomist, and pathologist; he explained the cause of one of the most painful affections of the heart, and advanced far in his investigations respecting the diseases of the lymphatic system, and the most numerous and extensive disorganizations to which animals are liable.

This sketch regards chiefly his character as a medical philosopher. It touches not those other qualities which entitle him to distinction. Had he done no more than what has just been specified he might justly have claimed an elevated station among his professional brethren; and had he been permitted to pursue to their full extent the various researches which he had commenced I am persuaded that there are few of our profession, in any age, who would have surpassed him as a promoter of useful knowledge. Many of his physiological and pathological views are still very imperfectly understood. When their full import is felt it will doubtless be acknowledged that he possessed a genius than which few more enlightened, or yet more humble, ever adorned the science of medicine.

It is time that justice should be rendered to him in this matter: he had acquired the most substantial title to respect as a medical philosopher, and to affection and veneration as a man, on grounds quite distinct from those which will carry his loved and honoured name to the latest age.

He did not labour for personal objects; he was careless, therefore, of his intellectual offspring, and often permitted others to appropriate what belonged to himself. He never ostentatiously put forth his claims to public distinction, and it has on that account been supposed by many that he had none to advance. Sir Isaac Newton was well nigh deprived of the merit of his discovery of fluxions by a corresponding state of mind: his friends stepped forward and secured to him the renown which was his due. Those who knew Dr. Jenner have a similar duty to perform; and to prove that, all important as have been the results of vaccination, he had other pretensions which ought not to be overlooked in any just estimate of his character.

CHAPTER IV.

EARLY HISTORY OF VACCINATION.

In order fully to unfold the progress of this discovery, and to develope the history of Dr. Jenner's mind, whilst meditating upon the probable issue of his investigations and carrying them on to their final accomplishment, we must cast our eye back for a season over that part of his life which has been already recorded. Little has been said about vaccination, because it seemed desirable that the various facts and incidents respecting it should be collected together in the shape of a continuous narrative, unbroken by details which do not immediately relate to that interesting subject. It has been stated that his attention was drawn forcibly to the nature of cow-pox whilst he was yet a youth. This event was brought about in the following manner.—He was pursuing his professional education in the house of his master at Sodbury: a young country-woman came to seek advice; the subject of small-pox was mentioned in her presence; she immediately observed, "I cannot take that disease, for I have had cow-pox." This incident* rivetted the attention of Jenner. It was the first time that the popular notion, which was not at all uncommon in the district, had been brought home to him with force and influence. Most happily the impression which was then made was never effaced. Young as he was, and insufficiently acquainted with any of the laws of physiology or pathology, he dwelt with deep interest on the communication which had been casually made known to him by a peasant, and partly foresaw the vast consequences which were involved in so remarkable a phenomenon. He was the more stimulated to meditations of this sort by frequent opportunities of witnessing the ravages of small-pox; and by retaining the most vivid and painful recollections of the severe discipline which he himself had not long be-

* An incident analogous to that above recorded is mentioned in one of Dr. Jenner's note-books of 1799, in the following words:—

"I know of no direct allusion to the disease in any ancient author, yet the following seems not very distantly to bear upon it. When the Duchess of Cleveland was taunted by her companions, Moll Davis (Lady Mary Davis) and others, that she might soon have to deplore the loss of that beauty which was then her boast, the small-pox at that time raging in London, she made a reply to this effect; that she had no fear about the matter, for she had had a disorder which would prevent her from ever catching the small-pox. This was lately communicated by a gentleman in this county, but unfortunately he could not recollect from what author he gained this intelligence."

fore passed through, preparatory to his inoculation for that disease. "There was," to use his own words, "bleeding till the blood was thin; purging till the body was wasted to a skeleton; and starving on vegetable diet to keep it so." The possibility of averting such evils could not arise in a mind like Jenner's without possessing it fully; and he resolved to let no opportunity escape of acquiring knowledge on so important a subject. How judiciously, how perseveringly, how successfully, he fulfilled this early resolution will be seen as we follow him through his various examinations and experiments.

The suggestions of inexperienced minds are sometimes treated with less respect than they merit. Probably, considering the average distribution of intellect and talent among mankind, this caution is becoming and prudent. But we have examples enough in many different departments of knowledge to prove that wisdom and genius in their purest and best estate do, at times, consent to dwell in youthful breasts.

Newton had unfolded his doctrine of light and colours before he was twenty: Bacon wrote his "Temporis Partus Maximus" before he attained that age: Montesquieu had sketched his "Spirit of Laws" at an equally early period of life; and Jenner, when he was still younger, contemplated the possibility of removing from among the list of human diseases one of the most mortal that ever scourged our race. The hope of doing this great good never deserted him, though he met with many discouragements; his notions having been treated with scorn and ridicule by some, and with indifference by almost all.

As has already been stated, Jenner went, in 1770, to prosecute his studies under Mr. Hunter. Among other subjects of interest which he carried with him from the country, and which he repeatedly mentioned to his teacher, was that of cow-pox. Mr. Hunter never damped the ardour of a pupil, by suggesting doubts or difficulties: on the contrary, as was usual with him on all occasions when the matter in hand admitted of being brought to the test of experiment, he advised that *trial* should be made, and that accuracy and faithfulness should guide the investigation. In cases of this kind he would say "Don't *think*, but *try*; be patient, be accurate." In language such as this he incited all who came within the sphere of his influence to cultivate their art; and his own example most fully accorded with his precept. It does not appear that his mind was ever so fully impressed with the probable consequences of the successful elucidation of the subject of cow-pox as Jenner's was. This, perhaps, is not to be wondered at, considering the extent of his occupations, and the great number of original and important pursuits which fully engrossed his attention. He certainly, however, made known Jenner's opinions, and the traditions in Gloucestershire, both in his lectures, and to his friends in conversation; and other lecturers, on his authority, mentioned them to their pupils.

It was constantly Jenner's habit, from the time that his mind was first awakened on the subject, to endeavour to stimulate all his professional friends and acquaintances to apply themselves to its investigation, because it was interesting as a branch of natural history, and moreover promised to bring with it knowledge most valuable to man. The state of feeling of those medical men in the district where he resided, who had heard of the reported virtues of the cow-pox, and who had also occasionally met with occurrences which seemed to corroborate the popular rumour, will have told the reader the sort of difficulties that Jenner met with, and will likewise prove that it required a mind possessed of all the firmness of purpose which he enjoyed, to induce him to persevere in his pursuits. "We have all heard (they would observe) of what you mention, and we have even seen examples which certainly do give some sort of countenance to the notion to which you allude; but we have also known cases of a perfectly different nature,—many who were reported to have had the cow-pox, having subsequently caught the small-pox. The supposed prophylactic powers probably, therefore, depend upon some peculiarity in the constitution of the individual who has escaped the small-pox; and not on any efficacy of that disorder which they may have received from the cow. In short, the evidence is altogether so inconclusive and unsatisfactory that we put no value on it, and cannot think that it will lead to any thing but uncertainty and disappointment." Observations such as these Jenner encountered from the commencement of his inquiries. They were often repeated while he was carrying them on, and some of them were even brought forward to weaken his claims to honour and gratitude, after he had refuted all objections by his perseverance, and by his consummate address and patience had divested the question of all its difficulties and obscurities, and given to vague, inapplicable, and useless rumour, the certainty and precision of scientific knowledge.

It was not till some years after his return from London that he had an opportunity of examining into the truth of the traditions respecting cow-pox. This was about the year 1775, and corresponds with the period specified by him in his tract on the Origin of Vaccine Inoculation. There is an apparent diversity in the account there given of the period at which his attention was excited to the subject, and that which has been delivered above. The difference, however, is extremely slight, and arises more from the absence of some of the minute facts which I have mentioned as characteristic of the first dawning of the inquiry in his own mind, antecedent to the actual prosecution of it by distinct and personal observations, than from any real discrepancy in the statements.

In his brief account of the Origin of Vaccine Inoculation, he evidently did not consider the details, which it has been thought proper here to give, necessary at that time to be laid before the public. He was not so much describing the history of his own mind, as the history of his efforts, as a medical man, in the investigation which then occupied so much of the public attention.

I am fully entitled to say this from repeated personal communications with Dr. Jenner, from the concurrent testimony of many of his friends, and from the internal evidence which the statement discloses. He could not have mentioned the cow-pox in London in 1770, had it not been a subject of meditation with him before that time. Moreover, in one of his last conversations with an intimate friend, but a few months before his death, he in a particular manner specified the occurrence at Sodbury, and spoke the story of the subsequent progress of his mind in a manner so engaging, so simple, and so humble, that it made an impression never to be effaced.

It was not till 1780 that he was enabled, after much study and inquiry, to unravel many of the perplexing obscurities and contradictions with which the question was enveloped, and which had impressed those who knew the traditions of the country with the opinion that it defied all accurate and satisfactory elucidation. In the month of May of the year just mentioned, he first disclosed his hopes and his fears respecting the great object of his pursuit, to his friend Edward Gardner. By this time Jenner's mind had caught a glimpse of the reputation which awaited him, but it was still clouded by doubts and difficulties. He then seemed to feel that it might, in God's good providence, be his lot to stand between the living and the dead, and that through him a plague might be stayed. On the other side, the dread of disappointment, and the probability of failing to accomplish his purpose, restrained that eagerness which otherwise would have prompted him prematurely to publish the result of his inquiries, and thereby, probably, by conveying insufficient knowledge, blight for ever his favourite hope.

He was riding with Gardner, on the road between Gloucester and Bristol, near Newport, when the conversation passed of which I have made mention. He went over the natural history of cow-pox; stated his opinion as to the origin of this affection from the heel of the horse; specified the different sorts of disease which attacked the milkers when they handled infected cows; dwelt upon that variety which afforded protection against small-pox; and with deep and anxious emotion mentioned his hope of being able to propagate that variety from one human being to another, till he had disseminated the practice all over the globe, to the total extinction of small-pox. The conversation was concluded by Jenner in words to the following effect:—"Gardner, I have entrusted a most important matter to you, which I firmly believe will prove of essential benefit to the human race. I know you, and should not wish what I have stated to be brought into conversation; for should any thing untoward turn up in my experiments I should be made, particularly by my medical brethren, the subject of ridicule—for I am the mark they all shoot at."

This caution respecting concealment sprung from no selfish or unworthy motive: it was a suggestion which came into his mind at the moment of making the communication to his friend, and arose entirely from the little sympathy which he had experienced in his previous efforts to excite his professional brethren to co-operate with him, or to treat the matter which so much engaged his own feelings with the consideration it so well deserved. He by no means continued to act on the principle of concealment. Far otherwise; for it will presently appear that vaccination was often the subject of his own correspondence, and also that of his friends.

Some of the facts which I have presented on the authority of Mr. Gardner were given by him in evidence before a Committee of the House of Commons, when a most extraordinary and preposterous attempt was made to strip Dr. Jenner of his well-earned reputation, and to deprive him of his not less merited reward.

While deliberating on the subject of vaccine inoculation he made some experiments regarding the nature of swine-pox (as it is vulgarly called). From facts elsewhere detailed it has been found reasonable to conclude that this disease, as well as the common variolæ and the variolæ vaccinæ, had one common origin and were, in fact, varieties of the same affection. The circumstance which I am now to mention affords a strong corroboration, and moreover forms a striking incident in the history which I am endeavouring to unfold.

In November 1789, he inoculated his eldest son Edward, who was then about one year and a half old, with swine-pox matter. The progress of the disease seemed similar to that which arises from the insertion of true small-pox matter when the disease is very slight. He sickened on the 8th day: a few pustules appeared; they were late and slow in their progress, and small. Variolous matter was carefully inserted into his arms at five or six different periods, without the slightest inflammation being excited in the part.

On Thursday, April 7th, 1791, variolous matter was again inserted by two small incisions through the cutis. 9th, Evidently inflamed. 10th, An efflorescence of the size of a shilling spread round the inferior wound. 11th, The incision assumed a kind of erysipelatous elevation: the efflorescence much increased. 12th, These appearances much advanced. 13th, A vesicle, containing a brownish fluid, and transparent, about the size of a large split pea on the superior incision, the inferior about twice as big; the surrounding parts affected with erysipelas. The erysipelas extended to the shoulder, and then pretty quickly went off. The child showed no signs of indisposition the whole time.

March 1792. E. Jenner was again inoculated: the matter was taken from a child that caught the disease in the natural way, and had it pretty full. It was inserted fresh from the pustule. The same evening an inflammation appeared round the incision, which, at the end of twenty hours, increased to the diameter of a sixpence, and some fluid had already been collected on the lips of the scratch, which the child had rubbed off.

Many years elapsed before he had an opportunity of completing his projected experiments in vaccination, and he encountered numerous difficulties in carrying on the preliminary part of his inquiry. In the first place, he had found from his own observation, as well as from that of other medical gentlemen in the county, that what was commonly *called cow-pox*, was not a certain preventive of small-pox. This fact damped, but did not extinguish, his ardour. By prosecuting his investigation a little farther, this difficulty was obviated. He discovered that cows were subject to a variety of spontaneous eruptions on their teats; that they all were capable of communicating sores to the hands of the milkers; and that whatever sore was so produced, was called, in the dairy, cow-pox. This was satisfactory information and removed one great difficulty, and suggested a distinction between these diseases,—

one being called by him the true, the other the spurious, cow-pox ; the former possessing a specific power over the constitution, the other not. This impediment was scarcely removed before another of more formidable aspect arose. He learned that there were well authenticated instances to prove that when the *true* cow-pox broke out among the cattle at a dairy, and was communicated to the milkers, even they had subsequently had small-pox. Tidings of this kind, which seemed to render farther investigation useless, checked for a season his fond hopes; but resistance and difficulty only augmented his energy, and he resumed his labours with redoubled zeal. The result was most happy, and enabled him to take that great step in the progress of his inquiry, without which none of its anticipated advantages could have been realized.

On the former occasion he discovered that there were two distinct affections, both vulgarly denominated cow-pox; that one gave protection against small-pox, that the other did not: but when he found that what he concluded to be the *true* cow-pox itself could not be depended on, he felt much perplexed. Most men would, at this stage, have abandoned the investigation in despair. It was not so with Jenner. He conceived that in such cases there must be some ascertainable cause for the deviation from the ordinary effects of the disease. It occurred to him that the virus of the cow-pox itself might have undergone some change whereby its specific virtues were lost; that, in its deteriorated state, it may have been capable of producing a local disease upon the hand of the milker, but no such influence upon the constitution as is requisite to render the individual unsusceptible of variolous contagion; so that the same cow might one day communicate a genuine and efficacious preventive, and the next, nothing but a local affection which could exert no beneficial influence whatever on the constitution. This most ingenious and forcible reasoning, supported by analogies drawn from the well-known properties of the virus of small-pox itself, received an ample confirmation from experience, and was the basis on which some of the fundamental rules for the practice of vaccination were founded. It was ascertained that it was only in a certain state of the pustule that virus was afforded capable of imparting to the constitution its protecting power; that matter taken after this period might excite a local disease, but not of such a sort as to render the individual proof against the effects of variolous contagion.

Having proceeded thus far, all doubts respecting the propriety of attempting to propagate the cow-pox by means of inoculation were at an end. The causes of failure in the casual dissemination of the disease were ascertained, and his chief care was to avoid them in attempting to propagate it by artificial means. A long period elapsed before he had an opportunity of putting his theory to the test of experiment; but he continued to collect information from all quarters. He carried, in 1788, a drawing of the casual disease, as seen on the hands of the milkers, to London and showed it to Sir E. Home and others. The subject was occasionally canvassed among the medical men there: it was (as has been already noticed) mentioned in lectures; and Dr. Adams thought fit to draw the attention of his brethren to it in one of his publications. Mr. Cline, in reference to this fact, has the following remark in one of his letters to Dr. Jenner:—

"I am very glad to learn that you are prosecuting your inquiries on the cow-pox, for it is a most interesting and curious subject. All that Adams had heard of the disease was from me."*

But this is not all: Dr. Jenner not only wrote and spoke about the subject himself, but he encouraged his friends to do so likewise. I present a striking proof of the liberality of his mind in this respect, in the subjoined extract of a letter from Dr. Haygarth, written in answer to one from Dr. Worthington, an intimate friend of Jenner.

"Your account of the cow-pox is indeed very marvellous: being so strange a history, and so contradictory to all past observations on this subject, very clear and full evidence will be required to render it credible.

"You say that this whole rare phenomenon is soon to be published; but do not mention whether by yourself or some other medical friend. In either case, I trust that no reliance will be placed upon vulgar stories."

* St. Mary Axe, London, August 11th, 1796.

"The author should admit nothing but what he has proved by his own personal observation, both in the brute and human species. It would be useless to specify the doubts which must be satisfied upon this subject before rational belief can be obtained.

"If a physician should adopt such a doctrine, and much more, if he should publish it upon inadequate evidence, his character would materially suffer in the public opinion of his knowledge and discernment."*

The subject of small-pox had in a particular manner engaged the attention of Dr. Haygarth, and he had published a work intended to promote the extermination of that disease. The opinion expressed by this eminent and learned physician is such as would have been delivered by most persons in his situation, who were not fully apprised of the strong evidence which had been accumulated in favour of the virtues of cow-pox.

The facts concerning the origin of this affection were investigated with as much patience, and passed through as severe a scrutiny as any of those which regarded the nature of the disease itself. Jenner was quite as communicative, also, on the one subject as on the other. His nephew George Jenner, in the year 1787, went into the stable with him to look at a horse with diseased heels.—"There," said he, pointing to the horse's heels, "is the source of small-pox. I have much to say on that subject, which I hope in

* Chester, April 15th, 1794.

due time to give to the world." Again, writing to a friend some years after this period (1794), he observes :—

"Our friend ———, at our last meeting, treated my discovery of the origin of the cow-pox as chimerical. Farther investigation has convinced me of the truth of my assertion, beyond the possibility of a denial. Domestication of animals has certainly proved a prolific source of diseases among men. But I must not anticipate : you shall have a paper."*

Subsequently to this letter, and before the publication of his Inquiry in 1798, he made many experiments in order to demonstrate the connection between the grease and the cow-pox ; but difficulties of a nature not easily overcome interfered with his success. Not long after this period he enjoyed the sublime gratification of seeing a more important part of his doctrine completely substantiated ; and the probability, therefore, of rendering his labours subservient to the most beneficent, the most interesting of all human objects, placed beyond a doubt. Hitherto he had only observed the casual disease, and investigated its laws : it yet remained to be proved whether it was possible to propagate the affection by artificial inoculation from one human being to another ; and thereby, at will, communicate security to all who were liable to small-pox. An opportunity occurred on the 14th of May, 1796, of instituting

* Berkeley, 1794.

this experiment. Matter was taken from the hand of Sarah Nelmes who had been infected by her master's cows, and inserted by two superficial incisions into the arms of James Phipps, a healthy boy of about eight years old. He went through the disease apparently in a regular and satisfactory manner; but the most agitating part of the trial still remained to be performed. It was needful to ascertain whether he was secure from the contagion of small-pox. This point, so full of anxiety to Dr. Jenner, was fairly put to issue on the first of the following July. Variolous matter, immediately taken from a pustule, was carefully inserted by several incisions, but no disease followed. He communicated the event to his friend Gardner, in the following letter.

Dear Gardner,

As I promised to let you know how I proceeded in my inquiry into the nature of that singular disease the Cow-Pox, and being fully satisfied how much you feel interested in its success, you will be gratified in hearing that I have at length accomplished what I have been so long waiting for, the passing of the Vaccine Virus from one human being to another by the ordinary mode of inoculation.

A boy of the name of Phipps* was inoculated in the arm

* The 14th of May is an annual festival in Berlin to commemorate the day on which Jenner made this experiment; and at the meeting of the medical men, held for this laudable purpose in 1819, the official returns made by vaccinators from the different departments gave an account of 307,596 persons vaccinated in 1817 in the Prussian dominions, which did not include

from a pustule on the hand of a young woman who was infected by her master's cows. Having never seen the disease but in its casual way before; that is, when communicated from the cow to the hand of the milker, I was astonished at the close resemblance of the pustules, in some of their stages, to the variolous pustules. But now listen to the most delightful part of my story. The boy has since been inoculated for the small pox which, as I ventured to predict, produced no effect. I shall now pursue my experiments with redoubled ardour.

Believe me yours, very sincerely,

Berkeley, July 19, 1796. EDWARD JENNER.

Were I to fix upon any period in the life of this admirable man that was more full than another of deep and intense emotion, more elevated by anxious and benevolent hopes, more absorbed with generous and ardent wishes for the complete success of a scheme fraught with great and disinterested benefit to his fellow men, I would mention that portion of it which we have now been contemplating. The situation in which he then stood seldom had a parallel in the history of our race. No invention ever

the total number; as from several of the departments the returns for that year had not yet arrived. This number, far exceeding the births that can have taken place during that time, shows a great advancement towards making the practice universal; and when this is once accomplished, the list of annual vaccinations, compared with the births and burials, will precisely indicate the state of the country in regard to the liability of its inhabitants to take the small pox."—See *Cross* on the Varioloid Epidemic at Norwich.

before promised to have such an immediate and extensive influence upon the lives of his fellow-creatures; no discovery elaborated by the patience, or skill, or science of man was ever calculated, in any comparable degree, to produce such consequences as that which at this period centred in the heart of Jenner. It was a heart above all others capable of rejoicing in the possession of such a secret. This was not, like most of the results of human labours, more rich in the prospect of future advantage than in present good. He, at that period, had it in his power to impart knowledge the advantages of which might be rendered as manifest and palpable as they were universal. It is pleasing to know that the state of his mind corresponded with the solemnity and magnitude of the occasion. There was a great struggle within him how to conduct himself. His natural benevolence would have stirred him up to act with some degree of precipitancy, to give to the world all that he had learned on the matter before his knowledge was complete, because he felt that the death of every one who became a victim to small-pox might almost be said to have been caused through his neglect. But in this, certainly one of the most trying emergencies that ever occurred in the life of any man, he was enabled to conduct himself with all the prudence, all the generosity and caution that befitted an individual to whom such high things were committed. He was not led away by selfish feelings, neither was he elated by pride nor vain glory, nor hurried beyond propriety by over-eagerness and zeal: he maintained the humility, the simplicity, and the disinterestedness of his character on this, as on every other occasion. It was his custom at this time to meditate much as he rambled in the meadows under the castle at Berkeley. He has left us a picture of his feelings, at this period, full of interest and full of piety:—

"While the vaccine discovery was progressive the joy I felt at the prospect before me of being the instrument destined to take away from the world one of its greatest calamities, blended with the fond hope of enjoying independence and domestic peace and happiness, was often so excessive that, in pursuing my favourite subject among the meadows, I have sometimes found myself in a kind of reverie. It is pleasant to me to recollect that these reflections always ended in devout acknowledgments to that Being from whom this and all other mercies flow."

From the period just mentioned till the spring of 1798, Dr. Jenner's researches were intercepted by the disappearance of cow-pox from the dairies. It again showed itself, and he had an opportunity of pursuing his inquiries. He repeated his inoculations with the utmost care, and then prepared his work for the press. It was his intention that it should have first appeared before the public in the Transactions of the Royal Society; but this design was abandoned, and the work appeared as a separate publication.

Before bringing it out he was extremely desirous of proving, by direct experiment, the truth of his opinion as to the origin of cow-pox, which at that time only rested upon circumstantial evidence. Being foiled in his hopes of seeing more of that disease in its casual form in the dairies he made many efforts, in 1797, to generate it from the heel of the horse. In reference to these experiments he wrote, on the 2d of August of that year, to a friend in the following terms:—

"The simple experiment of applying the matter from the heel of the horse, in its proper state, to the nipples of the cows, when they are in a proper state to be infected by it, is not so easily made as at first sight may be imagined; after waiting with impatience for months in my own neighbourhood, without effect, I sent a messenger to Bristol, in vain, to procure the true virus. I even procured a young horse, kept him constantly in the stable, and fed him with beans in order to make his heels swell, but to no purpose. By the time the Pamphlet goes to a second edition, I hope to be able to give some decisive experiments."

Notwithstanding the patience and accuracy of his researches, and the very interesting results to which they led, he continued, before he gave them to the public, to deliberate on his projected work with the most solemn and anxious care. This mature and laudable consideration is manifest in every sentence of the treatise, and it as much redounds to his honour as a philosopher as it does to his modesty as a man. Knowing, as we do, the magnificent results that have arisen from that small work, beyond all doubt the most extensively beneficial that science ever contributed for the welfare of man, it may be interesting to the reader to learn the manner in which the author, in his private correspondence with his friends, sometimes wrote about it.

"I have shown a copy of my intended paper on the cow-pox to our friend Worthington, who has been pleased to express his approbation of it, and to recommend my publishing it as a pamphlet instead of sending it to the Royal Society."*

The gentleman mentioned in the above extract was not the only one to whom he submitted the manuscript. His friends Gardner and Hicks were also often consulted about it; and, finally, before it was sent to the press it was accurately and faithfully scrutinized by a select number of his particular associates, at Rudhall near Ross in Herefordshire, the seat of Thomas Westfaling, Esq. They all felt deeply interested in the investigation; they all saw that a matter of so much moment ought to be canvassed with the greatest care; for the dearest interest of their fellow-creatures, as well as of their own affectionately loved friend, was involved in it. The party present on this occasion were Mr. Westfaling, Dr. Worthington, Mr. Paytherus, and Mr. H. Hicks.

These gentlemen listened to all the details with

* June 1797.

jealous ears; they sat in judgment on the work, and did honestly and kindly acquit them of their duty. Their judgment approved; their most benevolent feelings were gratified; and it only remained for them to applaud their friend, who then stood before them in a situation more truly interesting than they could express, and to urge him on his path by encouraging him in his purpose of opening, for the benefit of all, that stream of life and health which he had been permitted to discover.

It was a special honour to have been associated with Jenner on such an occasion. The mind, in dwelling upon occurrences of this kind, naturally seeks for parallels in the histories of the lives of eminent men in other times. But the situation of Jenner scarcely admits of illustrations of this sort: he seemed to hold in his hand one of the "gates of death," and to him it seemed to be given to close it.

When Columbus, by his judicious study of cosmography, anticipated the discovery of another hemisphere; when Newton beheld the host of heaven yielding up the secret of their movements to his patient and sublime researches; when Bacon, in the well-founded reliance on his almost superhuman powers, took a flight over the heads of men and with perfect confidence looked forward to a far distant age for the blossom and the fruit of that intellectual seed which he had so abundantly scattered,—the inward gratification derived from the consciousness that truth and wisdom were to be imparted through them to their fellow-mortals, and that the ultimate result would be felt in beneficial consequences to every class of society, doubtless imparted a joy and satisfaction to their souls of the most gratifying description. When Harvey, too, by the successful issue of his labours had accomplished the most remarkable discovery that up to his time had been made in medicine, and thereby enjoyed the happiness of knowing that improvements in all the practical parts of his profession must reward his perseverance and make up for the neglect and injustice of his contemporaries, we cannot doubt that he experienced that inward peace and happiness which his jealous detractors could not intermeddle with.

But if discoveries are to be estimated by their power of ministering to the benefit of man, which of all those that have most distinguished their authors can be compared with that of which we are now treating?

I cannot allude to the event commemorated above without bespeaking the reader's patience, while I dwell for a short time on the character of one of the gentlemen who was present at this interesting deliberation. Of the others I cannot write here; for, as they are still among us, truth itself might wear an appearance of flattery distasteful, I am sure, to their nature, and foreign to mine. But, unhappily for all those who knew him, this restraint does not apply to Thomas Westfaling. He expired suddenly at Bath in 1815; and though I may now speak of him as my heart dictates, I am unable to say what is worthy of him. As the friend and adviser of Jenner his character will not be uninteresting to any; but as the friend and lover of every thing good he has claims to the consideration of all. For myself, I should be unworthy of the office which I now attempt to execute were I not to indulge in paying a passing tribute of love and veneration to a character which bore conspicuously the marks of rare intellectual endowments and most generous and engaging virtues, and was in an especial manner endeared to myself by many acts of kindred sympathy and friendship during a most trying period of my own life.

All matters having been duly arranged the Inquiry was published about the end of June, 1798. The dedication to his friend Dr. Parry of Bath bears date the 21st of that month. The work was printed in the quarto form and very little exceeded seventy pages.

The first few paragraphs unfold the author's opinions respecting the influence that the artificial and luxurious habits of civilized life may have had in rendering man liable to disease; and in pursuing the same train of reasoning he is led to believe that the domestication of animals, not originally intended for our associates, may have inflicted upon our race maladies of various kinds and multiplied the number of our ills.

These notions are with much simplicity and elegance stated in the following sentences:—

"The deviation of man from the state in which he was originally placed by nature seems to have proved to him a prolific source of diseases. From the love of splendour, from the indulgences of luxury, and from his fondness for amusement he has familiarised himself with a great number of animals, which may not originally have been intended for his associates."

"The wolf, disarmed of ferocity, is now pillowed in the lady's lap.* The cat, the little tiger of our island whose natural home is the forest, is equally domesticated, and caressed. The cow, the hog, the sheep, and the horse are all, for a variety of purposes, brought under his care and dominion."

"There is a disease to which the horse, from his state of domestication, is frequently subject: the farriers have termed it *the grease*. It is an inflammation and swelling in the heel from which issues matter possessing properties of a very peculiar kind that seems capable of generating a disease in the human body (after it has undergone the modification that I shall presently speak of) which bears so strong a resemblance to the small-pox that I think it highly probable it may be the source of that disease."

Together with this statement of the author's opinion respecting the origin and nature of the disease

* The late Mr. John Hunter proved by experiments that the dog is the wolf in a degenerated state.

the Inquiry contained twenty-three cases, detailed at length, illustrative of the progress of the infection. The first sixteen cases were examples of the casual disease; the rest were the result of inoculation. One among the former is of an individual who having been infected from the heel of the horse, afterwards completely resisted small-pox contagion. Among the latter is mentioned the case of his second son, Robert Fitzharding Jenner, an infant eleven months old; and of several other children who were vaccinated on the 12th of April, 1798, with matter taken from the arm of Hannah Excell. It is particularly specified, "that Robert Jenner did not receive the infection."

I beg the reader to notice this circumstance; for many years afterwards a calumny of a very injurious nature was propagated respecting his abandonment of vaccination, because under danger of an urgent nature he found it indispensable to have this child inoculated with small-pox.

The important information contained in the part of the Inquiry just noticed, concluded by some modest and most sagacious remarks regarding alike the main object of the investigation, and others which were interwoven with it. He considered his assertion "that the cow-pox protects the human constitution from the infection of small pox" proved by the facts adduced. With regard to the opinion that the source of the infection of cow-pox "is a peculiar morbid matter arising from the heel of the horse" he believed that though it had not been completely proved by actual experiments made under his own eye, it nevertheless was supported by evidence sufficiently strong to establish it.

When the Inquiry was printed he imagined that the matter secreted in the heel of the horse required to be modified by passing through the system of the cow, in order to afford it the peculiar protecting powers which it evinced when it appeared in the shape of what is vulgarly called cow-pox on the hands of the milkers. In the infancy of the investigation this was a most natural conclusion; but subsequent trials proved that the *equine* matter, which had *never* undergone any change from passing through the constitution of the cow, exhibits all the characters of, and affords all the security which can be obtained from vaccine matter strictly so called.

The last paragraph of this interesting Inquiry indicates a strain of modest feeling respecting the probable issue of the author's labours, which cannot now be perused without exciting the utmost respect and admiration. The reader, in the preceding narrative, has witnessed the painful agitation of his mind while contemplating in secret the mighty aim which filled his soul. He had satisfied himself by direct and careful experiment of the truth of all the main facts, and of course saw clearly that his most sanguine expectations were likely to be completely realized. Under such circumstances the following subdued and unpretending sentences are eminently deserving of notice :—

" Thus far have I proceeded in an inquiry founded, as it must appear, on the basis of experiment in which, however, conjecture has been occasionally admitted, in order to present to persons well situated for such discussions objects for a more minute investigation. In the mean time I shall myself continue to prosecute this inquiry, encouraged by the hope of its becoming essentially beneficial to mankind."

Before the publication took place Dr. Jenner had repaired to London for the purpose of exhibiting the cow-pox, and of demonstrating to his professional friends the accuracy of his delineations and the truth of his assertions.

He left Berkeley with Mrs. Jenner and his daughter on the 24th of April, 1798. They slept the first night at Cirencester; next day they proceeded to Benson, and the following afternoon they arrived in Pall-Mall, where they dined with Mrs. Jenner's relative Mr. Ladbroke. Dr. Jenner remained in London till the 14th of July; on that day he quitted it, and arrived in Cheltenham the same evening.

I am thus particular in specifying the dates, which I have ascertained by a reference to his journal, because they are connected with a remarkable fact in the history of vaccination. It will scarcely be believed, that with all his efforts and those of his friends, he was unable during the period of nearly three months that he continued in the metropolis, to procure one person on whom he could exhibit the vaccine disease. I remember he often stated that his patience had been exhausted on that occasion, and that he had actually quitted the capital without having accomplished the object of his journey; but it was not till lately I discovered that he had so much cause for feeling disappointed. The tardiness and distrust evinced on the present occasion formed a striking contrast to the eagerness and zeal with which persons of all ranks, without knowledge and consideration, rushed to the adoption of a practice that was offered to them by unskilful hands, and by which consequences of a most disastrous nature ensued. Of course it was to be expected that a practice, involving such unexpected and uncommon results, should, on its being first mentioned, excite great doubt and surprise; but it is nevertheless a strange circumstance, that the author of that practice, a man known in the highest circles of medical science as worthy of all credit, and as an accurate and enlightened observer, should have been unable, notwithstanding the proofs which his Inquiry contained of the safety and importance of vaccination, to prevail on one individual to submit to the operation during his stay in London.

Some of the virus which he carried with him was consigned to Mr. Cline, who, in the end of July, inserted it by two punctures into the hip of a patient. There is a curious fact connected with this first interesting and successful exhibition of the *variolæ vaccinæ* in London. I mention it because it shows, together with the other circumstances just detailed, the extreme caution with which some of the early trials were conducted. The patient on whom Mr. Cline operated had some affection of the hip joint, and it was thought that the counter-irritation excited by the cow-pox might prove beneficial to the disease. This was the reason that the virus was inserted on the hip; and for the same reason it was intended to convert the vaccine pustule into an issue, after it had passed through its proper course.

I state these facts on the authority of repeated personal communications from Dr. Jenner; they are substantiated by the statements which are recorded in his journal, and are also corroborated by two letters from Mr. Cline to Dr. Jenner, which I think it right to annex in this place.

Extracts from Journal of 1798.

"That the matter of cow-pox, like the small-pox matter, may be preserved without any diminution in its active qualities is evinced by the following experiment.

"Mr. Cline inoculated a child with matter that had been taken from the pustule on the arm of Hannah Excell (see page 39 pamphlet) when in a limpid ichorous state, and dried by exposure to the air, after being preserved three months on a quill in a seal. The following is the result:—

COPY OF MR. CLINE'S LETTER.

Lincoln's-Inn Fields, 2*d Aug.* 1798.

"The cow-pox experiment has succeeded admirably. The child sickened on the seventh day; and the fever, which was moderate, subsided on the eleventh day. The inflammation extended to about four inches diameter, and then gradually subsided without having been attended with pain, or other inconvenience. The ulcer was not large enough to contain a pea, therefore, I have not converted it into an issue as I intended.* I have since inoculated him with small-pox matter in three places, which were slightly inflamed on the third day, and then subsided.

"Dr. Lister, who was formerly physician to the Small-pox Hospital, attended the child with me, and he is convinced that it is not possible to give him the small-pox.

"I think the substituting of cow-pox poison for the small-pox promises to be one of the greatest improvements that has ever been made in medicine: for it is not only so safe in itself, but also does not endanger others by contagion, in which way the small-pox has done infinite mischief. The more I think on the subject the more I am impressed with its importance.

With great esteem I am, dear Sir,

Your faithful servant,

HENRY CLINE."

* This boy was brought to town on account of some disease in the joint of the hip. Mr. C. therefore inoculated near the part, with the view of exciting inflammation, and subsequently of forming an issue. E. J.

"With the intention of proceeding with the experiments, Mr. Cline took matter from the pustule, and with it inoculated three other children; but on none of these did it take any effect."

"I have observed that the matter of cow-pox appears to lose its powers of infection after it ceases to be limpid. Probably it might have passed the bounds of perfection when Mr. Cline made his second experiment."

HENRY CLINE, ESQ. TO DR. JENNER.

Lincoln's-Inn Fields, 18th August, 1798.

"MY DEAR SIR,

"Seven days since, I inoculated three children with cow-pox matter, and I have the mortification of finding that the infection has not taken, and I fear I shall be entirely disappointed unless you can contrive to send me some fresh matter. I think it might come in a quill in a letter, or inclosed in a bit of tin-foil, by the same conveyance, or in any other way that may be more convenient.

With much esteem, I am, dear Sir,

Your faithful servant,

HENRY CLINE."

Mr. Cline having failed to propagate the disease from the first case of successful vaccination which occurred in London and Dr. Jenner having at that time no fresh lymph to transmit, it was not in his power to gratify the anxious wishes of the many professional men who now eagerly sought an opportunity of witnessing the progress of the affection, and of putting its alleged prophylactic powers to the test. It might have been expected that under such circumstances his brethren would have patiently awaited the result of further observation before they ventured to question the accuracy of the statements and to impugn the authority of Jenner, sustained as they were by many facts brought forward with minute and jealous attention to every thing that was necessary to gain them credit if true, and to secure their speedy refutation if false. There were indeed many, and those too of the most learned and respectable, who immediately did justice to the merits of Jenner, and cordially and thankfully acknowledged the many important consequences which were involved in the subject that he had so ably and so modestly brought before them. Others, confident in their own knowledge and trusting too much to the dim light of their own understanding, did not hesitate on very slender grounds at once to deride the doctrine and condemn the practice. Better things might have been anticipated from the gentlemen who distinguished themselves in this ignoble opposition.

Mr. Cline, perceiving at once from the success of his first trial what incalculable blessings were connected with the diffusion of the practice, with just and becoming regard for the welfare of Jenner wished his personal advantage to keep pace in some degree with the benefits which he had it in his power to impart to mankind. He therefore immediately advised him to quit the country and to take a house in Grosvenor Square, and promised him 10,000*l.* per annum as the result of his practice. In this opinion Mr. Cline was supported by the authority of an extremely accurate observer, and a most competent judge of such matters, the late Sir W. Farquhar. All these splendid prospects of wealth and distinction could not move Jenner. His sentiments on this occasion may be gathered from a letter, which he wrote to a friend who had also suggested to him a similar course.

Cheltenham, September 29th.

It is very clear from your representation that there is now an opening in town for any physician whose reputation stood fair in the public eye. But here, my dear friend, here is the rub. Shall I, who even in the morning of my days sought the lowly and sequestered paths of life, the valley, and not the mountain; shall I, now my evening is fast approaching, hold myself up as an object for fortune and for fame?—Admitting it as a certainty that I obtain both, what stock should I add to my little fund of happiness?

My fortune, with what flows in from my profession, is sufficient to gratify my wishes; indeed so limited is my ambition and that of my nearest connexions, that were I precluded from future practice I should be enabled to obtain all I want. And as for fame what is it? a gilded butt, for ever pierced with the arrows of malignancy. The name of John Hunter stamps this observation with the signature of truth. However, this I promise you, that as soon as my engagements here cease, you shall see me in Town.—In my last letter I told you how much I was perplexed; my perplexity really amounts to agitation. On the one hand, unwilling to come to town myself for the sake of practice, and on the other, fearful that the practice I have recommended may fall into the hands of those who are incapable of conducting it, I am thrown into a state that was at first not perceptible as likely to happen to me; for, believe me, I am not callous to all the feelings of those wounds which, from misrepresentation, might fall on my reputation; on the contrary, no nerves could feel more acutely; and they now are actually in a tremor from anticipation.

How very few are capable of conducting physiological experiments! I am fearful that before we thoroughly understand what is cow-pox matter, and what is not, some confusion may arise; for which I shall, unjustly, be made answerable. In the first place, instances will occur where those who have truly had the disease shall be subjected to the common process of inoculation, inflammation, vesication, and even pus will appear on the wounded part. The axilla will show that the lymphatics have been active and the system may even, in a very limited degree, feel the consequence. What would the enemies to the improvement of science say to this? I leave you to answer this question. But the very same thing has happened again and again to those who have had the small-pox; and do not those (nurses for example) who are much exposed to the contagion of small-pox—

(The remainder of this letter is unfortunately lost.)

Who can read this prophetic letter without admiring the sagacity of the writer? Who can know the sufferings which he actually endured from the very causes he had thus anticipated, and not lament that such a man should have been so injured?

I shall close this chapter with a few of the communications made to him, respecting his first publication, by men of high character and extensive medical information. These are chiefly valuable as marking their opinions at an early stage of the investigation.

DR. PERCIVAL TO DR. JENNER.

Manchester, Nov. 20th, 1798.

DEAR SIR,

About a fortnight ago, I received your very obliging letter, accompanying an "*Inquiry into the Causes and Effects of the Variolæ Vaccinæ.*" This truly valuable work I had before read with much interest, not only on account of the novelty and importance of the subject but as being the production of one whom I highly esteem. To receive it from your hands, therefore, I need not assure you was peculiarly acceptable to me; and I beg to return my best acknowledgments for this mark of your kind remembrance and attention.

The facts which you have adduced incontestably prove the existence of the cow-pox and its ready communication to the human species. But a larger induction is yet necessary to evince that the virus of the *variolæ vaccinæ* renders the person who has been affected with it secure during the whole of life, from the infection of the small-pox. You have opened, however, a new and most productive region of investigation; and I hope you will continue and enlarge your researches, and incite others to engage in the same laudable pursuit.

As soon as I had perused your work, several months since, I wrote to my excellent friend Dr. Haygarth (late of Chester, now of Bath) urging him to engage in a correspondence with you on his favourite and benevolent plan of exterminating the small-pox, of which your discovery points out a more probable mean than any which has yet been proposed. Permit me to express a wish that you would confer with Dr. H., either personally or by letter, on this interesting subject.

Mr. Simmons, an ingenious surgeon of this town, has inoculated a human subject with the ichor issuing from what is termed the grease in horses; but the fluid introduced, though eight punctures were made, neither occasioned inflammation nor eruption; yet the same child was soon afterwards inoculated with success for the small-pox. I showed your letter to Mr. Simmons, who desires me to present his compliments, and to state to you, that he has now engaged a herd of cows, and is busily employed in making such experiments as your publication has suggested. It is very remarkable, that the cow-pox has been hitherto unnoticed in Cheshire, which is not less a dairy country than Gloucestershire, and where the office of milking is performed also by men and maid servants indiscriminately.

* * * * * *

DR. HICKS TO DR. JENNER.

Bristol, October 3d, 1798.

DEAR JENNER,

I thank you for your friendly letter, the subject of which before I received it has lately been much in my mind. I am in doubt in what form to bring it forward. Dr. Beddoes has solicited me to publish it in a volume which he is about to publish at Christmas under the title of Contributions to Physical and Medical Knowledge. But, perhaps, it might excite the attention of the public more particularly if it were given to it separately.—What is your opinion?

Your book I have taken care to get read by many, as I

have recommended it to our public library, and to several private societies. It is in general much approved. Dr. Beddoes says it will do you much credit. I do not see that you need hesitate to accept of the invitation given you to inoculate with the cow-pox, convinced as you are that it will secure the person so inoculated from ever being infected with the small-pox.

I should be very much obliged to Mr. Shrapnell for his observations on what was vulgarly called the pig-pox. I wish you had been able to have communicated the cow-pox to the cow by means of inoculation from a greasy horse s heel,—your work would have been then more complete and satisfactory. But, as the inquiry will be prosecuted by you, you will of course make some experiments of that nature.

* * * * * *

From Francis Knight, Esq. to Dr. Jenner.

My dear Sir,

I hasten to make my best acknowledgment for your kind recollection of me in your late publication, which was left at my house a few days ago while I was absent with a patient in Yorkshire. I have read it over with much satisfaction; and, from a long residence in the dairy part of Wiltshire as well as in Gloucestershire, know the facts to be well supported; at least it was a general opinion among the dairy-men that those who had received the cow-pox were not susceptible of the variolous disease. The cow-pox pustule is very familiar to my eye, and I am quite charmed with the delineation of it in your plates. You have opened to the world a very curious field of investigation, and it is too interesting a subject to die with the day. No one can be so well qualified to pursue this subject as yourself yet I shall, in a more humble line, seek every opportunity of making experiments that may establish a point of so much importance to the world.

I am not anxious to know how or what the change is which the animal economy undergoes: it is sufficient for me to have proof that a lighter disease may be uniformly substituted for a greater one. I shall feel infinitely obliged by any new communications or directions that you may have to make, and in my next Gloucestershire visit shall eagerly seek the opportunity of acknowledging in *propria persona* how much I think the world indebted to your researches. Perhaps it may fall in your way to accommodate me with some fresh cow-pox matter. I know some people of fashion who are well disposed to let me make the experiment on some of their children.

Dear Sir,

Your obliged and faithful

Humble servant,

FRANCIS KNIGHT.

Clifford-street, Sept. 10,
1798.

CHAPTER V.

OPINIONS OF DR. JENNER RESPECTING THE VARIOLÆ, AND VARIOLÆ VACCINÆ—ILLUSTRATIONS DRAWN FROM THEIR LITERARY AND MEDICAL HISTORY.

HAVING now traced the progress of Dr. Jenner's observations respecting cow-pox it may not be unimportant, with a view to illustrate his opinions, to endeavour to find out if ancient records and traditions discover any affinity between that disorder and the small-pox, as it affects the human subject. Independent of the historical information which such an inquiry must elicit it claims attention from its intimate connexion with those questions which have most agitated the public mind respecting the origin of the cow-pox itself, as well as of its prophylactic virtues.

These considerations must be my apology for entering so much at length into a discussion which to the general reader may appear somewhat tedious. It is moreover my duty to state that this examination, undertaken to elucidate an obscure part of the most interesting of all pathological subjects, has afforded me the true gratification of finding at every step additional reason to admire the accuracy of Dr. Jenner's observations, and it has enabled me to bring forward a great weight of unprejudiced and impartial testimony to support his doctrines.

He always considered small-pox and cow-pox as modifications of the same distemper; and that, in employing vaccine lymph, we only made use of means to impregnate the constitution with the disease in its mildest, instead of propagating it in its virulent and contagious form, as is done when small-pox is inoculated. The name which he gave to the former, Variolæ Vaccinæ, sufficiently and strongly indicates his sentiments on this head. Different individuals have questioned the propriety of this name; but had they been aware of the facts about to be enumerated they would have seen that none more appropriate could have been adopted.

Many writers have imagined that the epidemic diseases which affect the human race are peculiar to our species, and have no influence on the inferior animals; and they have been not less decided in the opinion that the diseases of other animals are not communicable to man. The disclosures which have been made by the history of the Variolæ Vaccinæ have shown that both these opinions are erroneous: and other proofs of a more convincing nature will, hereafter, be adduced.

Without laying greater stress upon the following

facts than they can reasonably bear, I think it will be admitted that they sustain these propositions:—

First, that an eruptive disease *common* both to man and to the inferior animals has been known in different ages, and in different countries; and that the descriptions given of this eruptive disease by various writers accord so completely with those acknowledged to be characteristic of small-pox, as to render it highly probable that this disease actually existed at a much earlier period than that usually assigned to its origin.

Secondly, that as there are numberless writers who have described the small-pox in man, so there are others of established name and reputation, who have treated of a similar eruptive and pestilential disease as existing in various countries and in different times among the inferior animals, but especially among cattle; that to this disease they have unhesitatingly applied the name of VARIOLA; and actually recommended such treatment as experience had proved to be useful when that disease attacks man.

Should these propositions be established they will go far to prove that the variolæ of men and of the inferior animals are essentially and originally the same; and that from their first appearance to the present hour they have existed under various modifications. I am prepared to expect opposition to the opinion respecting their simultaneous origin; but whatever sentiments may be entertained on that subject, the facts connected with it are well worthy of consideration, and can scarcely fail to give a degree of confidence in the inoculation for the variolæ vaccinæ which that practice, unfortunately for mankind, has not yet acquired.

Although it is not, now, usual to refer to the Sacred volume in matters of medical history I feel that, on the present occasion, I should be unable to deal with this subject as it deserves were I to abstain from taking the strong ground it affords for corroborating events and opinions narrated by historians and naturalists. In this, therefore, the earliest historical record in the world we have a distinct announcement that there was one eruptive disease common to man and to beast. The number of similar instances mentioned in profane writers demonstrates that such phenomena were not peculiar to one era, but have been observed in different ages and countries ever since their first occurrence.

As every event connected with the origin and progress of small-pox has derived additional interest and importance from the singular phenomena exhibited in the history and properties of cow-pox, it will be right to examine more minutely into the descriptions above referred to.

The reader may be aware that Philo, the learned Jew, wrote, in the first century, a work on the Life of Moses; and on that part of the Book of Exodus which describes the plagues inflicted on the Egyptians he has introduced a paraphrase or commentary on the words of the inspired historian. That portion of the commentary which refers to the ninth chapter of Exodus, 9th and 10th verses, delineates the plague of "*boils and blains.*" The Greek words of the Septuagint, synonimous with our translation, "Boils breaking forth with blains," are ἕλκη φλυκτιδες ἀναζέουσαι, ulcera pustulis ex fervore ebullientibus. Philo's enlarged description runs thus: Ἔπειτα κονιορτὸς αἰφνίδιον ἐπενεχθεὶς, ἀνθρώποις τε καὶ ἀλόγοις ζώοις ἀγρίαν καὶ δυσαλθῆ κατὰ τῆς δορᾶς ἁπάσης ἕλκωσιν εἰργάζετο, καὶ τὰ σώματα εὐθὺς συνῴδει ταῖς ἐξανθήσεσιν, ὑποπύους ἔχοντα φλυκταίνας, ἃς ἐτόπασεν ἄν τις ἀφανῶς ὑποκαιομένας ἀναζεῖν, ἀλγηδόσι τε καὶ περιωδυνίαις, κατὰ τὸ εἰκὸς, ἐκ τῆς ἑλκώσεως καὶ φλογώσεως πιεζόμενοι, μᾶλλον ἢ οὐχ ἧττον τῶν σωμάτων τὰς ψυχὰς ἔκαμνον, ἐκτετρυχωμένοι ταῖς ἀνιαις. Ἓν γὰρ ἄν τις ἀπὸ κεφαλῆς ἄχρι ποδῶν συνεχὲς ἕλκος ἐθεάσατο, τῶν κατὰ μέλος καὶ μέρος διεσπαρμένων, εἰς μίαν καὶ τὴν αὐτὴν ἰδέαν ἀποκριθέντων.*

The late Dr. Willan, in his "Dissertation on the Antiquity of Small-pox" quotes some of these words of Philo, and he appeals to the passage to prove that it contains a lively and accurate description of small-pox. As Philo wrote in the first

* Φίλωνός τοῦ Ἰουδαίου, &c. Vol. II. pp. 100, 101. Περὶ Μωσέως. Ed. Mangey. "Coortus repentè pulvis in homines et bruta illatus, tetris et incurabilibus pene ulceribus totam cutem obserebat. Itaque statim tota corpora, efflorescente passim eruptione, in tumore erant, purulentis scatentia pustulis, quas existimasses subter occultè suffervescentes ebullire : dolore autem et cruciatu, ut æquum est, propter exulcerationem et inflammationem oppressi, magis adeo aut certe non minus animis laborabant, angoribus confecti. Quippe a capite ad pedes ulcus continuatum cerneres, iis quæ per membra et artus sparsa erant, in unam et eandem speciem confusis."

century, Dr. Willan very fairly contends that this disease must have been known as a specific malady in his time. Of the accuracy of Dr. Willan's opinion respecting the nature of the disease described by Philo no competent judge can doubt; but it is a singular fact that this description did not merely refer to a "malady known at the time Philo wrote," but to one known many centuries before. His words apply to the plague of boils and blains, as recorded by Moses. It is, therefore, evident that if Philo's account be descriptive of the small-pox, it carries back the antiquity of that disease, not to the first century, but to the much more remote period of nearly fifteen hundred years before the Christian era. And it is not unimportant here to remark that the histories and traditions of the Eastern nations, particularly the Chinese and Hindoos, refer the commencement of this disease to a corresponding epoch.

The interpretation given by Philo, explicit and distinct as it is, derives confirmation from the remarks of the learned Scheuchzer, in his "Physica Sacra." I subjoin his version of the Sacred text, with his comments thereon.*

* "Et dixit Dominus ad Moysen et Aaron, Tollite plenas manus cineris de camino, et spargat illum Moyses in cœlum coram Pharaone.

Sitque pulvis super omnem terram Ægypti, erunt enim in hominibus ac jumentis ulcera et vesicæ turgentes in universa terra Ægypti.

The original Hebrew is said to be still more expressive of the eruptive character of the disease; and, of consequence, goes to strengthen the opinion that *that disease* was the small-pox.

Tuleruntque cinerem de camino et steterunt coram Pharaone, et sparsit illum Moyses in cœlum, factaque sunt ulcera vesicarum turgentium in hominibus et jumentis.

Nec poterant malefici (Magi) stare coram Moyse propter ulcera, quæ in illis erant, et in omnibus Ægyptiis.

Altius rursus in graduali pœnarum scala scandit justissimus Deus, Ægyptios affligens apostematibus et ulceribus, (Hebraice) *ulcere germinante inflationes.* Non videntur inflationes hæ ulcerosæ *bubones* vel *carbunculi pestilentiales*, sed tumores inflammatorii cum vesicis vel pustulis in cute elevatis, sero acri urente plenis; fuit hoc malum, quod in Peste perrarum, commune hominibus et jumentis, molestum potiusquam lethale: nil legimus de strage magna vel hominum vel pecorum: imo vero, quod rursus in Peste rarum, obambulare poterant Ægyptii; Magi stetisse quidem se videntur coram Pharaone, sed non constiterunt, scilicet absque insigni doloris sensu et signo."

The following observations of this writer refer to the *bubonic* plague, and show that the distinction between the above disease and the pestilence is strongly contrasted in the Scriptures.

"Commoda heic se offert occasio explanandi specialius diram, quæ Pestis nomine venit, Luem. Prima fit mentio morbi truculentissimi Pestis, Exod. v. 3. 'Et dixerunt, Deus Ebræorum vocavit nos, licet nobis quæso ut eamus iter trium dierum in desertum, et sacrificemus Domino Deo nostro, ne forte accidat nobis Pestilentia aut Gladius.'

Id imprimis urgeo argumentum, quod ex textu citato desumi potest. Allegat Moses in conspectu infensissimi regis pro argumento dimissionis obtinendæ persuasorio Pestem, morbum Ægyptiis certe haud ignotum, æque cognitum ac est bellum, malum mundo coævum. Quis est quæso qui sibi persuadere

Whatever may be thought of this opinion, it is fit to observe that the eruption which constituted the plague of "boils breaking forth with blains upon man and upon beast throughout all the land of Egypt," does not appear, from the Sacred text, to have been removed from the sufferers, as some of their other inflictions were, by Divine interposition; but was left to its natural course, and thus may have been propagated through successive generations of mankind, as we know has been the case with small-pox.

Let us now endeavour to prove from other authorities that diseases similar in their nature have affected man and brutes in common from the earliest periods of profane history.

Though the testimony of Homer, as a poet, is not quite unexceptionable in matters of this kind, it is fully adequate to prove the antiquity of the belief that man may participate in the distempers of the brute creation.

possit, loqui populi Israelitici procuratorem, imo Dei oratorem extraordinarium, de re prorsus regi et Ægyptiis ignota, quam certè risu explosisset rex a veri Dei cultu, et populi Israelitici petito alienissimus?

Imo vero ex propositione oratoris inferre licet, *Pestem* quam in proscenium ducit, morbum fuisse maxime notum, Ægypti endemium, et forte tunc vel in Ægypto, vel in vicinia jam grassantem, quandoquidem profectionem populi proponit tanquam remedium abigendæ pesti idoneum. Malum leve quoddam nec curasset Rex, nec proposuisset vel comminatus fuisset Moses. Produci debebat argumentum, quo dari posset supplicationi pondus."

"On mules and dogs th' infection first began;
And last the vengeful arrows fix'd in man.*"

The instances of a similar description recorded by the Roman historians are numerous. Livy, in particular, mentions many such: some of his accounts are so striking that they must not be omitted in this place.

Though he makes mention of "Pestilentiæ," in a cursory way, in the early annals of Rome, we do not find any worthy of being referred to till the year 290 U. C. (Anno ante Christum 464.) This historian's words are: "Grave tempus et forte annus pestilens erat urbi agrisque, nec hominibus magis, quam pecori; et auxere vim morbi terrores populationis pecoribus agrestibusque in urbem acceptis." He adds, "Ministeriaque invicem ac contagio ipsa vulgabant morbos."

The next great pestilence recorded in Livy occurred in the year U. C. 300. (A. C. 454.) "Duo simul mala ingentia exorta, fames pestilentiaque, fœda homini, fœda pecori. Vastati agri sunt: urbs assiduis exhausta funeribus; multæ et claræ lugubres domus."

In the year 317 U. C. (A. C. 437.) "Pestilentia eo anno aliarum rerum otium præbuit: magna tamen clades in urbe agrisque, promiscue hominum pecorumque pernicie, accepta."

Of the following year he says, "Eo anno vis

* Οὐρῆας μὲν πρῶτον ἐπώχετο, καὶ κύνας ἀργούς·
Α'τὰρ ἔπειτ' αὐτοῖσι βέλος ἐχεπευκὲς ἐφιεὶς,
Βάλλ'· αἰεὶ δὲ πυραὶ νεκύων καίοντο θαμειαί.

morbi levata; neque a penuria frumenti, quia ante visum erat, periculum fuit."

In L. 4. c. 30. U. C. 325. (A. C. 429.) we find: "Defectus alibi aquarum circa torridos fontes rivosque stragem siti pecorum morientium dedit: scabie alia absumpta: vulgatique contactu in homines morbi, et primo in agrestes ingruerant servitiaque."

In the year 354 U. C. (A. C. 400.) "Tristem hiemem, sive ex intemperie cœli, raptim mutatione in contrarium facta, sive alia qua de causa, gravis pestilensque omnibus animalibus æstas excepit."

It would be easy to swell the list of pestilences noticed by this historian in the course of his work; the above-mentioned, however, afford ample testimony to the position that epidemic and epizöotic distempers are very intimately connected, if not on many occasions closely allied.

Again, in the years 434. 3. he gives an account of a pestilence raging extensively and mortally among men and cattle; the latter, he says, were affected with scabies, an eruptive disease, in describing which, Vegetius adds, "Scabies jumentis periculum generat, contagiosa namque est et transit in plures; cutem populis æstuans prurigo pervadit."

Of this epidemic Livy goes on to say that the diseases of the cattle were transferred, by contact, to slaves and labourers; but soon after, the dreadful pestilence fixed itself on the inhabitants of the city itself.

Of the year 396 A. C. Livy's summary is:—

"Tristem hiemem gravis pestilensque *omnibus animalibus* æstas excepit, insanabili pernicie." He describes the epidemic of the year 277 A. C. thus:—"Urbem adfecit pestilentia, pecudes hominesque communi strage corripiens; sed præcipue gravidarum abortibus formidabilis."

Another severe epidemic is described by the same historian, attended with some circumstances well worthy of notice. It raged among horned cattle as well as among men. U. C. 576. (A. C. 178.) "Delectus consulibus eo difficilior erat, quod pestilentia, quæ priore anno in boves ingruerat, eo verteret in hominum morbos. Servitia maxime moriebantur. Eorum strages per omnes vias insepultorum erat;—ne liberorum quidem funeribus Libitina sufficiebat. Cadavera, intacta a canibus ac vulturibus, tabes absumebat; satisque constabat nec illo, nec priore anno, in tanta strage boum hominumque vulturium usquam visum."

Orosius, lib. 4, in the year U. C. 477. gives us this account: "Gravis pestilentia urbem ac fines ejus invasit; quæ, cum omnes, præcipue mulieres pecudesque corripiens, necatis in utero fœtibus, futura prole vacuabat." That of 482 U. C. also was probably a continuation or revival of the former. Oros. lib. 4.

Dionysius of Halicarnassus, in his "Roman Antiquities," furnishes us with, at least, presumptive proofs that, amongst the numerous pestilential epidemics wherewith, from time to time, Rome was visited, an eruptive fever of a peculiar kind occasionally showed itself; thus, for instance, in that fatal λοιμος *(pestis)* which raged in the reign of Tarquinius Superbus. The historian mentions some of the striking symptoms incidentally, whilst relating the mission of Tarquin's two sons, Aruns and Titus, in company with Junius Brutus, to consult the Delphic oracle respecting the cause and cure of this very alarming plague "ὑπερ του λοιμου."—"For there raged over his whole kingdom a certain unusual disease, among *girls* and *boys*, under which many *died* (or many *bodies* were *corrupted*, as the Greek also signifies), but the sickness was most severe and difficult of cure in *pregnant* women, destroying the mothers, together with their infants, in crowds."*

There occurs another still more remarkable description of an epidemic pervading the Roman territory, (in the year U. C. 290. and A. C. 464): it commenced among the inferior animals, and then extended its ravages to man. This circumstance is strongly asserted both by Livy and Dionysius Halicarnasseus, but the latter states it more minutely and vigorously, thus:—"The following year, Lucius Fabius and Pub. Servius Priscus having commenced their government, the Romans did not en-

* Κατέσκηψε γάρ τις ἐπὶ τῆς ἐκείνου βασιλείας οὐκ εἰωθυῖα νόσος εἰς παρθένους τὲ καὶ παῖδας, ὑφ' ἧς πολλὰ διεφθάρη σώμαλα, χαλεπωτάτη τὲ καὶ δυσίατος εἰς τὰς κυούσας γυναῖκας, αὐτοῖς βρέφεσιν ἀποκτείνουσα τὰς μητέρας 'ν ταῖς ἀγοραῖς.—Vide Dionys. Halicar. Antiquit. Rom. Lib. IV. p. 196.—Lutetiæ, R. Stephan. offic.

gage in any work of a warlike or political nature worth mentioning, as they were grievously affected with a pestilential disease, more severe than ever attacked them before; it first assailed the *horses at pasture*, and the herds of cattle; after these it raged among the flocks of goats and sheep, and destroyed nearly all the quadrupeds; then it seized on the shepherds and agricultural labourers; and passing through the entire region, fell upon the city. Indeed, so great was the multitude of attendants and slaves, and poor population destroyed, that it was not an easy matter to ascertain the number. At first the dying were carried away in heaps on waggons, and those who expired of whom no account at all was taken were thrown into the stream of the river running by. Of the senate, a fourth part was estimated to have perished; amongst whom were the two chiefs, and many of the leading men of the state.

"This disease began about the kalends of September, and lasted through that whole year."*

* Though I have given in the text an accurate translation, yet I think the original much more expressive, and therefore transcribe part of it here.

Ῥωμαῖοι—ὑπὸ νόσου κακωθέντες, ὡς οὔπω πρότερον, λοιμικῆς, ἣ τὸ μὲν πρῶτον ἵππων τε φορβάδων καὶ βόων ἀγέλαις προσῆλθεν, ἀπὸ δὲ τούτων εἰς αἰπόλια καὶ ποίμνας κατέσκηψε, καὶ διέφθειρεν ὀλίγου δεῖν πάντα τὰ τετράποδα, ἔπειτα τῶν νομέων καὶ γεωργῶν ἥψατο, καὶ διελθοῦσα διὰ πάσης τῆς χώρας, εἰς τὴν πόλιν ἐνέπεσε. Id. Lib. IX. p. 459.

In this passage we have a most positive and distinct statement that this pestilence, of whatever nature or kind it might

Herodian relates, that in the reign of the emperor Commodus, a dreadful pestilence attacked the whole of Italy; but it raged more violently in Rome itself, inasmuch as that city abounded in its own proper population, as well as with a confluence of strangers from all parts of the world; whence followed a great destruction, both of beasts of burden and men ("πολλή τέ τις φθορὰ ἐγένετο ὑποζυγίων ἅμα καὶ ἀνθρώπων.") The historian, after relating the retreat of the Emperor from Rome, by the advice of his physicians, to Laurentum, to avail himself of the salubrity of its air, and its coolness, adds, "The disease, however, still advanced to a greater height, attended by a great mortality among men, and all the domesticated animals." *

It would not be difficult to increase the catalogue of pestilential diseases affecting both men and beasts at or about the same period; but as that point has been already sufficiently established, it will be better to bring forward such documents as seem more especially to illustrate the history of small-pox itself; and then to prove by unquestionable evidence that *that*

have been, commenced in the equine and bovine tribes of animals; from them it passed to the goat and sheep kind; then to the shepherds and farm-servants; and last of all made its way into the capital; having first traversed the whole Roman territory. Livy records the same epizöotic and epidemic disease in terms not materially different; and to which I have also referred.

* Πάντων τε ζώων τοῖς ἀνθρώποις συνοίκων.—Lib. I. cap. 12.

disease has at various times affected the inferior animals. I shall, then, have accomplished the object I proposed in commencing this disquisition; and its connection with Dr. Jenner's discovery of the Variolæ Vaccinæ will thus be rendered apparent.

Perhaps the earliest authentic account of any eruptive disease subsequent to that which is mentioned in Exodus, and commented on by Philo, is that given by Thucydides, in his history of the Peloponnesian war. This truly classical and elegant writer has given a delineation which carries with it all the authority of accuracy and truth. The symptoms which he enumerates resemble very strongly those of small-pox, whilst he is silent on those which characterize the true plague, such as buboes, parotids, and carbuncles. It, therefore, is fair to conclude that this latter disease is not that which he witnessed, as it is impossible to believe that a writer who is distinguished above most others by the minuteness of his descriptions, should have omitted those above-mentioned, had they existed. He states that the disease proceeded from Ethiopia, and that it was brought into Athens from the Piræus. His entire description is peculiarly animated, and gives an appalling view of the ravages of the disease. As it is too long for insertion here, I shall select that part which points out the manner of attack, and the subsequent symptoms.

" Some persons from no ostensible cause, but on a sudden, being in health, were first seized with violent heats of the head, and rednesses and inflammation of the eyes; and as to the interior (of the head), the faux or throat, and tongue, were immediately bloody (εὐθὺς αἱματώδη), and the breath emitted, bad and fœtid. Next after these (symptoms) sneezing and hoarseness came on, and in a little while the disease (πόνος) descended into the chest with a violent cough. And when it settled in the stomach (καρδίαν) it both turned it, and all off-scourings of bile, that have been named by physicians, succeeded; and these with great distress, (ταλαιπωρίας,) molestia. The greater part were affected with a fruitless hiccup, accompanied with strong spasm; with some, having immediate interval of ease; with others, much later. And the body, to the outward touch, was not very hot, neither was it pale, but somewhat red, livid, effloresced with small pustules and botches (ἕλκεσιν)."*

Such is the description of the incipient and eruptive symptoms of this fatal epidemic: that they are not those which characterize bubonic plague, is, I think, very evident. In the quotation from the ori-

* Τοὺς δ' ἄλλους ἀπ' οὐδεμιᾶς προφάσεως, ἀλλ' ἐξαίφνης, ὑγιεῖς ὄντας, πρῶτον μὲν τῆς κεφαλῆς θέρμαι ἰσχυραὶ καὶ τῶν ὀφθαλμῶν ἐρυθήματα καὶ φλόγωσις ἐλάμβανε, καὶ τὰ ἐντὸς, ἥ τε φάρυγξ καὶ ἡ γλῶσσα, εὐθὺς αἱματώδη ἦν, καὶ πνεῦμα ἄτοπον και δυσῶδες ἠφίει· ἔπειτα, ἐξ αὐτῶν πταρμὸς καὶ βράγχος ἐπεγίγνετο, καὶ ἐν οὐ πολλῷ χρόνῳ κατέβαινεν ἐς τὰ στήθη ὁ πόνος, μετὰ βηχὸς ἰσχυροῦ· καὶ ὁπότε ἐς τὴν καρδίαν στηρίξαι, ἀνέστρεφέ τε αὐτὴν, καὶ ἀποκαθάρσεις χολῆς πᾶσαι, ὅσαι ὑπὸ ἰατρῶν ὠνομασμέναι εἰσὶν, επῄεσαν, καὶ αὗται μετὰ ταλαιπωρίας μεγάλης. Λύγξ τε τοῖς μὲν, μετὰ ταῦτα λωφήσαντα, τοῖς δὲ, καὶ πολλῷ ὕστερον. καὶ τὸ μὲν ἔξωθεν ἁπτομένῳ σῶμα, οὔτ' ἄγαν θερμὸν ἦν, οὔτε χλωρὸν, ἀλλ' ὑπέρυθρον, πελιδνὸν, φλυκταίναις μικραῖς καὶ ἕλκεσιν ἐξηνθηκός. L. II. c. 49.

ginal will be found, if I mistake not, as accurate an account of the leading symptoms of variola as could possibly be expected from any historian not medical. The sudden attack, without any obvious exciting cause; the strong heats of the head; the erythema and inflammation of the eyes; the state of the pharynx, tongue, and fauces; the particular fœtor of the breath; the sneezing, and hoarseness, and violent cough, combined with distress (πονος) in the chest; let these be taken in conjunction with the symptoms which succeed, and which denote the eruptive stage of this epidemic fever, *(especially as pointed out in the last sentence of the Greek quotation)* and we have, if I mistake not, a very lively description of a varioloïd disease.

Another part of Thucydides' narrative worthy of attention is that wherein he states that even after the disease had continued so long as to have reached its height, "the body was not wasted, but bore up against the distress beyond all expectation, so that many perished on the ninth or seventh day, having still a degree of strength (τι δυνάμεως) remaining." Now this we know is often remarkably the case in variolous fever. The historian further states that many who escaped beyond this period of the disease were attacked with a strong ulceration of the bowels, and along with this, a diarrhœa which could not be restrained, by which many were at last destroyed through weakness (ἀσθενεία). He proceeds to add that, if any one got over the greatest (most severe) symptoms, a seizure, in return, upon the extreme parts left its mark there; for it made its attack ἐς αἰδοῖα, and upon the extremities of the hands and feet; and many ultimately escaped with the loss of these, and of *their eyes* also.

Another remarkable circumstance attendant on this epidemic, as noticed by Thucydides, is the immunity from a second attack of the distemper; at least from a fatal one, on the part of those who had once got well over it; so fixed was this belief in themselves and others, that they were deemed "*fortunate* and *happy*." The mode of expression adopted by the historian, is at once significant and forcible. διὰ τὸ προειδέναι τε', καὶ αὐτοὶ ἤδη ἐν τῷ θαρσαλέῳ εἶναι· δὶς γὰρ τὸν αὐτὸν, ὥστε καὶ κτείνειν, οὐκ ἐπελάμβανε.

Contemporaneous with the pestilential epidemy of Thucydides we find from Herodotus the historian, and the epistles of Hippocrates, the father of physic, that a similar disease prevailed in the hosts of Artaxerxes; but whence this pestilence proceeded does not appear:—that it was very fatal to the army is evinced by Artaxerxes' very urgent letter for medical assistance, as given in the epistles of the Coan sage, who afterwards made it his boast to the Athenians that he would not lend his aid to cure the enemies of his country.

Though Hippocrates has not in any part of his works treated expressly on small-pox as a disease, *sui generis*, yet I think it will be manifest, on an examination of some passages in his writings, that he was not only acquainted with it, but has absolutely marked the *eruption* as charácterizing a species in his classification of fevers. These he thus enumerates in the sixth book of his Epidemics, article 17. Πυρετοὶ, οἱ μὲν δακνώδεες τῇ χειρί. οἱ δὲ πρηέες. οἱ δὲ οὐ δακνώδεες μὲν, ἐπαναδιδόντες δέ. οἱ δέ ὀξεές μὲν, ἡσσώμενοι δὲ τῆς χειρός. οἱ δὲ περικαέες εὐθέως. οἱ δὲ διὰ παντὸς βληχροί. ξηροί. οἱ δὲ ἁλμυρώδεες. οἱ δὲ πεμφιγώδεες ἰδεῖν δεινοί. οἱ δὲ πρὸς τὴν χεῖρα νοτιώδεες. οἱ δὲ ἐξέρυθροι. οἱ δὲ πελιοί. οἱ δὲ ἔξωχροι. καὶ τὰ ἄλλα τοιουτότροπα.

"Some fevers are pungent, (or *biting*) to the touch; some are mild; some are not pungent, yet increasing, (or giving out;) some are sharp or acute, but yielding to the hand; some are forthwith, (or quickly) very ardent; others altogether, (or throughout) feeble and arid. Some are salt, others *pustular, dreadful to behold*; some moist to the hand, (or touch;) some very red, others livid, some very pale, and others of the same kind."

The above may be fairly considered a brief summary of the different kinds of fever with which Hippocrates was acquainted; but let us hear his commentator Galen on these very passages; chiefly, however, on that with which we have most concern, and which contains the very remarkable expressions "pustular, dreadful to behold." Passing over his comments, judicious as they appear to be, on all those species which precede the salt and pustular, let us confine our attention to them. He (Galen) begins by saying, "Of these the arid is sufficiently manifest," (i. e. distinguishable by the touch ;)—" not so the salt, for salt is not distinguishable by the touch, but by taste alone." But what follows is less clear, where he says, " others pustular, dreadful to behold ;"—and here we find a different reading. In most of the copies, the words " to behold," being placed before " pustular," but in a few, the word " dreadful" being inserted after " to behold." Sabinus is the only commentator that I have found who was acquainted with this reading, and he is followed by Metrodorus, and all who have come after him down to *———, though in a very few copies I have found it simply " others pustular," without either " to behold," or " dreadful."

After exercising his critical acumen in the examination of the word πεμφιξ ; and after offering several senses, he concludes with the question, whether the passage should be taken with two expressions out of the three, thus πεμφιγώδεες ˉἰδεῖν. Some, he says, have removed the word ἰδεῖν. Others who have not expunged it, have interpreted it either of fevers with φλυκταίναι, or of such as put life in danger, (or affect life :) and he thinks it probable, that Hippocrates may be here speaking of fevers of the λοιμὸς kind; for he says, " as yet, he (Hippocrates) passes over this *one* sort of fever. Now that it is attended with φλυκταίναι we have the testi-

* Galen has not mentioned the name of the author to whom he alludes in this place.

mony of Thucydides, who tells us "the body to the outward touch was neither very hot, nor very pale; but somewhat red, livid, broke out (or effloresced) with small pustules and ulcers." Galen, after some discussion of the passage in Hippocrates, then proceeds, "The addition of '*dreadful to behold*,' according to the reading of Sabinus's copy, is good, for the πεμφιγώδεες are *dreadful* to look at. I have said this before, and now repeat it: for that they are said to be dreadful to look at with reference to us, if they are affected with (φλυκταίναι and ελκεα) pustules and ulcers: but with respect to the patients themselves, when they become delirious and fix their eyes on you in a dreadful manner, &c." At length he sums up, thus, "Wherefore it is most probable to conclude, that either the fever puffy to the touch, (πνευματώδης) or that of the λοιμὸς kind has been described by the word πεμφιγώδης: but if we add the *dreadful* δεινοί to the term πεμφιγώδεες the λοιμώδης *alone* is intended or meant: the difference of which fever, with regard to others is, that the *heat* is of a putrid nature, as in the λοιμὸς which now prevails, and which has continued for so very long a time. For which reason, those who labour under it do not appear hot and burning to those who touch them. though their *inside* is violently affected with a sensation of heat, as Thucydides has said. Galen then cites Thucydides at some length, including the passage formerly quoted by him: 'ὅτε δ' ἐν αὐτῷ φλυκταίναι γίνονται, καὶ ὁ Θυκυδίδης μαρτυρεῖ γράφων οὕτως.

"καὶ τὸ μὲν ἔξωθεν ἁπτομένῳ σωμα οὔτε ἄγαν θερμὸν ἦν, οὔτε χλωρὸν, ἀλλ' ὑπέρυθρον, πελιδνὸν, φλυκταίναις μικραῖς καὶ ἕλκεσιν ἐξηνθηκός."

It may be to the purpose here to state, that according to Hippocrates, Galen, and the other Greek writers, the terms λοιμὸς and λοιμώδης were applied to all pestilential epidemics of whatever class or kind: as will be best evinced by consulting their works.

The opinions of the eastern nations, with regard to the high antiquity of small-pox, have been already alluded to. In China it would appear to have been known from time immemorial; or at least so far back as the dynasty Tcheouè, about 1122 years before Christ. The Chinese name for the disease is Tai-tou, which means "venom from the mother's breast." Père D'Entrecolles, a Jesuit, states that he had read some Chinese books, which notice small-pox as existing in the earliest ages. Père Du Halde, and other Jesuit missionaries, inform us of the existence of a Goddess in the Chinese mythology, under whose superintendence this disease (small-pox) is peculiarly placed: and further, their learned men believe that it has existed in China for 3000 years.

In Hindoostan, if the Brahmins are to be credited, the small-pox is of the remotest antiquity. The Brahmins say that the Veda (one of their sacred writings, which they refer back to an era nearly as remote as that of Moses,) contains forms of worship and offerings to a female divinity, whose tutelar care is exercised over small-pox. A particular tribe of these priests lays claim to the knowledge and practice of inoculation from time immemorial.

What weight is to be given to this testimony, in the general argument for the antiquity of small-pox, must be left to individual judgment; I cannot but think, however, that so many concurring traditions and historical statements go nigh to establish a strong probability of the origin I have assigned to small-pox. That a communication subsisted between the ancient inhabitants of China and India, on the one hand, and those of Egypt on the other, is evinced, as well from ancient history as from a similarity of religious rites, from names of remarkable places, and from existing monuments, emblems of superstition, and temples for the worship of their common idols.

In connexion with this subject we find that at a later period the elder Pliny adds his testimony to the accumulated evidence of ages that many eruptive diseases were to be traced to Egypt, as their *fons et origo*. In speaking of elephantiasis he says "Egypti peculiare hoc malum," and in his observations on "lichen," another cuticular disease imported into Italy in the reign of Tiberius Cæsar, he makes this remark "Adveneruntque ex Ægypto genitrice talium vitiorum medici hanc solam operam afferentes, magna sua præda."

And of the plague itself he says, " Qua in re observatum, a meridianis partibus ad occasum solis pestilentiam semper ire."

It now remains to endeavour to trace a connexion between the eruptive diseases just described, and those which are known from history to have extended their ravages over the world subsequent to the Christian era. During the reigns of more than twenty Roman emperors many descriptions are given, and different symptoms enumerated by historians,—of accounts strictly medical we have few or none. Loss of sight in numerous cases, and still more frequently a severe inflammatory affection of the eyes of those attacked, is particularly noticed. This, together with the pustular eruption and the absence of buboes and parotids, would seem to distinguish the disease from the *pestis inguinaria*, or true plague; and with other symptoms to identify it with small-pox.

So frequent were the visitations of these epidemics, that mention is made of them in almost every reign from that of Domitian downwards.

In the eighth year of the joint reign of the Roman emperors, Marcus Aurelius and Lucius Verus, and of the Christian era 170, a most destructive pestilence raged throughout many provinces of the empire: of this Eusebius, in his Chronicle, thus speaks, " Lues magna provincias occupavit, Roma ex magna parte vexata."

This is the same pestilential eruptive fever which

Galen witnessed at Rome, and of which he says, (as has been already noticed) that it exactly resembled the λοιμος at Athens, as described by Thucydides.

Of this also Julius Capitolinus (in Lucio Vero) writes—" Fuit ejus fati ut in eas provincias per quas rediït Romam usque luem secum deferre videretur. Et nata fertur pestilentia in Babylonia, ubi de templo Apollinis ex arcula aurea, quem miles forte inciderat, spiritus pestilens evasit, atque inde Parthos Orbemque complesse."

In the third year of the reign of Gallus and Volusian, (A. D. 256.) a great "pestis" afflicted the world: it was said to have originated in Ethiopia; thence to have spread over the whole empire, and scarcely to have ceased within a period of ten years. This distemper raged throughout the whole of Egypt during these times, as Eusebius relates (in Chronico).* Africa, too, in its entire extent would appear to have suffered from this pestilential visitation, as we learn from Pontius in Vita St. Cypriani; and Gregory Nyssenus.

Cyprian in his pastoral letter to the Christians under his care, enjoins them to show the nature and power of their religious faith, by their brotherly love towards each other, and their good offices to their

* As I had not access to Eusebius' Chronicon in the original, the passages adduced are from the Ecclesiastical Annals of Baronius.—The same may be observed with respect to some other Greek authorities.

Pagan neighbours and fellow-citizens; and they obeyed his injunctions with the willing minds of martyrs, in assisting and rendering service even to their enemies.

Cyprian's address is entitled "De Mortalitate;" and in the form of a note to it is given a description of the disease.*

A. D. 263.—In the ninth year of the Emperor Gallienus, a pestiferous and deadly disease broke out at Alexandria in Egypt, of which Eusebius gives us some notices from the letters of

* D. Cæcil. Cyprian. de Mortalitate, sive lue mortiferâ. Vide not. p. 156. fol. Oxon. "Revera morbus iste epidemicus tam varia et erratica habuit symptomata ut ad pestis proprie dictæ typum minime responderet. Corporis vires solutus in fluxum venter eviscerabat. In faucium vulnera conceptus medullitus ignis exæstuabat. Assiduo vomitu intestina quatiebantur, oculi vi sanguinis inardescebant, quorundam pedes vel aliquæ membrorum partes contagio morbidæ putredinis amputabantur, debilitabatur incessus, obstruebatur auditus, cæcatus fuit aspectus. Teterrima hæc lues in Arabia primum nata, deinde in Ægyptum devecta, postmodum in Africam processit, et exinde per occidentalis Romani imperii partes grassabatur; furore Atheniensis illius, quam describit Thucydides æmulo, itinere consueto, et via quasi prætoria: ait enim Plinius L. 7. c. 50: Observatum à meridianis partibus ad occasum solis pestilentiam semper ire.

Imperantibus Gallo et Volusiano hoc malum contigit, testibus Eusebio in Chron. Oros. l. 7. et Zonar. Annal. l. 2. Porro pacis tempore hoc evenisse docet Euseb. l. 7. c. 22. ex literis Dionys. Alex."

This description of the disease is taken from Tertullian.

Dionysius, then Bishop of Alexandria, written at or about the festival of Easter. "In such a crowd of miseries as at present overwhelm us, neither at this time nor at any other, nay not even at that period which men deem of all other the most joyful, does it seem fit that a festival should be celebrated. For now all are full of lamentations; now all weep; now nought but sorrow and complaints occupy the whole city, on account not only of the multitude of those who have just died, but of those who are dying every day. As it is written of the first-born of the Egyptians, so now also a great cry is sent forth: for there is not a house in which there is not one dead."

After forcibly depicting the exemplary conduct of the Christians towards their fellow-sufferers, he contrasts it with that of their Pagan neighbours thus, "But the Gentiles acted contrarywise in every respect; for they forcibly drove out of their houses those who began to show signs of sickness; they deserted their nearest friends; they cast the half-living into the streets; they exposed the dead bodies without sepulture to be torn asunder by dogs; seeking to avert from themselves all participation, or as it were all communion of death; from which, however, it was evident they could not escape, notwithstanding all their devices."*

* In this, and other similar quotations, the reader is requested to bear in mind that the early writers did not sufficiently dis-

The historian further adds, "At non tantum Ægyptum hujusmodi adeo immani clade contigit exagitari, verum et aliæ orbis regiones aliis diversis atque atrocioribus affectæ sunt malis." And again, "Nam et pestilentia (speaking of the same year) tanta extiterat vel Romæ, vel in Achaicis urbibus, ut uno die quinque millia hominum pari morbo perirent."*

On the authority of Eusebius, we have the following account of a pestilence attended with eruptions and other symptoms strongly indicative of small-pox.

It occurred in the 6th year of the Emperor Maximinus, and A. D. 311, and is thus described. "Pestis deinceps oberrare cœpit; morbi etiam cujusdam novi et peregrini (exulceratio quædam, quæ æstus et fervoris proprio nomine anthrax, id est

criminate between the pestilence, strictly so called, and those pestilential epidemics, combined with eruption, of which so many instances have been recorded. There can, therefore, be little doubt that under the head of *pestis* and *pestilentia* both affections have been often confounded; and that the symptoms which are peculiar to each are frequently omitted. In the general argument, then, I am not disposed to lay any stress on those pestilences where eruptions are not mentioned by the historian. When such examples are brought forward, it is chiefly to pseserve the continuity of the narrative. They cannot, however, be entirely rejected, inasmuch as symptoms omitted by one writer, are stated by another as occurring in the same Epidemic. This remark especially applies to eruptions.

* Euseb. Ecclesiast. Hist. Lib. vii. cap. xxii.
Baronii Annales Ecclesiastic. Vol. 2. page 582-3.

carbunculus, appellatur) gravis et violenta impressio, qui per universum corpus pererrans, mortifera his qui eo perturbabantur, injecit pericula. Quinetiam quoniam in oculis potissimum ejus insidebat cruciatus, infinitos ferè viros cum conjugibus et liberos cæcos reddidit." It is quite evident that the anthraces, or carbuncles, here spoken of, could not be those which denote the true plague, as in that disease they are always local; and besides, the eyes are seldom or never injured.*

This epidemic continued, or revived, during the subsequent year (A. D. 312), attended with famine: of the effects of both Eusebius has drawn a most appalling representation. He thus expresses himself respecting the disease: "Pestis autem omnes domos integras et familias penitus depascebatur, et eos vel maximè, quas fames propter alimentorum

* The description of this epidemic, or, to speak more correctly, of an eruptive disease, which accompanied it, is, in the Greek, so strongly indicative of variola, that I am induced to quote it: Λιμός δ' ἀδέκηlος επισκήπlει, καὶ λοιμὸς επι τούτῳ· καί τινος ἑτέρου νοσημαlος, ἕλκος δὲ ἦν φερωνύμως τους πυρώδους ενεκεν, ανθραξ προσαγορευόμενον, επιφορὰ· ὃ καθ' ὅλων μὲν ἕρπων των σωμάτων σφαλερους ενεποίει τοῖς πεπονθόσι κινδύνυς. ὀ μὴν αλλα καὶ καlα των οφθαλμων διαφερόντως επιπλεῖστον γινόμενον, μυρίυς ὅσυς ἄνδρας ἅμα γυναιξὶ καὶ παισὶ, πηρους απειργάζεlο.

Now, if we take this as a true account of the symptoms of the *certain other* disease above-mentioned, I would simply ask what disease, other than small-pox, *that* could have been, which, attended with the pustular eruption cver the whole body, and fixing itself about the eyes, rendered thousands (μυρίυς) of men, women, and children blind?

Consult Euseb. Hist. Ecclesias. Lib. ix. cap. viii.

affluentiam quâ fruebantur, consumere et conficere non poterat;" and goes on to say, that the magistrates and prefects, and many others who, having power and authority, possessed a superabundance of every thing, as if the famine had designedly spared them that they might perish by the pestilence, underwent a very severe, and, for the most part, a hurried death. In this way whole families were cut off very rapidly; so that two or three bodies of the dead were carried out from one house for burial at the same time. All places, therefore, by-ways, markets, streets, overflowed with tears, sorrow, and wailings; nor could aught be witnessed save the most piteous lamentations and weeping. During the prevalence of this pestilential epidemic, the conduct of the Christians was worthy of their high vocation, forming a strong contrast with the demeanour of their heathen countrymen.

Eusebius thus simply but strikingly pourtrays it:—"Quo quidem tempore clara singularis Christianorum erga quemque et studii et pii animi indicia apud omnes increbuerint. Nam hi solum in tanto malorum cumulo suis rectè factis, et piis officiis misericordiam declarabant et benignitatem: quorum alii indies singulos mortuorum funeribus et sepulturæ (infiniti enim erant, quibus sepeliendis nemo curam adhibuit) diligentem navarunt operam: alii, multitudine eorum qui per totam civitatem fame urgebantur in unum coacta, omnibus panes dispertierunt: usque adeo ut hoc præclarum facinus per omnes homines constanti famâ et magna celebritate pervaderet; et singuli Christianorum Deum gloria et laude prædicarent, eosque solos et veros Dei cultores re et factis comprobatos faterentur."

A. D. 376 and 13 of the Emperor Valens, Baronius, on the authority of St. Ambrose and Paulinus Nolanus, states that a pestilence raged throughout Europe, attacking both man and beast: "Lues pariter boum atque hominum, cæterique pecoris: ut etiam nos, qui bellum non pertulimus, debellatis tamen pares fecerit pestilentia." Ambrosius comment. in Luc. lib. 9. cap. 21.

In the year 410 of the Christian era, famine and pestilence prevailed at Rome to so great an extent, that, as Zosimus (according to the version of Baronius) expresses it, "Omnia plena cadaveribus erant. Cumque non possent extra urbem sepeliri cadavera quod omnem exitum hostes observarent, urbs ipsa mortuorum sepulchrum erat." Zosimus, lib. 5.

A plague (pestis ingens) raged at Rome A.D. 467. but no particular account is handed down to us either of symptoms or mortality. Pope Gelasius, in a letter to Andronicus, says, "pestilentia tanta subrepsit ut toleranda vix fuerit." Vide Baron. vol. 6. p. 281.

In the reign of Zeno (A. D. 484.) pestilence and famine ravaged all the cities of Africa. Id. p. 427.

Famine and plague distressed Italy in the 13th year of Justinian (A. D. 539.); the pestilence, however, seems to have been the result of scarcity. The symptoms mentioned are those of inanition and emaciation—not of plague.*

Procopius relates that, in the 18th year of Justinian's reign (A. D. 544.) a great plague began to rage in the East, for which no remedy was found, except from God, from whom it proceeded, although (he says) "many arrogant persons vied with each other in commenting on the causes of (its) physiology, all of which were vain and inexplicable, and only deceiving with idle words, for this disease spared neither age, nor sex, nor place; but whence it had its beginning, and in what manner it destroyed its victims, I will relate. It began in Egypt, at Pelusium; hence it took in the whole world, continually advancing; it left no lurking-places, nor did it attack the same persons a second time."

Evagrius records the same pestilence described by Procopius in his "Persic war;" but he has added many circumstances, either passed over by Procopius, or differing from his account. This, however, is easily explained, for Evagrius did not commit his recital to writing till fifty years after the first appearance of the epidemic (A. D. 544) to which time the narrative of Procopius applies: since he states that the disease prevailed in Constantinople for only three months, and that in the beginning of it a few died; but at length, five thousand, and, still oftener, ten thousand were hurried off by it daily; so that many, even of the

* Procopius de Bello Gothico, Lib. 2.

wealthy class, their attendants and servants having already fallen victims, perished rather from want of persons to take care of them than by disease; and, for the same cause, remained unburied.

Evagrius assures us that this pestilence raged amongst mankind during fifty-two years, nor were its ravages confined to the East alone but extended throughout the world; it depopulated some cities entirely and often returned to the very same places it had before invaded; and assumed many symptoms different from those which had marked its early progress.

It began to prevail about two years after the capture of Antioch by the Persians under Chosröes, (A. D. 542) in some respects resembling the plague at Athens described by Thucydides, in other respects differing from that. It was, by common report, said to have originated in Ethiopia, and thence diffused itself over the world attacking different places successively, and sparing none of human kind.

Evagrius says that at the moment of writing his account of the disease, it had invaded Antioch for the fourth time; in this attack it deprived him of a child and grand-child; he himself having undergone the distemper on its first appearance, fifty-two years before.

In the year of Christ 558, the 32d of the Emperor Justinian, this fatal pestilence again attacked Constantinople, of which Agathias, as reported by Baronius, thus expresses himself, "Eodem anno, vere jam ineunte, derepentè et iteratò urbem pestilens morbus invasit et infinitam sustulit multitudinem: haud tamen prorsus cessavit ex quo tunc primùm anno Justiniani Imperii quinto irrepere hanc nostram regionem occepit." *

On this occasion Justinian thought fit to issue his imperial edict of advice to the people of Constantinople, calling on them to depart from the iniquity of their ways, and to invoke the pardon of the Almighty. After having pointed out to them various crimes of which they had been guilty the edict concludes thus against the obstinate offender, "Primùm quidem obligatus erit judicio Dei, post hæc et nostram indignationem sustinebit."

To point out to my reader the difference between many of those epidemics which have been adduced, and the ordinary plague, I shall here trespass somewhat longer on his indulgence by bringing forward a brief account of *that* disease as it appeared in Italy in the last year of Justinian, A. D. 565; and therefore contemporaneous in its visitation with the more general and dreadful pestilence just noticed as prevailing epidemically in one part or other of the empire during his whole reign. It will show also that the prominent symptoms of the bubonic plague were well known to historians; so far at least as not to be confounded with those which marked many of the other pestilential epidemics, identified as these are, in my judgment, with small-pox.

* Agathias, lib. 5.

Paulus Diaconus* thus describes the Pestis inguinaria :—

"In these same times (A. D. 565) a very great pestilence arose in the province of Liguria. For there suddenly appeared certain signs (quædam signicula,) through the houses, on the doors, vessels, clothes, which, on any attempts to wash them out, became the more apparent. But, after the completion of the year, little glands (glandulæ) like a nut or date began to grow in men's groins and other more delicate places; these swellings were immediately succeeded by an intolerable heat of fever; so that the sufferer was carried off on the third day. Yet if any one got over the third day he had hopes of living."

Paulus says nothing farther respecting the symptoms of this disease, but those he *has* mentioned are fully characteristic of true plague, and distinguish it clearly from an eruptive fever. His detail of its natural and moral effects is eloquently piteous and impressive.

The historic sketch of pestilential eruptive fevers has brought us to that era to which the first appearance of small-pox has been usually assigned, namely, A. D. 568 or 569 during the siege of Mecca by the Abyssinian army under Abrahah the viceroy—and his sudden retreat from Arabia is attributed to the breaking out of small-pox among his troops. This

* See Paul. Diac. de gest. Longobard. lib. 1. cap. 4.

event is ascribed, in the language of fable, to an assault of pebble-stones thrown from the beaks and talons of great birds on the Abyssinian army; but there is no necessity to have recourse even to figurative expression; since nothing can be more probable than that the pestilence, which continued through Justinian's time, met the Abyssinians at Mecca. In this view of the subject it is quite as probable that the Abyssinians were infected by the Arabians as the Arabians by them; nor is it at all improbable that both might have had the disease in their ranks. Bruce relates that at Masuah on the confines of Abyssinia he met with a MS. account of the war of the Elephant, in which the native author told nearly the same tale as the Arabian writers on the subject, adding, that at that time the small-pox was first observed in Arabia.

From the time of the cessation of the great and long-continued pestilence during the reign of Justinian, which in its course depopulated many cities and districts of the empire and extended its ravages through Europe, Asia, and Africa destroying in its course two millions of the human race, the accounts transmitted to us, from the dark ages, of pestilential diseases are so obscured by monkish fables and miracles that the frequent recurrence and dreadful mortality of these epidemics are almost the only facts of which we can be certain; the attendant symptoms are scarcely ever stated; the consequences are sometimes alluded to, as for instance, disfiguration of the person, and loss of limbs occasionally; but most frequently *blindness;* inasmuch as restoration of sight lost, during the pestilence, is often recorded as the miraculous effect of saintly interposition. From this period too (the close of the 6th century) historians began to designate the genuine plague by the name of Pestis Inguinaria, or Lues Bubonum; so that by this appellation a more accurate distinction than had previously obtained was drawn between true plague and other pestilential epidemics. Yet even after the time that the Arabian physicians began to treat *expressly*, and by name, of the variolous eruption, their descriptions are not very accurate, nor their histories satisfactory. Of this any one may satisfy himself by consulting Rhazes, as published by Mead in his 1st vol. Paris, 1757, edit.

If such then be the inaccuracy of the older medical writers professedly treating of a specific disease what ground have we to expect that historians and chroniclers shall be more precise in their descriptions of wide-spread epidemics.

Moore in his history of small-pox, though he combats the antiquity of that disease in Europe, makes a very just observation respecting the alleged silence of historians on the subject of its appearance. "The small-pox being included in the term pestilence explains satisfactorily why it is not named by the older writers; and also accounts for the very frequent occurrence of the plague in early times. In the old chronicles the plague is recorded to have visited

France eleven times in the 9th, and six or seven times in the 10th century. Some of these visitations were unquestionably the small-pox and measles. The pestilence of *fire*, of which horrible descriptions are given, may have been in some instances the small-pox." Of the justness and truth of these remarks I have no doubt; even as applied to a period later than the 10th century. Thus, as an instance, I shall transiently adduce one from Baronius who, in speaking of St. Martial's miraculous powers, attributes to them the stopping of a pestilence even after he was laid in his grave. Baronius relates it on the authority of a writer of the history of Aquitaine of which some fragments only remain: his words run thus: "His diebus (A. D. 1029) *lues gravissima* Semovicinos devoravit, *incendens* corpora et *exardescendo* devorans."—Now, however we may reasonably deny the miracle, we have no reason to question the reality of the pestilential epidemic; or the accuracy of the historian with reference to its general and more obvious character which certainly assimilates it rather with an ardent eruptive fever, such as small pox, than with true plague or lues inguinaria, a name ordinarily given, then and for ages preceding, to the bubonic pestis.

Accounts of pestilential fevers still more strongly marked in their resemblance to small-pox, and in *their essential difference* from plague, occur so frequently in the Ecclesiastical Annalists as to weary the reader. But this branch of the subject has been so amply treated of by Dr. Willan, and his references are so numerous, that I think it unnecessary to go into it more at large.* Enough, I believe, has been already said to draw attention to this very interesting topic.

Having brought down the history of eruptive pestilential diseases, as they affected both man and the inferior animals, to that period in which small-pox is on all hands admitted to have been recognized I am now to prove by the direct testimony of those who have written expressly on the diseases of cattle, that an epidemic small-pox, described under that specific name, has been often known to have raged amongst them.

Lancisi, in his treatise *De Bovilla peste*, par. iii. p. 142, asserts that the disease among the horned cattle which he describes, and which was epidemic in the papal territory in 1713-14, was similar to that which had occurred in Italy, nearly two centuries before (1514) and of which Fracastorius thus writes in Lib. I. de Contagione, cap. 12. "Referemus etiam insolitam anni 1514 contagionem, quæ in boves solùm irrepsit. Visa primo circa Foro-juliensem tractum, mox sensim, et ad Euganeos delata, atque indè in agrum nostrum. Abstinebat primò bos à cibo sine causa ulla manifesta; spectantibus autem in ora eorum bubulcis, asperitas quædam et parvæ pustulæ percipiebantur in palato et ore toto. Separare protinus infectum oportebat à reliquo ar-

See Willan on the Atiquity of Small-pox.

mento, alioquin totum inficiebat. Paulatim labes illa descendebat in armos, et indè ad pedes; ac quibus ea permutatio fiebat, sanabantur ferè omnes; quibus autem non fiebat, plurima pars peribat."

Lancisi, after referring to the above passage in Fracastorius, says, "It is, therefore, evident that this kind of pestilence was by no means unknown to our ancestors. The name which they gave it I shall now examine into, as I think even this not foreign to our purpose: for, though it be true that names are imposed not by any authority of nature but at men's pleasure, still such investigations should, by no means, be neglected." He then proceeds: "*Malis* igitur, ut a Gesnero et Aldrovando accepimus, armentorum ægrotatio illa vocatur, quæ cibi fastidio et cessante ruminatione se prodit. Duplex est apud illos *Malidis* genus, siccum et humidum. Illud narium atque oris ariditas consequuntur. Istud mucus et fluens ex ore pituita. *Quatuor* Græci assignant commemorantque *Malidum* species. (μάλις enim aut μαλίη Græcorum vox est;) ὑγρὰν primam vocant, scilicet humidam; ξηρὰν alteram, seu siccam; tertiam ἀρθρῖτιν, articularem videlicet; quartam denique ὑποδερματῖτίν, hoc est subcutaneum. Quatuor istas non casu, sed datâ operâ, recensui. Certum quippe mihi est eas omnes, si ab articulari discedas, in hujus temporis contagione se immiscere. Nam ut nihil de sicca humidaque dicam, quæ in dubium revocari nequeunt, cur subcutaneum excludamus? Nonne illam in omnium oculis ponunt et tabe ad cutem depositâ depilatæ, et horror pilorum, et armorum, cluniumque tremor, maculis denique et pustulis infecta cutis? Adeo ut quibusdam in mentem venerit cogitare boves non lue, ut nunc res est, sed *ipsis pustulis,* quas *variolas* vocant, interire."

Bernardinus Ramazzini, in his Constitut. Epidem. for the year 1690, after having entered at large into the subject so far as it respected the diseases of man, proceeds thus:—"Vim hujus male moratæ constitutionis experta quoque sunt cujuscunque generis Animalia, quæ in magno numero interiere; sicut autem in ea, quam describit Silius Italicus, quæ sicca fuit et præfervida vim primi sensere canes, ut in hac frigida et præhumida, prima clades pecudum fuit, quibus postquam per dies aliquot ægrotassent, veluti per crisim apparebant *variolæ* in capite et collo, ac plerumque ab iisdem excæcabantur, sic quæ ex vehementia morbi non perierant, tandem ex inedia contabescebant. Sues quoque turmatim suffocati moriebantur." To this account he farther adds, "Tubercula autem illa quæ in capite, collo, et cruribus pecudum visebantur, revera *variolas* fuisse licet profecto asserere, quando nec figura, nec colore, nec liquore in illis contento, nec magnitudine, nec modo quo solvebantur post suppurationem, nigra crusta superstite, quicquam a puerorum variolis discrepabant."

This description of symptoms may be the more strongly relied on, inasmuch as Ramazzini merely narrates what he witnessed then, and at a subsequent period in 1711. Of the year 1691 he observes, "Haud secus quam anno elapso magna fuit animalium strages, ac pecudum præcipuè, ita ut ovillus grex totus penè deletus fuerit."

I have already observed that the statements of this physician are the more to be relied on because he had no favourite theory to support by them, nay he was even adverse to a prevailing opinion of his Italian brethren that the diseases of cattle were communicable to men; yet with the integrity which should always scrupulously attend the medical historian he adds, at the close of his dissertation on the Epidemic Constitution of the year 1691, these remarkable words:—At profecto gravis hæc morborum tempestas omnino desiisset, ni Variolæ, quo magis ambientis calor tepescebat, eò ferocius sæviissent, cum non solum puellarem ætatem demeterent, sed etiam grandævos, ac præcipuè utero gerentes, quotquot hoc morbo laborarent. De Variolarum natura et causis, cum tot extant doctissimorum hominum scripta, nil addam nisi quod in hac Variolosa Constitutione, quæ circa autumni finem vim suam magis exercuit, facilius evaserint ii quibus nec detractus fuit sanguis, nec ullum administratum remedium, toto curationis negotio naturæ commisso." But it was in the course of the year 1711 that the destructive pestilence, since called Lues Bovilla, raged in Italy, and which Ramazzini has so strongly and so feelingly delineated. His introductory expressions would lose so much by translation that I shall present them to my reader in their original form. "Nemo non novit quam inopinatè, quam violenter Bubulum genus dira contagio pervaserit flammæ ad instar, quæ ope nulla humana consopiri, nedum restingui potuerit. Hæc primum quidem in Vicentino agro subobscure observari cœpit; mox in Patavinum transgressa apertè se prodidit, ac longè latèque effusa, usque ad ipsa urbis pomæria, tam magnam ac horrendam böum stragem edidit, ut tum rura tum civitatem mœrore ac metu complerit. Mœret rusticana plebs, imo attonita stupet dum ampla bovilia vacua ac deserta intuetur."

His detail of symptoms is also very forcible and clear. "The kind of affection which seemed to have declared exterminating war on the whole race of oxen was evidently a malignant, destructive, and (if you will) a pestilential fever commencing with chills, rigor, horripilatio, succeeded quickly by pungent, violent heat diffused over the whole body, with frequency of pulse, and accompanied by great anxiety and heavy panting, together with stertor, and, in the commencement of the fever, with stupor and a kind of lethargy; a continual flow of stinking matter from the mouth and nostrils; a most fœtid discharge from the bowels, and this at times bloody; loss of appetite and rumination was altogether destroyed; on the fifth, or sixth day pustules broke out over the whole body of the animal, and tubercles resembling *variolæ* in kind and appearance; death common to all, and in the same manner, about the fifth or seventh day; very few escaped, and these rather by chance than the efficacy of any remedies."

Our author then proceeds to inquire into the epidemic constitution of the air and season, and fully acquits them of having either caused, or contributed to the production of the disease: he says, "Aërem itaque et pascua ab hoc crimine, quòd hujus morbi in causa fuerint, absolvere fas erit." He goes on to state that "it now is sufficiently certain, and is related in *the public acts*, that from the crowds of oxen which the traders are in the habit of bringing from Dalmatia and the neighbouring countries, a single ox separated from the rest, and being found by a herdsman, and brought to the pastures of Count Trajan Borromeo, '*pro hospitio, eadem labe qua erat infectus hospites suos fœdavit;*' for having died a few days afterwards, by degrees the whole herd miserably perished by the same disease, one ox only excepted, in the neck of which a seton had been inserted. The same contagion, therefore, creeping by degrees in a short time has pervaded the whole extent of the territory of Padua; and then, at length having crossed the Po, threatens the same destruction to the inhabitants of Æmilia."

Ramazzini having thus established the fact of the importation of this eruptive plague among the herds of Lombardy proceeds to account for its wide-spread dissemination by means of its highly contagious nature. His words are "Quoniam autem morbosi seminii ea est indoles, ut facile sobolescat et in immensum se propaget, si in subjectum proprium incidat, non est quod miremur si hæc contagio tam longè latèque sese effuderit." He then touches on the opinion entertained, even at the time, that this distemper might, in some shape or other, be propagated to man. "Num autem hic epidemicus bovini populi morbus humanis corporibus labem aliquam affricare valeat, *ut nonnulli suspicantur*, non alienum est disquirere." After having discussed some very interesting and curious arguments on each side of this question, supported by historical facts, he concludes *against* the probability of the disease attacking the human subject, and sums up the evidence on his own side with this expression, "habemus igitur unde spes nostras alamus." Nevertheless he prudently recommends a strict attention to such measures as were in his opinion calculated not only to prevent the infection from acting on man, but to check its operation on the untainted cattle, very judiciously remarking—"Ubi enim de morbo contagioso agitur nunquam satis cavemus, dum cavemus."

With respect to the means to be employed in the case of the diseased cattle themselves, he treats more at large: his general plan, however, he expresses in a few words—"Crediderim itaque in curatione hujus malignæ febris ea methodo procedendum quæ a bonis medicis servatur in curanda Variolosa puerorum febre." By this short sentence, Ramazzini strongly evinces his opinion of the nature of this bovine pestilence.

It is worthy of observation, that the medical writers who saw and described the Variolæ among cattle, especially Lancisi and Ramazzini, have referred to the Latin poets for illustrations of the disease. Such authority is sufficient to justify any writer of the present day in making use of the language of Virgil, of Ovid, of Lucretius, or of Seneca, to elucidate any statement that he may have advanced. Before producing the quotations themselves, I would beg the reader to remark that as modern physicians have recognised in the delineations of the poet a real disease which they themselves witnessed, so in like manner the poets assimilate the disease which they paint to that left us by Thucydides, and refer to Egypt as its source,

"Nam penitus veniens Ægypti è finibus ortus."

Such coincidences are too striking not to arrest attention; in the general argument they can not and ought not to be overlooked; and they go not a little way to strengthen and to confirm the positions we have taken respecting the origin and progress of Variolæ.

Virgil, after a most pathetic description of an epidemic distemper which destroyed the domesticated animals, among which he enumerates oxen, horses, dogs, swine, and sheep, concludes thus:—

"Jamque catervatim dat stragem, atque aggerat ipsis
In stabulis turpi dilapsa cadavera tabo:
Donec humo tegere, ac foveis abscondere discunt.
Nam neque erat coriis usus: nec viscera quisquam
Aut undis abolere potest, aut vincere flamma:
Nec tondere quidem morbo illuvieque peresa
Vellera, nec telas possunt attingere putres.
Verum etiam invisos si quis tentaret amictus;
Ardentes papulæ, atque inmundus olentia sudor
Membra sequebatur. Nec longo deinde moranti
Tempore contactos artus sacer ignis edebat." *

Ovid is equally strong in his expressions, though not so minute in his description of symptoms.

"Strage canum primâ, volucrumque, oviumque, boumque,
Inque feris subiti deprensa potentia morbi.
Concidere infelix validos miratur arator
Inter opus tauros; medioque recumbere sulco.
Lanigeris gregibus, balatus dantibus ægros,
Sponte suâ lanæque cadunt, et corpora tabent.
Acer equus quondam, magnæ que in pulvere famæ,
Degenerat palmas, veterumque oblitus honorum,
Ad præsepe gemit, letho moriturus inerti."

"Omnia languor habet, silvisque, agrisque, viisque
Corpora fœda jacent: vitiantur odoribus auræ."

"Pervenit ad miseros damno graviore colonos
Pestis, et in magnæ dominatur mœnibus urbis." †

* Vide Georgic. lib. III. sub fin.

† Metam. lib. VII.

Ovid proceeds to depict the ravages of this malady when it affected men, almost in the very words that Thucydides has employed in his account of the Athenian pestilence. But this historian is, as it were, imitated or paraphrased still more closely in Lucretius' sixth book "De Rerum Natura."

"Nunc, ratio quæ sit morbis, aut unde repentè
Mortiferam possit cladem conflare coorta
Morbida vis hominum generi, pecudumque catervis,
Expediam."

I shall not follow the poet through the whole of his animated and accurate detail; but select such prominent symptoms as tend most impressively to characterize the disease.

"Principio, caput incensum fervore gerebant,
Et dupliceis oculos suffusâ luce rubentes.
Sudabant etiam fauces intrinsecus atro
Sanguine, et ulceribus vocis via septa coibat,
Atque animi interpres manabat lingua cruore,
Debilitata malis, motu gravis, aspera tactu:"

"Nec nimio cuiquam posses ardore tueri
Corporis in summo summam fervescere partem:
Sed potiùs tepidum manibus proponere tactum,
Et simul ulceribus quasi inustis omne rubere
Corpus, ut est, per membra sacer cùm diditur ignis."

"Quorum siquis, ut est, vitârat funera lethi
Ulceribus tetris, et nigra proluvie alvi;
Posterius tamen hunc tabes lethumque manebat.

Aut etiam multus capitis cum sæpe dolore
Conruptus sanguis plenis ex naribus ibat."

"Horrida pædore, et pannis cooperta, perire
Corporis inluvie: pellis super ossibus una,
Ulceribus tetris prope jam sordique sepulta.
Omnia denique sancta Deûm delubra replêrat
Corporibus mors exanimis, oneratáque passim
Cuncta cadaveribus cœlestûm templa manebant,
Hospitibus loca quæ complêrant æditentes."

Seneca, the philosopher and poet, enlarging on the circumstances of the λοιμὸς that ravaged Thebes in Bœotia, during the reign of Creon, as described by Sophocles in his 'Œdipus Tyrannus,' thus depicts the symptoms of the disease:

"O dira novi facies leti!
Gravior leto!—Piger ignavos
Alligat artus languor, et ægro
Rubor in vultu, maculæque caput
Sparsere leves; tum vapor ipsam
Corporis arcem flammeus urit
Multoque genas sanguine tendit,
Oculique rigent, et sacer ignis
Pascitur artus. Resonant aures
Stillatque niger naris aduncæ
Cruor, et venas rumpit hiantes.
Intima creber viscera quassat
Gemitus stridens."

On referring to Goëlicke's dissertation "De Lue Contagiosa Bovillum genus nunc depopulante," (1730), we find the more common symptoms described very much as they had already been pourtrayed by former writers, such as Lancisi, Lanzoni, Ramazzini, and others. From this close resemblance, he draws the necessary inference that the distemper was one and the same throughout the various regions it visited: he thus expresses himself on this topic:—"Ne quis vero existimet luem nostra inter pecora hodie grassantem alterius forte indolis esse quam quæ bovilia et armenta aliarum regionum devastavit, agedum, excutiamus tantum principes, quæ morbum istum comitatæ sunt, circumstantias, atque observatis per extispicia factis comparemus, morbumque per omnia sibi similem esse constabit." He then subjoins to his enumeration of the more ordinary appearances, "Observarunt nonnulli linguam admodum inflammatam, in cujus etiam superficie copiosas aphthas conspexerunt. Quædam istorum pecorum ophthalmia laborarunt. In quibusdam circa ambitum intestini recti ecthymata, carbunculos vel petechias representantia, in aliis vero vibices esse detectas. In aliis toto corpore *pustulæ* emicuere, quibus ruptis ichor tenax et sanguinolentus effluxit." Corresponding with this last symptom, is a remarkable passage in Lanzoni's detail: "In aliquibus pustulas sub cute effloruisse adeo ut nonnulli crediderint *variolis* boves esse adfectos." Goëlicke, as well as all the authors whom I have consulted, mentions two other circumstances worth noting; namely, "the disease attacked plump, healthy, and strong cattle, in preference to those that were lean and weakly; which latter, when affected, were more likely to survive:" and "quamplurimæ vaccæ gravidæ abortivere."

That this virulent and deadly disease was propagated by contagion or infection there can be no doubt: and indeed even those writers who have been disposed to ascribe its origin to "local causes," or to an "epidemic constitution of the atmosphere," do nevertheless admit its dissemination by contact or proximity.* Goëlicke, in his proëmium, on the authority of a German physician, John Kanoldus, (whose work I have not been able to obtain) derives the origin of the "Lues contagiosa," of which he treats, "from the confines of Tartary, through Muscovy into Poland, and thence partly towards the North, and partly to the South, this dire pestilence spread itself by degrees, and in that course into Livonia, Curonia, Prussia, Pomerania, Holstein, Holland, and thence to England itself; from England, partly into the Turkish empire, Hungary, Sclavonia, Croatia, and thence into Austria, Moravia, Stiria, Carinthia, Carniola, Bavaria, &c. &c.: it partly penetrated into Italy, and moreover into France, and even Spain, and thence at length again into Germany."

It may here be worth while to give Sauvages' character of this disease. He places it in the genus Dysenteria; though this flux of the bowels is evi-

* Goëlicke remarks, in common with many others, that the cattle who were once affected, and recovered, never again, "nisi rarissimis exemplis," were subjected to the same disease.

dently only symptomatic; and, at times, altogether absent.

"9. *Dysenteria pecorum. Malis humida, Lancisii de peste bovilla. Pestis bovilla, Ramazzini. Lues vaccarum Tubingensis.*

"Morbus hic totam successive Europam, et demum nuper Galliam devastavit, atque boves eripuit. Hunc observavi apud Helvios, Comitiorum jussui ut obtemperarem; boves ibi, in Occitania capræ et verveces infestabantur; oriebatur ab inappetentia, ruminationis cessatione, languore, boves passim in campis tristi et demisso capite vagabantur interdum ejulantes, salivantes, dein frigebant auribus, fremebant toto corpore, alvi fluxum successive cruentum et torminosum patiebantur, hinc colliquatio, fæces oleosæ et mucosæ cruentæ, paucissimi sanabantur obortis ad nares et caput *pustulis crustaceis*; et alopecia subsequente."

Sauvages, in the genus *Pestis,* would appear to describe under the species *carbunculosa* a disease nearly allied to, if not the same with, the *pestis bovilla* of Ramazzini.

"*Pestis carbunculosa. Charbon pestilentiel, pestis Indica, Carazzo* Gemelli Carreri. *Anthrax pestilens* Auctorum." Of this disease he says, "aggreditur rusticos qui carne vervecum anthrace peremptorum in cibum adhibent; pastores, laniones, lotrices qui eorum cadavera, lanam tractant."

He then subjoins this question—"An morbus ardentium, *mal des ardens*, qui olim in Gallia pestis instar strages horrendas edidit, fuerit pestis carbunculosa?" And yet afterwards, in his *character* of

Erysipelas pestilens, he adds, " Feu S. Antoine, Feu Sacre; ægri ardentes; ignis sacer, Mezeray hist. Franc. anno 1090. Mal des ardens en 1130 sous Louis VII. Mezeray."—and he farther says, " An ad erysipelas, an anthracem referenda sit hæc epidemia dubito."

Another species of pestis in Sauvages Nosology is, from the accounts of various writers on the Lues bovilla, intimately connected with that distemper, indeed it may be said to be identified with it, as occurring in the same epidemic. He thus defines it:

" *Pestis Glossanthrax,* Ligeri *de morbis jumentorum, domo rustica.*

" Boves, equos, mulos per Galliam, necnon homines plures Nemausi anno 1732 infecit et jugulavit hæc pestis, quæ ad linguæ radicem anthracem proferebat quo tota sensim lingua intra paucos dies exedebatur, hâc vero erosâ penitus, jumentum quod huc usque præter febrem et virium aliqualem prostrationem functiones suas aliquatenus obibat, illico moriebatur."

It is here worthy of remark, that Sauvages under the head of " Cynanche maligna" observes: " Hæc species ante triginta annos epidemica fuit circa Nemausem, maximè inter boves; atque etiam aliquot homines infecit."

As we descend nearer to our own time, we meet with increased accuracy of observation in writers of medical and natural history. Thus, in the course of the last century are to be found luminous and instructive accounts of various epizöotic diseases:

among others, Dr. Layard has in the Philosophical Transactions for the year 1780, part 1st, furnished us with an excellent description of an eruptive disease attended with fever, which affected black cattle in England, the progress and symptoms of which he watched with sedulous care, and has given with great discrimination. "The disease," he says, "among horned cattle is an eruptive fever of the variolous kind: it bears all the characteristic symptoms, crisis and event of the *small-pox;* and whether received by contagion, or by inoculation, has the same appearances, stages, and determination, except more favourable by inoculation, and with this distinctive and decisive property, that a beast having once had the sickness, naturally or artificially, never has it a second time.

"According to the several prejudices of different countries, various opinions have arisen of the nature of this sickness. Such as are averse to inoculation have obstinately refused to acknowledge it was similar to the small-pox in the human body, and have very idly asserted, that the only intention of declaring this contagion to be a species of small-pox, was purposely, and with no other view than to promote inoculation for the small-pox. Others have as positively declared it to be a pestilential putrid fever, owing to a corrupted atmosphere, and arising from infected pastures. But unfortunately for the supporters of this opinion, while the contagious distemper raged with the utmost violence on the coasts of

Friseland, North and South Holland, Zealand, and Flanders, there was not the least appearance of it on the English coast, from the North Foreland to the Humber, although the coast and climate are the same."

This destructive disease so vividly depicted by Dr. Layard's pen appears to have been first noticed in England, in the year 1745; and to have continued there through the three succeeding years. The contagion is said to have come from Holland, the horned cattle of which had been previously and were at the time, affected similarly. In a letter from Vicq d'Azyr, dated Paris, August 28th, 1780, to Dr. Layard, respecting the distemper among horned cattle in Picardy and the neighbouring provinces during the years 1779, 1780, are these words: "Il me paroit comme à vous que c'est toujours la même maladie qui a regnée depuis 1711, et qu'elle a de grands rapports avec l'eruption varioleuse." And in his "Precis historique" of the pestis bovilla as it appeared in Picardy, he says of the diseased cattle: "Quelques uns ont eu le cou couvert de boutons; et cette terminaison étoit ordinairement heureuse." From the preceding facts it is impossible to doubt that some of the inferior animals, as well as man, have been liable to small-pox in some of its worst forms. The result of Dr. Layard's experiments by inoculation has given an interest and a confirmation to this truth, which greatly increases its value. Had the disease which he observed in cattle been of a mild, instead of a malignant nature; and had he attempted to communicate it to man by inoculation he might, perhaps, have anticipated the great discovery which I am commemorating in the life of Jenner. He has, at all events, proved that the disease was communicable from one animal to another by inoculation, and that it was thereby rendered milder. When Dr. Layard wrote, it was of less importance than it now is to illustrate the connexion between the diseases of man and the inferior animals: no trials, therefore, were made to ascertain whether the *Variolæ* of man could be communicated to the brute, or *vice versa*. The discovery of the *Variolæ Vaccinæ* has fully established the latter point; and although attempts to demonstrate the former have failed in the hands of some, other investigators have been more successful.

M. Viborg, professor of the veterinary college at Copenhagen, says that he succeeded in communicating the human small-pox to dogs, apes, and swine; he asserts also that it has been proved by experiments at the royal veterinary college at Berlin, that the cow likewise receives the small-pox by inoculation.*

* Medical and Physical Journal, Sept. 1802, page 271.

CHAPTER VI.

SKETCH OF THE HISTORY OF VARIOLA, AND OF VARIOLOUS INOCULATION.

I TRUST that the propositions which the preceding discussion was intended to elucidate and establish have received confirmation. The evidence disclosed by the phenomena of the *Variolæ Vaccinæ* will, I doubt not, be found to corroborate in a very striking manner the historical proofs. To render that evidence still more complete we must trace the *Variola*, as it has been recognized in man, with somewhat greater minuteness. Having so done, and also given a sketch of the effects of variolous inoculation, the connexion of the whole argument with Dr. Jenner's discovery will be rendered more apparent.

In opposition to the views taken of the remote origin and subsequent progress of small-pox it may be argued, as has already been urged by Sydenham,

Freind, Mead, aud other celebrated physicians, that as no full or accurate description of small-pox, as a specific disease, has been given either by Hippocrates or Galen, we are fully warranted in concluding that no such disease was known to them, or to the ancient world. This appears plausible, at least; but is by no means convincing—for in addition to the arguments heretofore stated I would now remark that Aaron, the first acknowledged writer who has treated of small-pox methodically and by name, assigns the same cause for its production as for that of measles, and bubonic plague. Furthermore, Rhazes, who refers to the works of Aaron with high approbation, considers small-pox and measles as one disease; and expressly declares his belief that it was well known to Galen: indeed, in proof of this he quotes three passages from Galen's works, in each of which a particular morbid affection or disease is mentioned by Rhazes, which is asserted to be *variolæ* or small-pox. He even expresses great surprise that Galen has passed it over in a manner so slight or transient—especially when it must have been of such very frequent occurrence in his day; and when, also, there must have been such weighty reasons for opposing it by remedies, and he, (Galen) so diligent in investigating the causes of other diseases, and in finding out remedies for them. Rhazes, however, adds, as if in full confidence of Galen's zeal and accuracy: "But God knows whether he may not have treated of small-pox in some of his books, which have not yet come out in Arabic."* Yet from this very silence, as they call it, of Galen on this particular subject, Sydenham, and still more strongly Mead infer his entire ignorance of *variolæ*. Dr. Mead uses this strong language on the occasion: "Frustra enim sunt, qui ἀνθρακας, ἐπινυκτίδας, et consimilia in cute ἐξανθηματα, variolas nostras esse contendunt." "Ex Arabum igitur medicorum libris petenda est prima morbi hujus notitia. Horum facile princeps Rhazes circa annum æræ Christianæ 900 inclaruit."†

The explanation of this discrepancy in the opinions of medical writers on small-pox will, I believe, be found only in the views I have endeavoured to unfold respecting the antiquity of this disease; whence it will be seen that though the ancients, as well physicians as historians and poets, saw and knew it in its various forms and modifications yet from its mode of

* Vide Rhazes de Variolis et Morbillis apud Mead, Vol. I. p. 357.

† It is worth while here to remark, that in the tract of Rhazes, translated from an Arabic MS. at the express desire of Dr. Mead, and by him published as an appendix to his own treatise on small-pox, the term Variola is that adopted by the translator to express the original Greek word or epithet employed by Galen to designate the disease, which is pronounced by Rhazes to have been small-pox. The translator of Rhazes' work into Greek has denominated it Περὶ λοιμίκης, the identical epithet used to denote pestilential epidemics among the Greeks from Thucydides, downwards: he uses λοιμὸς in his account of the pestilence at Athens.

attack, its concomitant fever, its malignant and highly contagious nature, but above all its terrific mortality, they blended and confounded it with other desolating pestilential distempers; considering and treating of it as such: yet notwithstanding this error and confusion, many of them have described some of the characteristic symptoms (the eruptive) of Variola, quite as accurately as more recent authors who have written expressly on that disease as distinct from bubonic plague. We have also much reason to think that the ancients, under the term of Ignis sacer, have frequently depicted the eruption of small-pox; and that the ecclesiastical and monkish chroniclers since the Christian era have, under the same name, or under the designation of pestis, given us accounts of epidemic small-pox.

After the presumptive, indeed I would almost say the positive, proofs of the existence of small-pox in various cities and districts of Europe anterior to any, even the earliest, period usually assigned to its appearance by medical writers of the last two centuries, I should deem it an idle task to dwell much longer in tracing its history through the remaining ages of ignorance and mental darkness; a rapid glance through these times of superstition may suffice.

The first recorded case of small-pox or Variola, under that peculiar, and now appropriated, name, is probably that of Elfrida daughter of our English Alfred and wife of Baldwin the Bald, Earl of Flanders: the date of her illness is generally fixed as A. D. 907: she recovered.

The next case to be met with in the monkish annals is that of the same Elfrida's grandson Baldwin, who was carried off by fatal small-pox in early life A. D. 961. Thus more than half a century passed over without any historical notice of Variola, under that special designation, though not a doubt can now be reasonably entertained that the disease had, in the interval, ravaged many parts of Christendom.

I mention these cases chiefly for the purpose of suggesting to my reader how very little of accurate medical statement he is to look for, or to expect, in the European history of small-pox, even after the disease was fully recognized by the Arabian physicians as distinct from ordinary or genuine plague.

It was probably about this period that the name of Variola, with its connate or derivative terms, was adopted by the Latin family of Europe, whilst the Teutonic or Saxon assumed that of pocca, or pock, or little pouch; the former most likely derived either from varus, a pimple, a spot, or from varius, speckled, spotted. But though the disease had thus obtained a more distinctive name than those by which it had been previously designated, or rather confounded with others, such as λοιμὸς or λοίμικη among the Greeks; *pestis, pestilentia,* lues, among Latin writers; still its characteristic or peculiar nosological symptoms continued ill-defined even among the Arabian physicians prior to Rhazes, who quotes their authority with his approval. Aaron of Alexandria, in the beginning of the 7th century; Bachtishua, physician to the Caliph Almansor, in the 8th century; John of Messue, at the close of the 8th or beginning of the 9th century; Isaac Judæus, in the 9th century; Serapion, at the close of the 9th century,—though they all treat expressly on small-pox, yet their respective descriptions of the pathognomonic signs are not very accurate or complete, their theories, as to its origin or cause, absurd or ridiculous, and their plan of cure injurious. None of them even hint at its being a *new* disease, though it must have been in their days a most widely diffused and fatal affection. * Aaron assigns as its efficient cause, *adust blood*, and *bile*, in common with plague and all other malignant fevers. "Et generantur omnes (eruptiones malignæ) ex malo sanguine adusto cum cholera."

† Isaac the Jew has descanted expressly "De febre variolarum, quæ *fere omnibus* accidit."

‡ Johannes Serapion treats of it among impost-

* Consult Rhazis Continentis, lib. xviii. cap. 8.

† Lib. Febrium Isaac Israel. cap. V.

‡ In the subsequent abstract of the medical writers on small-pox, I have to acknowledge my large obligations to the History of Small-pox, by James Moore, Esq., and to Dr. Valentin, of Nancy, in his History of Small-pox. Both of these gentlemen entertained the sincerest friendship for the author of vaccination, and I am sure that they will never consider their labours as misapplied when employed to elucidate his doctrines.

humes in his work "De Plegmone, et Almessere (Erysipelas) et Igne Persico, et Variolis." Yet this same author classes small-pox among fevers in a chapter, "De cura febris causatæ a variolis.*"

Thus far we do not find that the Arabic writers much excelled their Greek and Latin precursors or contemporaries, either in their theories or delineations of this disease. Indeed Hali Abbas (A. D. 980) has improved somewhat in the descriptive part, and he seems to have had a glimpse of the contagious nature of Variola.

In the following century Avicenna has given a view of small-pox and measles more clear and distinct than any of his predecessors. He describes the pustular eruption more accurately, and distinguishes between the confluent and distinct kind; and seems to consider the measles as different from small-pox, in a degree at least; as he takes notice of the weeping eye and the pneumonic inflammation in the latter disease. Both he pronounces to be very contagious: and he, like Rhazes, observes that *small-pox may affect the same individual twice.*† Nearly contemporary with Avicenna in Asia, lived Avenzoar in Spain. He has added little to the knowledge imparted by his forerunners in the history of small-pox.

Constantinus Africanus flourished towards the end of the 11th century. He was undoubtedly a

* Lib. Joh. Serapionis in Lat. tradit. Venet. 1558.

† Avicennæ lib. 4. 1. cap. 10.

learned physician, considering the age of ignorance in which he lived. Though well acquainted with the Arabic he wrote in the Latin language, and settled ultimately in Italy. He treats of small-pox and measles as one disease; and has not increased our knowledge of either. He, however, revived the knowledge of Greek medicine, and introduced an acquaintance with that of the Arabians.

Little additional light is let in on the history or treatment of small-pox from the beginning of the 12th to the middle of the 17th century, though we have a long list of medical writers, as Averrhoës and Albucasis, both Arabic physicians, and among the last of that school.

Next follow Francis de Pedemontium, Arnaud de Villeneuve, and others who treated of small-pox and measles under the title of Anthrax and Carbuncle, conceiving them to be of the same pestiferous nature, which, as I have already pointed out, was the opinion of the older historians and physicians: these close the 13th century.

Nearly contemporaneous lived Gilbert, in the reign of our first Edward. His Compendium of Medicine is the earliest medical production that England can lay claim to; yet his work is not original, with respect to small-pox and measles, since he has copied from the Arabian writers;—as did also his English successor John of Gaddesden, physician to Edward II. He was a servile copyist of the Arabic school, and his only dogma worth recording with regard to small-pox is "Aliquando variolæ bis hominem invadunt."

During the 15th century astrology and alchemy occupied so powerfully the minds of the learned that the medical history, symptoms, and treatment of small-pox could not be expected to receive either elucidation or improvement, even from such great names as Fracastorius and Paracelsus. But towards the end of this century, or early in the following one appeared Fernelius, who becoming Professor of Medicine at Paris, may well take his rank as the restorer of that science. Though he was a believer in astrology and magic yet he possessed uncommon talents and brought industrious application to the study and practice of his art, and to the acquisition of natural knowledge. He does not make mention of either small-pox or measles by name, but includes them among pestilences, under the terms of Exanthema and Ecthyma.*

Fernelius rejected the theories of the Arabian writers on small-pox, and made some approximation to the doctrine of the operation of contagious effluvia: his mode of practice in this disease is not exactly known, but his treatment in ardent fevers appears to have been wonderfully judicious, for his time.

In the 16th century Forestus returned to all the absurd notions of the Arabians respecting the *causes*

* Fernelii Opera, lib. ii.

of "*variola et morbilli;*" and followed the hot and stimulating mode of treatment.

He observes that the same individual may have the small-pox twice "id quod verum esse experientia rerum magistra testatur: ut et nos in nostro filio, qui bis variolas habuit, licet puer; et in aliis multis ægris observavimus."*

It is apparent from what has been said that it was a general belief among the early writers on small-pox that the disease might affect the same individual twice. Dr. Mead, without due inquiry, promulgated an opinion of a different kind, and from his time till within these few years, that opinion was so blindly followed that when small-pox did occur a second time it was set down as chicken-pox, swine-pox, or some other eruptive disease. When the friends of vaccination were compelled to prove that the small-pox did occur a second time it was believed that they drew on their imagination for such facts. The antivaccinists who made this charge showed themselves as little acquainted with the real history of small-pox as they were with that of cow-pox. One very remarkable example of the kind alluded to, namely, the death of Louis the Fifteenth at the age of sixty-four, from small-pox, after having undergone that disease at the age of fourteen years, made a great sensation throughout Europe. Without loading this work with a long list of medical authorities I may mention that more than one hundred

* Petri Foresti Opera, lib. vi. obs. 43.

and thirty different writers may be named who have recited examples of such an occurrence.

Mercurialis and Sennertus, in the sixteenth and early part of the seventeenth century, have added but little to the records of small-pox, though they both improved its mode of treatment by recommending a cooler regimen; and to abstain from opening the pustules, or otherwise meddling with them. These writers declare that the disease may attack the same person twice. This fact is also asserted of their own knowledge by Willis and Diemerbroëk, neither of whom, however, has much increased our historical information respecting this disease.

During the seventeenth century Kircher, Sylvius, and several other continental writers have treated of small-pox, inventing new theories, or modifying or adopting the old: but it would be worse than idle to follow them in the labyrinths of vain speculation, for Sydenham arose at this time, imbued with the spirit of experimental philosophy which he had learned from the mighty mind of Bacon, the first of men uninspired.

Sydenham discarded all theories respecting small-pox: he separated this disease from measles, with which it had been always blended by authors from the revival of medical learning under the Arabian physicians to his own time: and he gave such an accurate description of symptoms, and adopted such a judicious plan of cure in both diseases that he left little room for improvement in either, till the introduction of inoculation, which so far mitigated small-pox to the individual undergoing that treatment as almost to have disarmed a fearful pestilence of its terrors. But, though by this salutary practice individual welfare was consulted and, in most instances, obtained, still the public safety was endangered, nay more than endangered; for there cannot be a doubt that, in this way, a loathsome and destructive malady was kept alive and widely spread abroad, in all its baneful activity, among those classes of society who either would not, or could not avail themselves of the disarming power of inoculation.

The views and treatment of small-pox by Hermann Boerhaave were not far different from those of Sydenham. He (Boerhaave) considered the disease very much as an inflammatory fever; epidemical, however, and alone proceeding from a peculiar contagion exhaling or thrown out from the bodies of persons infected by, or labouring under, the disease; perhaps it was from a consideration of this kind that he was led to believe that a specific might possibly be found to overcome the matter of this contagion, or to cure the disease when formed. He suggested a combination of antimony and mercury. Could he have looked into futurity, and known by anticipation the sure preventive since traced out so admirably by the unwearied sagacity of Jenner, how must his best feelings have been gratified!*

* As Boerhaave's aphorism containing the conjecture here alluded to has been put forward, apparently to lessen the merit

Succeeding writers on the small-pox, as Freind, Mead, and others immediately prior to the introduction of inoculation, follow very closely the track marked out by Sydenham, and where they have deviated from it have not added to his positions, nor improved on his descriptions.

of Dr. Jenner in the discovery of vaccination, I feel it right as his biographer to remark in this place that Boerhaave's prospective view of an antidote extended, and could only extend, to some pharmaceutical preparation, or physical remedy to be used *internally* for the purpose of destroying or correcting the virulence of the contagious poison whatever that might be: and I will further add that by no possible interpretation can his words be made to show the most remote allusion to any external means to be discovered as preventive. I shall now give, in proof of this assertion, the aphorisms which refer to this subject in Boerhaave, merely premising that his view of it led to the combination of mercury and antimony in *Sutton's* celebrated powders for small-pox.

Aphorism 1388. "Cognito hoc morbi statu (1380 ad 1387) indicatio oriri videtur hæc primo; ut, stimulo inflammatorio ablato, sanetur status præsens et impediatur ulterior ejus progressus, et proinde caveatur futura suppuratio, gangræna, &c."

Aphor. 1389 "*Stimulus* videtur auferri posse correctione per specifica ita dicta, vel methodo universali antiphlogistica."

Aphor. 1390. "Correctio specifica niti debet invento *remedio* opposito illi veneno contagioso, quod tam parvâ mole susceptum reliqua parit ut effecta."

Aphor. 1391. "Quale (1390) invenire posse comparatio historiæ antidotorum, et indoles hujus mali, faciunt sperare, et ad indagandum impellit summa hinc futura humano generi utilitas."

Aphor. 1392. "In stibio et mercurio, ad magnam penetrabilitatem arte deductis, nec tamen salina acrimonia nimium corrosivis sed bene unitis, ut quæramus, incitat aliquis horum aliquando successus."

We have now arrived at that period in the history of small-pox when this disease was about to undergo a most salutary modification, through the determined and persevering good sense of an accomplished woman;—I say a woman, for though it is very true that the practice of inoculating or engrafting the small-pox had been published in England so early as 1714 by Dr. Timoni, of Constantinople; at Venice, in 1715, by Pylarini; and in the same year in London by Mr. Kennedy, a surgeon, who had been in Turkey, yet it is certainly to the influence and example of Lady Mary Wortley Montague we are indebted for its introduction and adoption in England; and for its consequent diffusion through Christendom. Her ladyship's first letter to England on the subject is dated 1st April, 1717: and soon after she caused her own son Edward, then a child, to be inoculated at Constantinople by Mr. Maitland, surgeon to the British Ambassador Mr. Wortley, her husband.

On her return to England in 1722 Lady Mary had her daughter inoculated by the same Mr. Maitland, this child being the first known subject of the new practice in civilized Europe; the second was a child of Dr. Keith, a London physician, who had witnessed the progress of inoculated small-pox in Lady M.'s daughter. This was creditable to the Doctor, and must have given a fresh impulse to the practice: yet it made very slow advances, although the then Princess of Wales, with the King's consent, caused two of her own daughters to be inoculated. It was ascertained that the accounts of success and immunity at Constantinople were not fully borne out by facts; and in England the inoculated small-pox was found in some cases not to turn out by any means so mild an affection as had been represented by the advocates for the new practice, and it was occasionally fatal. During the first eight years after its introduction into England the number of persons inoculated amounted to 845 only and, of these, 17 died of the disease: nearly as one to fifty. Though this was a greatly diminished mortality compared with *that* attendant on natural small-pox yet it impeded the progress of inoculation, which had still to encounter violent opposition from many quarters. Some over-zealous men from a religious feeling (as it is to be hoped they thought) opposed the novel practice of inoculation as an attempt, at once impious and unavailing, to counteract the visitations of an all-wise Providence; asserting that in the case of adults, who assented to the voluntary infliction of a fatal disease on themselves, the crime was that of suicide, but in respect to children it was "horrid murder of the little unoffending innocents." Even clergymen preached from their pulpits in this style of argument (if so it might be called). Some went so far as to pronounce inoculation an invention of Satan himself—and its abettors were even charged with sorcery and atheism. These things would scarcely obtain credence were it not that similar arguments and assertions have been employed against vaccination itself. One sermon, by a reverend rector of Canterbury, is dated so lately as 1753.

Justice to the learned body of the clergy of that day requires it to be mentioned that many of them, with the Bishop of Worcester (Dr. Maddox) at their head, maintained publicly the innocency of the new practice; nay more, they defended its adoption on the score of Christian duty. The opposition of medical men was temperate though ill-founded, save in those who said, and said truly, that by inoculation the small-pox, however milder and safer it might be thus rendered to the individual subject of it, must necessarily be more extensively propagated, from its contagious nature, and its fomes kept alive and active for ever.

These objections carry with them great weight, the last especially: nevertheless the great body of the medical profession was decidedly favourable to inoculation.

Notwithstanding all this the practice made little progress in England for nearly twenty years; indeed, about the year 1740 it had almost fallen into neglect, or disuse. In Scotland and Ireland it was still more unsuccessful; and in Germany it could scarcely have been said to have been tried, except at Hanover very partially.

Such was the hopeless state of the cause of variolous inoculation in Europe when accounts highly favourable were brought from both Americas, and from the West Indies. In one of these islands (St. Kitt's) a planter (as is related by Dr. Mead) inoculated three hundred slaves without the loss of one; in Carolina the loss was stated as one in one hundred; and in South America it was reported as still less. These accounts revived the practice in England; so that in 1746 the Small-pox Hospital in London was founded for the purpose of inoculating the poor, and of keeping the patients distinct from the general population of the city.

This judicious regulation was afterwards departed from by inoculating all persons who applied at the hospital, and thus the disease in its natural form was disseminated far and wide throughout the community.

It was confidently, though mistakenly, asserted by the advocates for inoculation about this time (1752) that the number of deaths from small-pox, in the bills of mortality, had diminished by one-fifth since the practice of inoculation had been adopted: this statement, however, was erroneous; for, in fact, they exceeded those of any former year; amounting to 3538, in 1752. In 1754 the College of Physicians in London declared their fullest approbation of the practice of inoculation. Had they accompanied this with a recommendation to keep the infected apart from the non-infected they would have more fully performed their public duty.

In Scotland the progress of inoculation was so slow that, so late as the year 1765, not more than

6000 persons had received it: and there the mortality was as one to seventy-eight, of these persons.*

At this time also its advancement on the Continént had been very tardy and limited. In Germany attempts had been early made to introduce inoculation but they did not succeed to any extent, perhaps in consequence of the opposition of Van Swieten and De Haën; and in France the practice was condemned by the faculty of medicine at Paris. This brought it into disuse for thirty years at least, till M. de la Condamine, and the celebrated minister Turgot (in 1755) espoused the cause with energy and zeal; the nobility, also, led on by the example and countenance of the Duke of Orleans who had caused his own son and daughter to be inoculated, gave effect to the recommendations of the statesman, and the man of letters; and soon the practice spread through France. It was, however, in 1763 prohibited by royal authority, in Paris, in consequence of small-pox having raged as an epidemic in that city attended with such dreadful mortality as to have aroused the notice and fears of the police and of the parliament, who on investigation found that the infection was multiplied and diffused by means of inoculation. This is a melancholy truth; and at all times formed the most forcible, perhaps the only strong objection to inoculation.

* See an account of inoculation in Scotland, by A. Monro, M. D.

It is asserted that in Spain,* where the practice was scarcely ever admitted, small-pox has caused less mortality in proportion to the population than in any other country in Europe. If documents prove this assertion (and I have no doubt of its truth) it had probably been better for mankind that the practice of inoculation had never been adopted, in the partial way it has been, in Europe and without the proper restrictions and the entire seclusion and separation of the infected from the uninfected.

The Suttonian method, as it has been called after its author Daniel Sutton, was certainly a great improvement both as regarded individual recovery and public safety. In 1768 the Empress Catharine of Russia invited Baron Dimsdale to Petersburgh; and submitted herself and her son Paul to his treatment by inoculation. This high example, of course, spread the practice among the Russian nobility, and with it diffused the natural small-pox and all its attendant evils and mortality. So fatal had that disease become in Russia throughout its immense extent that Sir Alexander Crichton, the imperial physician, states a calculation from which it appears that, previously to the adoption of vaccination, every seventh child born in Russia died annually of small-pox.†

* See Moore's History of Small-pox, p. 287.

† Moore's History of Small-pox, p. 286.

CHAPTER VII.

DR. JENNER'S OPINION RESPECTING THE ORIGIN OF SMALL-POX AND COW-POX—ILLUSTRATIONS OF THAT OPINION—PROOFS OF ITS ACCURACY.

WE have now brought the history of the Variolæ to a period not far distant from that in which the disorder incident to cows began to attract attention in England, from its reputed virtues as a preventive of small-pox; and have arrived at that point which, I trust, will give an interest to the preceding investigation that it could not have acquired as a mere historical disquisition.

It was quite an unlooked for, and at first almost an incredible thing that a disorder immediately derived from one of our domestic animals should exert an influence so powerful and so beneficial on the human frame. But if it should appear that the disease incident to man and to beasts had one common origin; and that an analogy, close and well-defined, may be traced in their subsequent history and progress we shall have obtained evidence to explain pathological facts which are of the utmost value to mankind.

From what has been already adduced it is clear that a fatal pestilential eruptive disease, common to man and the inferior animals, has been known from the earliest period of authentic history: that the same, or at least a disease somewhat similar, continues to exist in various regions of the earth, often attended with great mortality: that it appears to have undergone various modifications in respect to virulence; and to be susceptible, by artificial communication, of still greater modification.

Should it appear that the views which I have attempted to illustrate rest upon a solid foundation, they will tend, I would hope, to give a stability to the practice of vaccine inoculation which was not formerly experienced. They will also explain how sheep, or horses, or any other animals may be subject to the disease as well as cows or oxen: that it is not a poison peculiar only to one variety, but may be found and propagated among many.* It need not

* To corroborate the preceding views I select the following passage from one of Dr. Jenner's manuscripts.—" Our domesticated animals are subject to a variety of eruptive diseases: the horse, the cow, the sheep, the hog, the dog, and many others. Even poultry come in for their share. Again, there certainly must be a reason why the term *chicken* is annexed to a species of pock, which infests the human skin. In the province of Bengal the poultry are subject to eruptions like the small-pox, which become epidemic and kill them by hundreds. The Europeans to stop its depredations have even tried the effects of in-

therefore excite surprise that *matter* capable of producing the genuine pustules should be found in the horse, as it unquestionably has been in this country and elsewhere; or that the disease should make its appearance among sheep, as it is reported to do in Persia, and in goats in other countries. That the goat also is subject to an eruptive pustular disease, similar in its nature to the Variolæ Vaccinæ, is proved by the following extract from a letter of Mr. Dunning, surgeon, of Plymouth-Dock, addressed to Dr. Jenner some time in the year 1804, "My late correspondent at Madrid is now in England, and has been lately favoured with the following from Professor Heydeck:—

"Madrid, March 9th 1804.—I am not able to send you, at present, our observation on the goat-pock subsequent to the 8th of June last, because it is not finished yet; for the king has ordered in September last that all the children in the Foundling-house, and those who are in the Desamparados should be inoculated with the goat-pock, which did its effects; we are now employed in the contra-proofs, and after every thing is finished, shall send the whole process to you for the inspection of your medical friends and Dr. Jenner; and as I am at present *on another discovery, not less useful*

oculating their chickens. The East Indians have only one name for the small-pox, and this disease, *gootry*. The origin of the term chicken-pox, then, no longer remains mysterious. Whether the poultry in our own island feel the same complaint I do not know."

as the goat-pock, I shall give also an account of its results in my next letter." Mr. Gunning adds, "I wrote to the Professor about three weeks ago, told him that his discovery had excited very much the attention of the medical world in England, and more immediately Dr. Jenner's, and urged him to forward his further observations with all the expedition in his power, and that I would transmit them to you."

I cannot *here* refrain from again adverting to a fact stated by Sauvages under the head of his *Dysenteria Pecorum*—(the *pestis bovilla* of Lancisi and Ramazzini; and the *Lues Vaccarum Tubingen.*) "Morbus hic totam successivè Europam, et demum nuper Galliam devastavit, atque boves eripuit. Hunc *observavi* apud Helvios Comitiorum, *boves* ibi, in Occitania *capræ* et *verveces* infestabantur."

As the existence of the Variolæ Vaccinæ in the dairies of England would seem not to have been of very long duration I think there is good ground for believing that the disease, as originally noticed by Dr. Jenner in Gloucestershire, was the endemic or local remains of the more general or epizöotic disease which prevailed in many parts of this island, at the period when Dr. Layard wrote.

This opinion is strengthened by the following well-authenticated facts. Soon after the publication of Dr. Jenner's discovery it was found that cow-pox existed in several counties of England—in Devon, Dorset, and Somerset, in Hampshire, Buckinghamshire, Middlesex, Wiltshire, Staffordshire, Norfolk, &c. in all, eighteen counties. Accounts were also received of the existence of the Variolæ Vaccinæ among the cattle of Lombardy, Holstein, and other regions where the pestilential eruptive disease, denominated *pestis bovina*, had previously raged, as recorded by Ramazzini and other medical writers. At a later period it was also found in Persia, both in cows and sheep, as appears in a letter from Mr. Bruce, English Resident in Bushire, (11th vol. of the Edinburgh Medical and Surgical Journal, p. 270). It was discovered also by Don F. X. Balmis among the cows of the valley of Atlixco, near the city of Puebla des los Angeles; in the neighbourhood of Valladolid de Mechöacan where the Adjutant Antonio Gutierrez found it; in the district of Calabozo in the province of Caraccas, by Don Carlas de Pozo Physician of the residence, and by Humboldt in the Peruvian Andes.

In the Report of the Central Committee of vaccination in Paris for the year 1821-22 it is stated that the disease has likewise been found in the vicinity of Clairveaux, by M. Janier Dubry; and to this list Ireland ought to be added on the authority of Dr. Barry, of Cork. From these facts it may very fairly be inferred that the Variolæ will hereafter be found among cows in other parts of the world.

That this disease has already been met with in regions so remote from each other is a proof that it cannot depend on mere local circumstances, either for its origin, or propagation; and that, like small-pox itself, it has pursued its victims through every clime and in every season. I cannot but feel that this circumstance, coupled with the facts already mentioned, adds great weight to the opinion as to the affinity between the two affections.

This opinion, drawn as it is from an impartial consideration of well-authenticated historical documents, comes with much force to support the sentiments of Dr. Jenner grounded on his own observations. He always, as has been said, considered small-pox and cow-pox as modifications of the same distemper; and, that in employing vaccine lymph, we only made use of means to impregnate the constitution with the disease in its mildest, instead of propagating it in its virulent and contagious form; as is done when small-pox is inoculated. Had his views been deliberately examined, and had the consequences to which they clearly lead been kept in mind by the different writers who have distinguished themselves by their zeal in the vaccine controversy, much of their bitterness might have been forborne, and many needless and irritating discussions avoided. The fact having been admitted, as with reason it might have been, that there was such an affinity between small-pox and cow-pox, it would have at once appeared reasonable to expect that the same general laws should govern both affections with regard to their prophylactic powers. Thus, if there were different degrees of security afforded by small-pox against the recurrence of that disease, it was quite natural to expect that similar modifications might be found in the protecting virtues of the Variolæ Vaccinæ.

These views are not at all affected by the doctrine which Jenner promulgated respecting the origin of Variolæ Vaccinæ from the heel of the horse. Although there is now no room for any doubt that the Variolæ may be thus derived yet it is probable that the grease, as it is called in the horse's heel, is only the mode in which the disease commonly exhibits itself in that animal. The following remarks of Dr. Jenner himself will best illustrate this statement.

"The skin of the horse is subject to an eruptive disease of a vesicular character, which vesicle contains a limpid fluid, showing itself most commonly in the heels. The legs first become œdematous; and *then* fissures are observed. The skin contiguous to these fissures, when accurately examined, is seen studded with small vesicles surrounded by an areola. These vesicles contain the specific fluid. It is the ill management of the horse in the stable that occasions the malady to appear more frequently in the heel than in other parts; I have detected it connected with a sore on the neck of the horse, and on the thigh of a colt."

This doctrine I hold to be substantially true; for it has been established by unquestionable evidence that matter from the horse *does* produce a pustule similar in appearance to the vaccine; and likewise possessing the same protecting power; and *that*, without having passed through the constitution of the cow. This fact, though it prove the identity of the diseases, does not prove that they both originated in the horse; but it goes strongly to confirm the view of the simultaneous origin of the affections in question. When Dr. Jenner published his *Inquiry* he was not aware of the independence (if I may so speak) of the properties of the *matter* of *grease;* that is to say, he did not believe that it could excite in the human constitution an influence like that of vaccine matter, without previously undergoing certain modifications in the system of the cow.

It seems certain that there are, at least, *four* animals, namely the horse, the cow, the sheep, and the goat which are affected with a disorder communicable to man; and capable of securing him from what appears to be a malignant form of the same disease. It is moreover proved by direct experiment that other animals are capable of receiving the vaccine disease by inoculation; and that matter taken from pustules so produced affords the genuine cow-pox in man. The animals on which these experiments have been tried are the dog, the goat, the she-ass, and the sheep. The fact as regards the dog was ascertained by Dr. Jenner: with respect to the other animals I rest on the authority of Dr. Valentin, of Nancy, who made his experiments in 1801–2.

When we place these facts in conjunction with

those already stated on the authority of professor Viborg, of Copenhagen, it is impossible not to find strong corroboration of the doctrine of the similarity of small-pox and cow-pox. Some incidents whicn occurred at the Small-pox Hospital in London, soon after the publication of Dr. Jenner's first work, infallibly lead to the same conclusion. The blunders which unhappily occasioned these incidents had for a time a most injurious effect on vaccination: it will, therefore, be some consolation if we can now extract from them any thing to give confidence in the practice, and to further its dissemination. By a most unfortunate mistake some of the very first trials that were made in the Metropolis were conducted in the Small-pox Hospital, under the direction of the late Dr. Woodville. The place chosen for these trials, as well as the manner in which they were carried on, reflect no credit on the projectors. What might have been anticipated soon took place, the vaccine and variolous virus were commingled, and an eruptive disease was thus produced milder in its nature than ordinary small-pox, but very different from the benign *solitary* pustule which characterizes the Variolæ Vaccinæ. Dr. Jenner immediately detected the error, and its cause; but the persons who fell into it did not abandon it till they had propagated the evil over many parts of Europe. The consequence was that eruptions, in greater or smaller proportion, followed the inoculations which were performed with the vitiated virus.

But the fact of highest importance, ascertained from these trials, was that even this *matter* after repeated re-inoculations lost much of its virulence; ceased altogether to produce eruptions; and at length became almost assimilated to the true vaccine character. This will be found recorded in a subsequent page, in a letter from Dr. Jenner to Lord Egremont.

Still further to elucidate the preceding observations let us advert to some of the facts connected with the natural history of the Variolæ Vaccinæ. Though they have not appeared recently in a fatal or malignant form among the cattle, it is certain that they sometimes proved a severe and troublesome disease to those who caught it from the cattle, insomuch that an experienced surgeon anticipated little benefit from propagating it by inoculation because as *he* had seen it among the dairy people, it was almost as severe as the inoculated small-pox. As, therefore, the Variolæ Vaccinæ sometimes assume the character of small-pox under one of its modifications, so the latter under certain circumstances approximates in its nature the mildness of the former. After a series of inoculations with true variolous matter it has often been observed that the severity of the symptoms and the number of the pustules gradually diminish till only *one* is to be seen, at the point of insertion; and that this pustule though it may have excited no constitutional indisposition, provided it has regularly gone through its course, protects the individual from subsequent attacks of small-pox.

This fact did not escape the observation of Dr. Jenner; in reference to which he has remarked in one of his memoranda "Here then we see the cow-pox and the small-pox acting similar parts: and that in either case the virus may steal, as it were, imperceptibly through the constitution, and give no signal of its presence."

As connected with this subject it is not irrelevant here to remark that Dr. Adams, physician of the Small-pox Hospital in London, succeeded in producing a benign form of variola attended with scarcely any eruption of pustules and little or no constitutional affection; and this species of small-pox he considered capable of being rendered fixed and permanent. He selected such patients as came under his care with a mild kind of small-pox, which occasionally showed itself in London. Dr. Adams denominated this the Pearly-pox, from the pearl-like appearance of the attendant eruption; the pustules were few, distinct, and filled with limpid matter not unlike that in the cow-pox vesicle, to which indeed, according to him, this variolous pustule bore no distant resemblance. The constitutional symptoms, he also says, were very trifling, and he expressly declares it as his decided belief that he had succeeded in giving this innoxious form of small-pox permanence and uniformity.

The ancient *Thessala*, too, so celebrated for successful inoculation at Constantinople more than a century ago, was careful in taking the infectious matter of small-pox from such children only as had the disease in a mild, distinct form; thereby insuring (as she thought, and as is said to have been the case) a similarly safe affection in the person inoculated by her. She also remarked that *one* regular pustule, proceeding through its natural course uninterruptedly, gave full protection.

Although Dr. Jenner's opinion respecting the origin of cow-pox is comparatively of little moment when contrasted with the important consequences arising from the successful practice of vaccine inoculation itself, it is, nevertheless, necessary whilst investigating his character as a philosopher, to show that even in this, which was considered a wild speculation, he proceeded with his usual caution and discretion. The fact that the disorder in the cow originated from the horse had not been proved by direct experiment when he published his *Inquiry*; yet the evidence on which this doctrine rested was so complete as to entitle it to much attention. He himself has thus stated that evidence.

"1st. From its being the fixed opinion of those who have been in the habit of attending to cows infected with this disease, for a great number of years.

"2dly. From its being a popular opinion in this

great dairy-country, and from the cautions the farmer observes when he has a horse with a sore heel.

"3dly. From observing, in almost every instance, that the appearance of the cow-pox at a farm was preceded by some disease of a horse at the same farm, which produced the discharge of some fluid from the skin.

"4thly. From having attempted, in vain, to give the small-pox to the son of a *farrier* who had had sores and a fever from dressing a diseased horse.

"And 5thly. From the peculiar appearance of the pustule, and its disposition to run into an ulcer in the arm of the boy who was inoculated with matter taken from the hand of a man who received the infection from dressing a slight spontaneous sore on a horse's heel."*

Mr. Tanner had the merit of proving the truth of this doctrine: he succeeded in communicating the disease to the cow by inserting some liquid matter taken from the heel of the horse. This

* I cannot refer to this incident without calling the reader's attention to the modest and diffident manner in which the author speaks of a fact which was well nigh conclusive as to the truth of his doctrine. A beautiful representation of the pustule in its advanced stage is given in the second plate of the "Inquiry." The character of the pustule is so correct as to excite some surprise that it has been so little attended to. If it had, his opinions would, doubtless, have been treated with more respect.

produced on the teat of the cow a complete vaccine pustule. "From handling the cow's teats," he observes, "I became infected myself and had two pustules on my hand, which brought on inflammation and made me unwell for several days. The matter from the cow, and from my own hand proved efficacious in infecting both human subjects and cattle."

This experiment was in complete contradiction to some which had been instituted by Mr. Coleman at the Veterinary College, London, in November 1798, as well as by Mr. Simmons, who published in the same year, "An Account of some Experiments on the Origin of the Cow-pox."

In the month of April 1800, some observations were made by Mr. Lupton, a respectable surgeon of Thame, Oxfordshire, which were communicated to Dr. Jenner by Sir Christopher Pegge, and printed in the Medical and Physical Journal of that year, and in a very striking manner tended to confirm Jenner's opinion. The same thing was still more conclusively established by the work of Dr. Loy, entitled "Some Observations on the origin of Cow-pox," and published in 1801.

It was proved by him that the matter taken from the heel of the horse communicated an efficacious preventive of small-pox to the human subject, without previously passing through the cow. An incident corroborative of this truth occurred to Dr. Sacco, at Milan, in the year 1803. He had taken up an opinion adverse to that entertained by Jenner; but with the candour which becomes a scientific inquirer he immediately announced his error, and bore testimony to the accuracy of his friend.

Milan, le 25 Mars, 1803.

Monsieur,

J'étais depuis long temps occupé à faire des expériences sur le *grease* pour confirmer votre opinion sur l'origine de la Vaccine. Jusqu'au commencement de cette année je n'avais jamais pu rien obtenir. La lecture du petit livre de Mr. Loy m'encouragea à repeter une autre suite. L'hiver de cette année ne pouvait pas être plus abondante de grease à cause de la quantité de l'eau qu'il avait, et par consequence de la boue dans les chemins: ainsi presque tous les chevaux souffraient le grease. Mon domestique en fut attaqué au deux avant-bras par cinq boutons pansant un de mes chevaux qui avait le grease; il ne m'en a averti que quand les boutons passaient en exsiccation: celui m'encouragea de plus à continuer mes tentatives. J'ai inoculé plusieurs enfans, plusieurs vaches avec le virus qui sortait du grease à differentes époques, mais toujours inutilement. Un cocher se présenta à l'hospital pour se faire visiter d'une éruption qu'il avait sur les mains. On connaît de suite que c'était vaccine prise en traitant les chevaux, qu'effectivement il pansait. Il fut conduit à l'hôpital des enfans trouvés où on fit quelques inoculations: il vint le même jour chez moi, et je fis neuf inoculations sur autant d'enfans, et de plus j'ai inoculé les pis d'une vache. Trois de ces enfans ont contracté une éruption toute pareille à la Vaccine. La vache n'a point pris. J'ai fait des autres inoculations avec la matière prise de ces enfans, et c'est déjà la quatrième génération que se reproduit avec la même effet comme le Vaccin. J'ai déjà inoculé plusieurs de ces individus avec la petite verole, mais sans aucun effet. C'est donc bien sûr et consenté que le *grease est cause de la Vaccine*, et on pouvait bientôt changer denomination en *équine*, ou en ce que vous croyez mieux. J'ai aussi enfin obtenu avec le virus de grease inoculé sur six autres enfans deux boutons tous semblables aux Vaccins. Je continue mes observations. Il-y-a tout pour s'assurer qu'enfin nous aurons du grease le virus pour se mettre à l'abri de la petite verole sans passer aussi par *l'intermedium* de la vache. J'espère que cette nouvelle preuve pourra ôter les doutes qu'il-y-avaient encore sur l'origine de la Vaccine. Je publierai les résultats de ces expériences sur un code doctrinal de vaccination, au quel j'ajouterai une planche illuminé de grease. J'espère que vous aurez reçu les médailles par Mr. Woodville, à qui je me pris la liberté d'addresser le paquet pour vous le faire obtenir avec certitude. Je renouvellais mes remercîmens pour les livres que vous m'avez envoyés, avec les regrets aussi de n'avoir pas reçu aucune de vos lettres. Je comte à cette heure plus de 25 mille inoculations faites par moi seul.

Je vous prie, mon très-estimable collegue, de me donner quelque nouvelle avec quelqu'autre enseignement sur cette matière; mais sur tout honorez-moi de votre réponse.

Très humble serviteur,

LOUIS SACCO, Med. Chirurg.

A curious piece of rustic history communicated to Dr. Jenner by Lord St. Asaph corroborates the doctrine respecting the origin of the Variolæ Vaccinæ

LORD ST. ASAPH TO DR. JENNER.

SIR,

When I read, between four and five years ago, your publication respecting the cow-pox there was one point only which appeared to me not sufficiently proved: namely, that the disease was in its origin that which is known by the name of the grease in horses: and that it was conveyed from the one animal to the other in the manner therein asserted.

I therefore made some inquiries, in the part of Suffolk where I reside, with a view to discover whether the cow-pox was at all known in a county where the milking of the cows is performed by *women only*. For this purpose I applied to Mr. Harwood of Battisford Hall, a very respectable and experienced farmer; and particularly asked him if he remembered to have had a horse with greasy heels: and if he had remarked any such consequences ensuing from it as those which I then related to him. He assured me that he did not recollect any thing of the kind: and further told me that *he positively disbelieved* the fact which I doubted.

A few weeks after this conversation he called upon me and informed me that he had very unexpectedly obtained some information on the subject, concerning which I had before questioned him. He then told me that Samuel Nunn and his wife (who had both been servants to him and had quitted his service a few years before, on their marriage) had made him a visit; and that, talking over past occurrences, the wife, in order to ascertain the date of some event by that of another more strongly impressed on her memory, mentioned the time when all the cows had sore udders, and when she herself suffered so much from a violent eruption on her hands and arms, the marks of which were still visible. On hearing this Mr. Harwood (recollecting the questions I had put to him) asked the husband whether he remembered if any horse in his stable had *at that time* greasy heels. The man replied that he certainly did: and named a particular horse that was so diseased for a long while. Mr. Harwood then inquired whether he ever went to milk the cows. The man's answer was, "Why to be sure, sir, that was not my business: but I was *then* courting my wife; and sometimes when I had finished my work I went to help her in hers."

I have further learned that S. Nunn was not himself infected with the disease which he thus conveyed from the horse to the cows, and that every cow in the dairy, consisting of upwards of twenty, had it. And this is the only instance that I have been able to discover of cows in my neighbourhood having been thus affected.

Such is the case which you desired me to transmit to you: and which I have thought that I could not so clearly do in any other manner as by relating to you that in which it came to my knowledge. The details into which I have entered may preclude the necessity of adding any remarks. At least I have endeavoured to render sufficiently obvious to every one, who shall have the patience to peruse my narrative, that the facts were communicated to me by an avowed disbeliever in the truth of your discovery, and obtained from those who never had heard of Dr. Jenner and of vaccine inoculation: one of whom indeed had had the disease, but without knowing it, or its name. I flatter myself that such ignorance will be in a short time no where to be found: and that in this instance, at least, we shall not "see nations slowly wise" in availing themselves of the invaluable advantages of the discovery: nor "meanly just" to the merits of the man who made it.

I am, Sir,

Berkeley Square, Your obliged and obedient servant,
April 25th 1803. St. Asaph.

Subsequently to the observations above recited many opportunities were afforded of verifying them, insomuch that Dr. Jenner was in the practice of using *equine* matter with complete success. He supplied myself and many of his medical friends from this source. He also transmitted it to Edinburgh, where it produced the genuine cow-pox.* A fact similar to that which occurred to Dr. Sacco was observed in Paris in 1812. A coachman who had not had small-pox, and who dressed a horse affected with the *grease*, had a crop of pustules on his hands, which resembled the vaccine. Two children were inoculated from these pustules, and the genuine vaccine was excited in both: from this stock many successive inoculations were effected, all possessing the proper character. A similar series of inoculations took place from another infant who was infected from one of the scabs taken from the pustules on the hand of the coachman.

It happened to me to see one case of this kind in the autumn of the year 1817. A young man in this neighbourhood, who had dressed a horse with the grease, had not less than fifty pustules on his

* See Thompson on the Varioloïd Epidemic, p. 316.

hand and wrists. They exhibited the true character of the Variolæ Vaccinæ when taken in the casual way. The pustules were too far advanced to permit of any experiments being made with virus taken from them. I cannot refrain from remarking in this place that as the disease, whether caught from the cow or the horse, is much more severe than when communicated by inoculation, so it likewise differs from the last in being sometimes what may be truly called an eruptive disease. Besides the case just specified I know of instances where the disease, when it has been caught from cows in the dairy, has produced pustules more extensively diffused over the body than in the case above-mentioned.

Whoever has attended to the reasonings in a preceding chapter respecting the affinity of the small-pox and the Variolæ Vaccinæ will, it is presumed, find in the facts just enumerated a strong corroboration of the opinions there delivered.

CHAPTER VIII.

DIFFERENCE BETWEEN VARIOLÆ AND VARIOLÆ VACCINÆ—OBSERVATIONS ON VARIOLOÏD DISEASES.

THE preceding investigations have, I would hope, thrown some additional light upon the history and character of the Variolæ Vaccinæ, as well as upon those of small-pox. The analogies which subsist between these two morbid affections have been rendered apparent; and their high antiquity, as well as the identity of their origin, has been supported by very probable evidence. My object in collecting this evidence, connected as it is with the labours and opinions of Jenner, was to secure for the practice of vaccination that confidence to which it is so fully entitled. In furtherance of this design I shall now proceed to place small-pox and cow-pox in contrast with each other; and thus to mark wherein the excellencies of the latter consist. This cannot be done without in some degree breaking in on the regular order of events, because the comparison would be incomplete unless the whole of the past history of vaccination were in some measure brought to bear upon the subject. I will, therefore, first take a rapid view of the mortality attendant on small-pox, and then show by a selection of unquestionable proofs that cow-pox inoculation is perfectly capable, if duly and vigilantly practised, of accomplishing all that its benevolent inventor anticipated. It will thus be seen that vaccination has already subdued small-pox as well in extensive regions as in limited districts; and, therefore, that by proper regulations its protecting power may be made available to the whole world.

From authentic documents and accurate calculations it has been ascertained that *one* in fourteen, of all that were born, died of the small-pox. This was the calculation even after inoculation had been introduced. Of persons of all ages taken ill of the small-pox, in the natural way, one in five or six died, whilst of those who had been inoculated one only in fifty died. These conclusions were drawn by Dr. Jurin from an examination of the London bills of mortality for a period of forty-two years.

In addition to this frightful mortality, alike observable in the icy regions of Greenland; under the burning sun of Asia and Africa; and in the temperate climates of Europe, I must mention the many evil consequences which resulted from the progress of the disease, even when it did not extirpate its victims. Of those who recovered very many were permanently disfigured, or deprived of eyesight; in others, many diseases such as scrofula, chronic ophthalmia, pulmonary consumption, &c. were called forth. Even under the most careful inoculation, and with every mitigation which art can insure, the disease is always troublesome and often dangerous; and seldom leaves the constitution without imparting a shock from which it, with difficulty, recovers.

I would hope that future ages may know this dreadful disease only as matter of history; and feeling, as I do, that the description of it by a physician may be liable to the suspicion of high colouring I subjoin one from the hand of the great and good Sir Matthew Hale. In his fourth letter to one of his grandsons, after admonishing him in his usual solemn manner regarding the feelings of piety and gratitude which his late illness ought to inspire, he thus proceeds:—

"First, therefore, touching your late sickness (small-pox) I would have you remember these particulars: 1st, The disease itself in its own nature is now become ordinarily very mortal, especially to those of your age. Look upon even the last year's general bill of mortality, you will find near two thousand dead of that disease the last year; and, had God not been very merciful to you, you might have been one of that number with as great likelihood as any of them who died of that disease. 2d. It was a contagious disease that secluded the access of your nearest relations. 3d. Your sickness surprised you upon a sudden, when you seemed to be in your full strength. 4th. Your sickness rendered you noisome to yourself and all that were about you; and a spectacle full of deformity, by the excess of your disease beyond most that are sick thereof. 5th. It was a fierce and violent sickness; it did not only take away the common supplies of nature, as digestion, sleep, strength, but it took away your memory, your understanding, and the very sense of your own condition, or of what might be conducible to your good. All that you could do was only to make your condition more desperate, in case they that were about you had not prevented it, and taken more care for you than you did, or could for yourself. 6th. Your sickness was desperate, in so much that your symptoms and the violence of your distemper were without example; and you were in the very next degree to absolute rottenness, putrefaction, and death itself."

The inestimable discovery of the properties of the Variolæ Vaccinæ has divested this picture of all its horrors. One of these properties is that it is an affection extremely mild in its nature and affords, when it has regularly passed through its stages, as complete immunity from subsequent attacks of small-pox as that disease itself does. By this property all danger either of death or of future ill effects are removed. But the property, of all others, which peculiarly distinguishes the Variolæ Vaccinæ from small-pox, and which would enable us to banish this disease entirely, is that *they* are not communicable by effluvia.

The contagious nature of small-pox renders all attempts to mitigate its severity of little avail as regards the interests or the safety of the community at large. The practice of inoculation, the greatest improvement ever introduced in the treatment of small-pox, although beneficial to the individual inoculated, has been detrimental to mankind in general. It has kept up a constant source of noxious infection which has more than overbalanced the advantages of individual security.

The advantages that have already arisen from the employment of the Variolæ Vaccinæ, though less extensive and complete than they might have been, are nevertheless of a magnitude so truly astonishing as to fill the mind with wonder and admiration. Of the millions of every age, condition, temperament, or constitution, who have passed through the disease I doubt if it has proved fatal in one single instance. The *alleged* fatal cases which occurred under Dr. Woodville at the Small-pox Hospital in London deserve no consideration, as these were examples of small-pox, not of cow-pox; and after the most mature examination I cannot but conclude, with the Commission for vaccination at Milan, "that if an infant does happen to die even during the course of vaccine inoculation, reason, experience, authority ought to tranquillize us; inasmuch as they sufficiently assure us that a like fate would have befallen without vaccination, which cannot bestow immortality."

How different are the sad records of small-pox! It is infinitely more destructive to human life than the plague itself. To designate its ravages, as regards the whole race of man, the eloquent Condamine forcibly exclaimed, in his pleading for the adoption of variolous inoculation in France, " La petite verole nous decime." It did *more*, when it attacked a savage or half-civilized people; whole tribes were swept away. The misery of the sufferers was increased by neglect and desertion; the nearest relatives and the dearest friends fleeing from the infected, as persons doomed by divine wrath to irrevocable death.

The records of historians and travellers, from the earliest ages to the present hour, teem with dismal details of this kind. In Europe, Asia, and Africa, as well as among the native tribes in North and South America, such awful visitations were very common. In Thibet, the capital was, on an occasion of this kind, deserted by its inhabitants for three entire years; nor was it re-peopled till it was supposed to have been "purged from the pestilence."*

In Ceylon its very appearance caused whole villages to be abandoned. In the Russian empire small-pox is reported to have been so malignant as to have cut off *two millions* of inhabitants in a single year.† At Constantinople the small-pox has, in many epidemics, proved fatal to one half of those infected.

Dr. Lettsom, of London, calculated that 210,000 fall victims to it annually in Europe; Bernouilli believed that not less than fifteen millions of human

* See Turner's Embassy. Also, Mackenzie's Travels in Canada.

† See Woodville on Small-pox, p. 292.

beings are deprived of life by small-pox every twenty-five years, that is, six hundred thousand annually. This dire disease is not only universal in its ravages, but so subtle is its influence and so insidious its attack as to render it impossible to prevent its approach, or to stay its violence, save by the general adoption of vaccination. In the space of seven years it is said to have been imported more than one hundred times into the British Channel; and in the year 1800, twenty times by the Channel Fleet alone.*

It has been proved by the records of the Institution for the instruction of the indigent blind that three-fourths of the objects† relieved had lost their sight by small-pox. I have already remarked that the practice of small-pox inoculation, though it has been salutary to individuals, has been injurious to the community. In addition to what has been said on this subject I have to observe that the late Dr. Lettsom delivered in, to the Committee of the House of Commons, a document formed from the yearly bills of mortality, by which it appears that in forty-two years, between 1667 and 1722, the average number of deaths occasioned by small-pox was to the whole number who died as 72 to 1000; but that in forty-two years after inoculation came into full use, namely from the year 1731 to 1772, the proportion was no less than 89 in 1000.

Sir Gilbert Blane brought forward a calculation

* See Ring on Cow-pox, p. 687. † Moore's Reply, pp. 64, 66.

made by Dr. Heberden, stating the numbers who died of small-pox in the last thirty years of the late century as 95 in 1000, while in the first thirty years the proportion was only 70 in 1000. Sir Gilbert adds that this is perhaps much more strongly exemplified in the country than in London: since there were many parts of the country in which, previously to the practice of inoculation, small-pox was not known during periods of twenty, thirty, or even forty years, so that a great many passed through life without ever having been affected with it in any way: whereas, at present, both from inoculation and from the free and extended intercourse between the most distant parts of the united kingdom, an adult person who has not had small-pox is scarcely to be met with or heard of.

The salutary power of vaccination in controlling this dreadful mortality having been ascertained different governments adopted different modes of promoting the practice; some by authoritative statutes, others by rewards and inducements of various kinds. Sweden and Denmark adopted the first method principally; as did also many of the German States. One of the best ordinances is said to have been that of Austria in 1808. Vaccination was likewise established by law in Westphalia and Saltzburgh. The French government issued similar ordinances for the Lower Rhine, and for Illyria. The colonial governments of Spanish America acted on the same principle and, so complete was the success, that very soon after the arrival of Balmis's celebrated expedition the small-pox was exterminated in the department of Venezuela. The same zeal was evinced in other parts of South America; and it was calculated in the year 1813 that the population of that extensive, yet thinly-peopled region, had received an annual augmentation of one million of lives, which but for the glorious discovery of Jenner would have fallen a prey to the small-pox.

The Juntas under whose direction these benevolent and judicious measures were carried on made it their business likewise to find out whether the Variolæ Vaccinæ existed among the cows in their country. After a diligent search they found it in the district of Calobozo, in the Caraccas.

Our Government in India endeavoured to entice the natives by all the arguments in their power; and it will be seen in another place that several "pious frauds" were practised by gentlemen connected with our establishments there to induce the Brahmins to give vaccination their sanction. The native inoculators were tempted by considerations of a different kind: a pecuniary reward, amounting to about five pounds sterling for every one hundred persons vaccinated, having been offered to them. It is to be feared that this allowance excited their cupidity more than their vigilance and carefulness in the practice it was intended to promote, their great object being to swell the numbers, rather than to give proper attention to the accuracy of their inoculations.

Vaccination was introduced into Vienna by Dr. De Carro, in May 1799. The effect of it was so great that in the year 1804 only two persons died of small-pox in that city, and these two cases were imported the one from Suabia and the other from a distant village. In the year 1812 it is stated in the Report of the National Vaccine Establishment, on the authority of Drs. De Carro and Sacco, that though the mortality from small-pox was formerly more considerable in proportion to the population in Vienna and in Milan than in London, it had become unknown in the two former cities for several years: Vienna had been free from this pest for five years, and Milan for eight years.

In a letter which I received from Dr. Sacco, dated at Milan, July 13th, 1824, he confirms the preceding favourable statement. He adds "that vaccination is carried on very extensively throughout the kingdom. Almost all the new-born children are annually vaccinated, so that we have now no fear of the small-pox. It is occasionally imported from the neighbouring states of Parma, Piedmont, &c. Such occurrences never fail to prove the efficacy of the preservative, for the disease never becomes epidemic." The result of the observations and experience of this indefatigable and most successful vaccinator is contained in the following sentence, "If all governments would exert themselves to procure the regular vaccination of the children born in their states small-pox would soon disappear, and with it the new eruption."

A letter which I received from Dr. De Carro, dated August 26th, 1826, contains the following gratifying information:—"As to the state of vaccination in Germany you may say, without entering into tedious particulars, that it is everywhere admirably conducted; that Government interferes without employing violent measures, by requiring, on every occasion, certificates of vaccination, without which no children are admitted into any school; without which they can never have any share in any public institution, nor be admitted to any religious sacraments; and without which even the burial is not regular, if they die of the small-pox.

"Nothing is so rare among us as a case of small-pox after vaccination. I do not believe that six such cases could be ascertained at Vienna: in short, vaccination has been brought to a regular system, from which very few individuals are exempted. I cannot speak so positively for other German States, but I have every reason to believe that it is everywhere as well conducted as in Austria, without any sort of medical or popular opposition."

Let us contrast these occurrences with some facts observed on a smaller scale in our own country. In the Royal Military Asylum for the children of soldiers vaccination was introduced by order of Government, at the time of its establishment in the year 1803. This institution, which contains more than eleven hundred children, from the period just mentioned to the year 1811, lost only *one* child by small-pox, and this casually arose from the child not having been vaccinated, in consequence of its mother having declared that it had had the small-pox in infancy. In the Foundling Hospital *no death* had occurred by small-pox since the introduction of vaccination in 1801. Every public institution, I believe, throughout the kingdom, where vaccination has been steadily adopted, will afford results not less satisfactory. If the practice can accomplish such benefits in a crowded city, which is never altogether free from variolous contagion, what might it not achieve, were it employed as it ought to be? In a climate infinitely more unfavourable than England, and where small-pox used to rage with dreadful severity, the disease was extinguished by the judicious regulations for the employment of vaccination: I allude to Ceylon. Between the years 1802 and 1810, under the vigilant direction of Dr. Christie, medical superintendent-general, 128,732 persons were regularly vaccinated. The consequence was that small-pox was unknown in any part of the Island from February 1808 till October 1809, when it was introduced from Quilon on the Malabar coast. It spread to a few individuals who had not been vaccinated; but by the removal of the infected, and the employment of vaccination, the progress of the disease was instantly arrested.

In the year 1803 the Governments of Sweden and Denmark so effectually enforced the practice of vaccination that small-pox was soon extirpated from these countries. They remained free from it for nearly twenty years. It was introduced into the latter kingdom by a traveller from Hamburgh.* It is said that a considerable number of persons were affected with the varioloïd disease, but not one death occurred among them. At Stockholm forty-nine cases of variola appeared, of which seven proved fatal; of those who died none had been vaccinated. In the Swedish provinces sixty cases of small-pox occurred, and thirty-five of varioloïd disease; of the former twelve were fatal, of the latter not one died. It is to be hoped that these untoward occurrences will induce both Sweden and Denmark to renew their efforts to diffuse vaccination universally among their people, and again secure for them that immunity from small-pox which they had so long enjoyed. The practicability of saving mankind from the ravages of this disease has been proved by what has occurred in these countries, and in other parts of the world. They have set an example which all nations ought to follow; and surely there is no one to which the example should more forcibly speak than to our own favoured land, wherein this most salutary discovery had its origin; and from which the small-pox should long since have been banished.

The manner in which vaccination has been conducted in Anspach in Bavaria has afforded results nearly as favourable as those derived from Denmark and Sweden. One fortieth of the population is vaccinated annually. This number is supposed to include almost all the children that live to the age of six months.

* See London Medical Repository for February 1826, p. 164.

In this district, which contains upwards of 300,000 inhabitants, only four deaths occurred from small-pox in the year 1809, and from that period up to the end of the year 1818 *not one* died of that disease. This account is the more striking and satisfactory when it is considered that during the years 1814, 1815, 1816, and 1817, small-pox prevailed epidemically in every part of the contiguous state of Wirtemburgh.* It ought further to be

* The decided effects of the salutary regulations enforced in Anspach would seem to have, at length, aroused the attention of the King of Wirtemburgh to a sense of his own interest and of his people's safety.

The mortality of the four years already mentioned as occurring in his kingdom where small-pox raged epidemically, and probably the observation that during the same period the adjoining dominions of Bavaria were almost entirely free from small-pox, (five deaths only having taken place from the *casual* introduction of that disease, in the course of eleven preceding years) the King of Wirtemburgh at length issued strong enactments in the year 1818, with a view to promote vaccination. The chief regulations are as follow: "Every child must be vaccinated before it has completed its third year, under a penalty *annually* levied on its parents so long as the omission continues; and if the operation fail, it must be repeated every three months until a third trial. No person to be received into any school, college, or charitable institution; be bound apprentice to any trade; or hold any public office, who has not been vaccinated. When small-pox appears, all those liable to take it must be vaccinated without delay; and, the operation not succeeding, it must be repeated every eight days to the third time, under a penalty. The superintendence of vaccination is limited to medical men, each of whom takes charge in a given district; and a fine is levied on all who undertake to vaccinate without added that in Anspach the deaths from small-pox in the year 1797, 1798, and 1799 were more than 500 annually, and in 1800 they actuallyrose to 1609.

In Prussia the deaths from small-pox before the introduction of vaccination were 40,000 annually. In the year 1817 they were under 3000, notwithstanding a considerable increase of population by an accession of territory. In the department of Breslaw, containing about 500,000 inhabitants, the small-pox was introduced in nine different places in the course of the year 1818. By insulating all who fell sick, and by the extensive employment of vaccination, the progress of the disease was immediately arrested; only twenty-eight persons caught it; but of these, six died. *

Dr. Casper has shown that since the general introduction of the cow-pox the mortality among children has been very much diminished; and this diminution

being duly qualified. The name of every child is to be enregistered the day after its birth; and if it die before vaccination, notice must be given. Provision is made for a supply of fresh ichor annually from the cow: and for vaccinating from *arm to arm*, the parties being recompensed for the time thus taken up. Variolous inoculation is prohibited when small-pox is not present; and when it is, the practice can only be done by a medical man, and under proper precautions of seclusion, &c., to prevent the disease from spreading: all expenses incident to the *isolation* and care of variolous patients, whether the disease has been taken by contagion or inoculation, is to be borne by the parents, unless the individual had previously gone through the cow-pox, or been *thrice* vaccinated without effect."

* See Cross on the Variolous Epidemic, pp. 244, 245.

has not been gained at the expense of an increased mortality from other diseases.

According to the researches of Black, Lüssmileh, and Frank, eight or nine per cent. of the human race were carried off by small-pox. Duvillard endeavoured to show that of 100 persons, only four reached the age of thirty years without having it; that one in seven or eight, who were affected, died; and that of those who were attacked in infancy only two-thirds escaped.

The effect of vaccination is clear and distinct, and is admirably exhibited in Prussia, which seems to have got the start of most other European kingdoms. In the French empire 12,857 persons died of small-pox in 1818 and 1819, which in a population of fifty-four millions gives one in every 4518; whilst in Prussia, in 1820 and 1821, the deaths from small-pox were 3137, which in a population of twenty-three millions, gives one in every 7204.

In Berlin alone, the average number of deaths for the twenty years immediately preceding the general introduction of vaccination in 1802, was 472—for the twenty years following, it was diminished to 175, although during that period the disease had been several times epidemic. Since the founding of the vaccine establishment in 1812, the annual mortality from small-pox has decreased to fifty; since 1817 it has been only twelve—and in 1821 and 1822 there was only one death each year. In 1823, during which year Hüfeland considers the small-pox to have been *epidemic* in Prussia, the *cases* of that disease in Berlin were 200, and five died; and he contrasts these numbers with those in the last epidemic, before the general introduction of the cow-pox, namely in 1801, when 16,000 were affected with small-pox, and of these 1646 died.

It is a lamentable fact that, notwithstanding these cheering and irresistible proofs of the virtues of vaccination, we should still have to record the dreadful ravages of small-pox in some of the most civilized countries of Europe; and with shame and sorrow be it said that England stands conspicuous among the number.

Small-pox was nearly as prevalent in London in the year 1825 as during any of the three great epidemics of the preceding century. The admissions into the hospital were more numerous than in any year since 1796, when, according to the bills of mortality, 3549 died of small-pox.

The numbers admitted into the Small-pox Hospital have only been exceeded on two other occasions during the last half-century, namely in the years 1777 and 1781, when the deaths, according to the bills of mortality, were 2567, and 3500 respectively.

In contemplating this very painful statement it is some consolation to know that vaccination, though it had not been permitted to effect all the good that it might have accomplished, has nevertheless been of signal benefit in staying the pestilence. Twelve hundred and ninety-nine persons perished in London, in twelve months; and that, too, after the means of eradicating this plague had been in the hands of the public for twenty-seven years. The physician, from whose report I am quoting these facts, supposes that, frightful as this mortality is, it probably would have amounted (taking into account the increased size of the town) to at least 4000, but for the employment of vaccination.

The proof that vaccination could do so much must fill every rightly-disposed mind with grief that it had not been *so* employed as to prevent altogether this waste of human life. This feeling is augmented when we consider that in the year 1824 the number of deaths was reduced to 725; and in 1818 they were only 421.

In another part of his report Dr. Gregory mentions that, during the last year, 419 persons were admitted into the Small-pox Hospital. Two hundred and sixty-three took the disease in the natural way without previous protection, of whom 107 died. Two had it subsequent to variolous inoculation, of whom one died; and 147 had small-pox, after real or presumed vaccination, of whom twelve died.

The latter part of this alarming statement attracted the notice of Mr. Secretary Peel, who issued orders to the board of the National Vaccine Establishment to consider the alleged fact of twelve deaths from small-pox after vaccination, as stated in the report of the physician to the Small-pox Hospital. The result of this examination was so satisfactory, as it regards the Variolæ Vaccinæ, as to leave no cause to doubt that these individuals had not been properly vaccinated.*

Vaccination seems ever to have been fated to suffer more in character from events within the walls of the Small-pox Hospital than from any other quarter: its atmosphere has always been unfriendly to the benign influence of Vaccina; but I trust that the inquiry which has taken place will counteract the ill effects that might have arisen had the statement of the physician remained unexplained. I hope it will not be thought out of place if I express an ardent wish that my professional brethren may be slow to publish fatal or other cases of small-pox after vaccination, until they have good grounds for believing that their patients had regularly and duly passed through the protecting process; and surely there is no reason to think that this had taken place in any of the fatal instances reported by Dr. Gregory; as a reference to his examination will fully evince.

It appears that the deaths from small-pox in Paris during the last year amounted to 1264.

In the twelfth arrondissement during October last it was found by M. Devilliers that of 215 deaths in private houses 91 were from small-pox, and none of these individuals had been vaccinated.

* See the letter of Sir Henry Halford, Bart. to H. Hobhouse, Esq. Under Secretary of State.—Medical and Physical Journal for May, 1826, pp. 436 and 7.

Those who *had* been, associated with impunity with the persons affected by small-pox; and in those establishments where no persons are admitted who have not been vaccinated the small-pox did not occur.

It appears from the report of the Committee of Vaccination presented to the Minister of the interior of France, in 1811, that from the 4th April 1804 to the end of 1811, the number of individuals vaccinated amounted to 2,300,937. This did not include the numbers vaccinated from the first establishment of the Central Committee, on the 11th May 1800; nor those which have occurred in private practice. The Minister of the interior, in his address introductory to the report itself, made the following striking declaration; he asserted that 1,400,000 children were annually born in the French empire; that formerly one million passed through the small-pox in that period, of whom 150,000 perished. In the year 1811 there were only 70,000 cases of small-pox, and of these 8500 died. Thus, he observes, nearly 150,000 lives are preserved annually to the empire; and a number at least as large snatched from the evils which follow small-pox.

From the documents furnished by the *Provincial* Vaccine Institutions throughout France by order of Government, a report was drawn up soon after this period by M. M. Berthollet, Perce, and Halle (in 1812) in which it is affirmed that of 2,671,662 persons *properly* vaccinated in France, only seven cases of subsequent small-pox appear, i. e. one in 381,666; and it is stated that well authenticated instances of *secondary* small-pox occurring after complete small-pox inoculation are *proportionally far* more *numerous:* and this report adds that in Geneva, Rouen, and several other large cities where the Jennerian method was fully adopted, small-pox is no longer known.

I do not know whether the vaccinations have continued to increase in France, but I rather fear that they have not. In 1811, the births are reported at 1,196,111. Vaccinations 702,218. The number of persons who had the small-pox is 68,711, of whom were disfigured or injured 4436; and died 8377. The report for the years 1821 and 1822 gives the following numbers, births 1,213,082; vaccinations 681,331; persons affected with small-pox 30,480; deaths from small-pox 6538; disfigured or injured by it 2432.

In consequence of the difference of the time, as well as of the number of departments referred to in the preceding reports, it is impossible to determine the relative proportion of vaccinations during the two periods; one fact, however, may be ascertained which seems to prove that vaccination has rather receded than advanced. In the first mentioned period the vaccinations were to the births as more than 3 to 5, in the last they were only as 5 to 9.

The influence of Variolæ Vacinæ on population has not yet been so clearly elucidated as the importance of the subject demands. It has recently attracted the attention of the Institute of France, and that learned body requested M. B. de Chateauneuf to inquire into it. He states in his memoir that the mortality in France among children is much less at present than it was in the last century, particularly from the time of birth to the age of five years. Before the introduction of vaccination the deaths within this range were in the proportion of 50 in 100. Since the introduction of vaccination the proportion has been reduced to 37 in 100. The number of children vaccinated since 1811, throughout France, has not risen above three-fifths, and in the capital it has only amounted to a seventh.* This neglect of the practice fully accounts for the great mortality from small-pox in Paris during the last year.

This part of the subject cannot be better concluded than in the words used by the Commission appointed (in 1825) by the Royal Academy of Medicine of Paris to examine into the state of vaccination.

"Is it reasonable to expect more protection from the vaccine than from small-pox itself?

"Let those who doubt the *efficaciousness* of cow-pox compare the small number of such as are said to have been attacked by small-pox after vaccination, with the important mass of twelve or fifteen millions vaccinated in France within the last five-and-twenty years, and who have altogether escaped from small-

* See Christian Observer, for January 1826.

pox; then must the instances of *secondary* attack appear as slight *exceptions* to a *general* rule. But an irrefragable proof of the preservative powers of the vaccine is to be found in the great public establishments of France, into which no person is received except such as bring with them a certificate of having had either small-pox or cow-pox. Into these establishments Variola has never made its way, as, for instance, the Polytechnic School, St. Cyr, Facultes de Droit et de Medecine, the different colleges of Paris, the Pensionnats de St. Denis, de Saint Germain. These different institutions, situated for the most part in the very midst of contagion, what could have preserved them unless vaccination?"

It appears to have been proved by the most incontrovertible evidence that the Variolæ Vaccinæ, as Dr. Jenner always maintained, are the mild species of small-pox. From this fact it is evident that the same general laws must govern both affections; and that, if there are different degrees of protection afforded by small-pox itself, similar gradations must be looked for in the protecting power of the Variolæ Vaccinæ. When Dr. Jenner published his "Inquiry" it had escaped the attention of medical men that small-pox occurred a second time as frequently as it has been proved to do, both by recent experience and past history. This mistake led, perhaps, to a too confident statement of the never-failing efficacy of vaccination; but after the most mature examination it cannot be doubted that this process, when duly gone through, does certainly afford as complete immunity from subsequent attacks of small-pox, as that disease itself can do. Indeed, if I were to be guided by my experience in this district, I would say that cases of failure have been so rare as to justify even a stronger statement of the protecting qualities of cow-pox: and it ought ever to be remembered, (in the words of Mr. Bryce*) "that there are on record more instances of persons suffering severely, nay fatally, from what was considered to be a second attack of small-pox, than from small-pox after what has been considered perfect vaccination."

The more the whole of this subject is considered and scrutinized the more needful it will appear to attend to all the minute peculiarities connected with the character of Variolæ Vaccinæ; and to obtain full assurance that they take place whenever we attempt to propagate that affection by inoculation.

These observations are introductory to a short consideration of the occurrences which happened during the varioloid epidemics already alluded to, that have recently prevailed in different parts of Great Britain and on the Continent, and which may be supposed to modify the preceding estimate of the virtues of vaccination.

It is not my design to enter upon the nosological question which has been agitated in Edinburgh with regard to these diseases. I cannot, however, but feel that the discussion intended to elucidate the

* See his letter, in Thomson on Varioloïd Diseases, p. 86.

natural and literary history of *variolous diseases*, together with the opinions entertained by Dr. Jenner, *does* forcibly lead to conclusions similar to those which have been so ably stated by Dr. Thomson.

It is agreed on all hands that the eruptive disease, which appeared in Edinburgh and other parts of Scotland, was produced by the contagion of true small-pox. When it occurred in those who had been vaccinated, it, in general, put on that modified character which is described by Dr. Willan * and, in short, resembled the disease which had been excited by that physician and by Dr. Woodville, when they inoculated with the matter of small-pox and cow-pox at the same time.

Its variolous origin is further proved by the following circumstances. The disease was produced in persons previously vaccinated by inoculating with true small-pox matter, and inoculation with matter from the modified disease in many cases produced regular small-pox in those who had not been vaccinated.

During this epidemic Dr. Thomson saw from June 1818, to December 1819, 556 cases. These he distributes into three classes. Of the first, 205 had neither small-pox nor cow-pox previously. Of the second, 41 had gone through the small-pox. Of the third, 310 had been vaccinated.

Of the first class 50 died, or nearly one in 4.

Of the 41 of the second class, which he himself

* See Willan on Vaccine Inoculation, p. 50.

saw, together with 30 others communicated by friends, in all 71, 3 died; one in 23 nearly.

Of the third class, one is reported to have died. The account of this case in the Edinburgh Medical and Surgical Journal for 1820 is somewhat different from that given by Dr. Thomson. It is there represented as an anomalous case, not of small-pox, but of peculiarly aggravated chicken-pox. It is, moreover, right to remark that the proofs of correct vaccination are not stated by Dr. Thomson.

I shall have another opportunity of stating Dr. Jenner's opinions on these occurrences; in the mean time the following sentences from Dr. Thomson's work deserve to be brought forward in this place. "It has been impossible to see the general mildness of the varioloïd epidemic in those who had undergone the process of vaccination, and the severity, malignity, and fatality of the same disease in the unvaccinated, and not to be convinced of the great and salutary powers of cow-pox in modifying the small-pox in those who were afterwards affected with this disease. Proofs cannot be imagined more convincing and satisfactory of the efficacy of the practice of vaccination and of the incalculable benefits bestowed upon mankind by its discoverer, than those I have had the pleasure of witnessing."*

Events similar to those which took place in Edinburgh happened at Millau near Montpelier in France, in the year 1817. M. Paugens, physician

* See Thomson on Varioloïd Epidemic, pp. 40, 108, 111.

there, produced in one child distinct, and in another, confluent small-pox, by inoculating with matter taken from eruptions in two other children, which eruptions he believed not to be chicken-pox, as some supposed, but small-pox rendered mild by previous vaccination. But whatever may have been the nature of this epidemic it did not prove fatal to a single individual who had undergone vaccination, though it carried off more than 200 unvaccinated children. According to M. Fonteneille's account, of 290 persons attacked with this disease at Millau, 55 died; almost all who recovered were deeply marked; and several lost their sight.

In the fatal epidemic which raged at Norwich in the year 1819, and which has been so accurately and ably described by Mr. Cross, he kept a regular register of the effects produced by the contagion in 112 families comprehending 603 persons. Of these, 297 had previously had small-pox, all of whom escaped; 91 had been vaccinated, of whom all, excepting 2 who had a mild affection, and one who had chicken-pox, were protected. * 200 who had neither had cow-pox nor small-pox took the latter; and 15 others who were in the same situation resisted the contagion altogether. It was proved,

* The protecting power of vaccination is very strongly evinced on a comparison of the event in these 200 cases with that which took place in the 91 vaccinated individuals, all of whom, save two, entirely resisted the variolous contagion to which they were exposed; and which it is evident was of the most virulent kind, from the registered statement of the 200

nevertheless, in the same epidemic that in several instances adults caught the small-pox who had at various times resisted the intimate and continued exposure to the contagion of that disease. He mentions two clear cases of regular small-pox, the one after the natural, the other after the inoculated disease. He also met with a few cases of modified small-pox subsequent to small-pox. His correspondents informed him of many cases of secondary small-pox, three of which are stated to have been fatal.

With a view to ascertain the protecting power of vaccination Mr. Cross searched for alleged cases of failure in every quarter of the city, and spared no pains in examining all such reports to the bottom. He found, in all, six cases wherein small-pox ran its usual course in persons who had been vaccinated. The progress of the cow-pox had been watched by medical men; all the cases save one had the usual cicatrix. Of these, two proved fatal; the one. being a confluent disease, on the eleventh day; th other on the eighth day, the pustules being intermix-

cases of small-pox occurring in the same families, with whom they were intermixed; thus given by Mr. Cross:—

Cases of small-pox, in 112 families.

Mild	75	46 deaths.
Severe	78	
Confluent	42	
Petechial	5	
	200	

Vaccinated persons 91, of whom 2 had mild small-pox, one had chicken-pox. No death occurred.

ed with many petechiæ. His conclusion upon the whole is that such fatal cases ought to have no weight against the practice of vaccination, when it is considered that 530 deaths occurred among a little more than 3000 persons who had not been vaccinated; whilst of 10,000 who had been vaccinated, and lived in the midst of a contaminated atmosphere, an immense proportion escaped all illness, a very small number only having eruptive complaints. It is also to be observed that regular and fatal small-pox had occasionally been met with in those who formerly had had that disease.

Facts such as are above recorded, when they are observed on a great scale, necessarily arrest the attention of medical men as well as of the public. Had Dr. Jenner's observations been as much regarded as their importance required many of these untoward events might have been foreseen and, possibly, prevented. So early as the year 1799, he had ascertained by the clearest evidence some of those points which the late fatal epidemics have confirmed. He showed, in the first place, that the constitution cannot by previous infection be rendered totally unsusceptible of the variolous poison.

In the next place he proved that by inoculating a person who had gone through the cow-pox, with variolous matter, it was possible to excite a local vesication from which virus was obtained capable of producing a mild, but efficacious, small-pox. He further maintained in his tracton the varieties and modifications of the vaccine pustule, that these varieties were such as to produce " every gradation in the state of the pustule from that slight deviation from perfection, which is quite immaterial, up to that point which affords no security at all." In this paper he republished some of his observations on the characters of the pustules taken from his instructions for vaccine inoculation. He afterwards adds that fluid taken from a spurious vaccine pustule can propagate and perpetuate its like, and, even if it be taken from a genuine pustule in its far-advanced stages, it is capable of producing varieties which will be permanent if we continue to employ it.

The deviations to which the remarks in this paper chiefly refer are those occasioned by the herpetic state of the skin ; but every deviation, from whatever cause it may have arisen, was considered by him of the greatest moment; and in all his published works, as well as in every private communication, he never failed to express his deep sense of the importance of the most scrupulous attention to that subject; and to the last he felt (and, I believe, most truly) that had his admonitions been received as they ought, had the phenomena connected with vaccine inoculation been studied by all who conducted the practice, a large portion of the failures would have been avoided. The following passages published twenty years ago express so forcibly these sentiments that I cannot withhold them; and now that their author is no more they may perhaps obtain greater notice than when he was among us.

"I shall conclude this paper by observing, that although vaccine inoculation does not inflict a severe disease but, on the contrary, produces a mild affection scarcely meriting the term *disease*, yet, nevertheless, the inoculator should be extremely careful to obtain a just and clear conception of this important branch of medical science. He should not only be acquainted with the laws and agencies of the vaccine virus on the constitution, but with those of the variolous also, as they often interfere with each other. A general knowledge of the subject is not sufficient to enable or to warrant a person to practise vaccine inoculation: he should possess a particular knowledge; and that which I would wish strongly to inculcate, as the great foundation of the whole, is an intimate acquaintance with the character of the true and genuine vaccine pustule. The spurious pustule would then be readily detected, whatever form it might assume; and errors known no more." *

It may not be without its use to connect the events which occurred in recent varioloïd epidemics with those which occurred at the commencement of vaccination. In London, at Petworth, Geneva, and Marblehead in the United States, a series of events took place illustrative of the disasters that arose out of inattention to Dr. Jenner s rules. The matter used at Petworth came originally from the Small-pox Hospital. It was employed in fourteen cases:

* See "Varieties and Modifications of the Vaccine Pustule." p. 13.

variolous eruptions were the result. The disease did not prove fatal in any of the inoculated; but an elderly woman who attended on them caught the disease and died. Her husband was affected by her, but recovered after severe suffering. In this history we have an epitome of what took place in some of the varioloïd epidemics. The vaccine virus sent from London was mingled with that of small-pox. This combination modified the small-pox in those who were inoculated, and produced a comparatively mild disease; but when it was communicated by effluvia to others the variolous affection predominated and assumed its worst character. This I take to be quite analogous to what happened when persons previously vaccinated caught the small-pox or varioloïd disease. In such it became modified and mild; but in those who had never felt vaccine influence it assumed when propagated, whether by inoculation or by effluvia, its most malignant form.

The disasters at Geneva and Marblehead were of a different kind. They arose from employing spurious matter. The disease excited in both instances had nothing of the true vaccine character, and afforded no protection to those to whom it was communicated. Who can tell how many of the alleged cases of small-pox after vaccination may not have been of this description?

It were easy to add numberless facts to those which have been accumulated in this chapter. But they all speak one language. Let Vaccination be effectively and universally employed and Small-pox must disappear; and with it, as Dr. Sacco says, "the new eruptions." They are only met with when small-pox has been permitted to show itself because vaccination has either been neglected, or partially and insufficiently performed.

The experience of the public institutions in Paris and London is quite conclusive on this point. The same fact is equally capable of proof whether we appeal to small districts, or to large kingdoms. There are many country parishes in England from which small-pox has been entirely excluded ever since the general employment of vaccination, even though the disease was prevalent in their immediate vicinity. In such situations both variola and varioloïd eruptions have been unknown.

Shall not these unquestionable statements arouse the attention of the community to secure all the blessings that are placed within its reach? Small-pox inoculation, I believe, is now abandoned by almost every respectable medical man. In the country it has fallen into the hands of the lowest and most illiterate of the people. Surely this ought not to be suffered. If qualifications are required for selling or dispensing medicines, the power to disseminate a poison, whose influence is not and cannot be confined to the individual who receives it, but may carry destruction through a whole district, ought not, certainly, to be left at the disposal of every ignorant person whose prejudice or cupidity may prompt him thus to sport with the lives of his fellow-creatures.

CHAPTER IX.

LIFE, AFTER THE PUBLICATION OF THE "INQUIRY," TO 1800. DISASTERS AT THE SMALL-POX HOSPITAL, AT PETWORTH, &c.

In endeavouring to elucidate and confirm the opinions of Dr. Jenner, and to demonstrate the salutary power of vaccination, we have been carried away from his personal history, and been obliged to deviate from the regular succession of events that took place subsequent to the publication of his *Inquiry.* The first opposition which he encountered, after this event, came from a quarter that might have caused misgivings in the mind of any one who was not thoroughly convinced that both his facts and reasonings were fitted to stand the test of the severest scrutiny. The celebrated Dr. Ingenhousz, distinguished both as a man of science and as a physician, came to visit the Marquess of Lansdown at his seat in Wiltshire, soon after the publication of the Inquiry. On the 12th of October, 1798, he addressed a long letter to Jenner on the subject of that work.

It is written respectfully, but at the same time with a degree of pomp and authority scarcely even befitting one who could subscribe himself Physician to the Emperor and King He had made it his business to inquire among the dairies of Wiltshire, where cow-pox sometimes prevailed, respecting its reported virtues as a preventive of small-pox; and he got, as might have been expected, such answers as Dr. Jenner had obtained at the commencement of his investigation. Some individuals, who had had what was called cow-pox, were subsequently affected with small-pox; and therefore, he said, all the security which you promise from the inoculation of the former is so neutralized by this testimony that you had better confess that you have been in error, and prevent the disappointment which must follow from ungratified expectations.

Jenner s discussions with his professional brethren at Alveston had fully enabled him to meet this objection; and the different steps in the investigation, by which he had proved when protection was to be expected and when it was not, had so completely armed his own mind against assaults of this nature that he had nothing to fear in the encounter. The station and character of such a man as Ingenhousz was nevertheless not to be neglected. He had made small-pox his particular study, and had been employed in inoculating some of the Imperial family of Vienna as well as of the Grand Duke of Tuscany, with this disease. Jenner pointed out to him in the most becoming manner the source of his misapprehension—that he had met with individuals who had confounded the spurious affection which sometimes appears on the hands of the milkers, and which affords no security at all, with that genuine and regular pustule which, when not disturbed in its course, gives the most perfect exemption from the attacks of small-pox.

As the early letters on this subject must have a peculiar interest, and will convey a better idea of the state of feeling than any description of mine, I will now, as on other occasions, illustrate my narrative by reference to documents of that kind.

Dr. Ingenhousz to Dr. Jenner.

Sir,

Having read with attention your performance on the Variolæ Vaccinæ, and being informed by every one who knows you that you enjoy a high and well-deserved reputation as a man of great learning in your profession, you cannot take it amiss if I take the liberty to communicate to you a fact well deserving your attention, and with which you ought to be made acquainted. I prefer this private method of conveying my information to any other which might expose you to the disagreeable necessity of entering into a public controversy, always disagreeable to a man so liberal-minded and well intentioned as your treatise indicates you to be.

As soon as I arrived at the seat of the Marquess of Lansdown, Bowood, near Calne, I thought it my duty to inquire concerning the extraordinary doctrine contained in your publication, as I knew the cow-pox was well known in this country. The first gentleman to whom I addressed myself was Mr. Alsop, an eminent practitioner at Calne. This gentleman made me acquainted with Mr. Henry Stiles, a respectable farmer at Whitley, near Calne, who, thirty years ago, bought a cow at a fair, which he found to be infected with what he called the cow-pox. This cow soon infected the whole dairy; and he himself, by milking the infected cow, caught the disease which you describe, and that in a very severe way, accompanied with pain, stiffness, and swelling in the axillary glands. Being recovered from the disease, and all the sores dried, he was inoculated for the small-pox by Mr. Alsop. The disease took place: a great many small-pox came out, and he communicated the infection to his father, who died of it. This being an incontrovertible fact, of which I obtained the knowledge from the very first man to whom I addressed myself, cannot fail to make some impression on your mind, and excite you to inquire farther on the subject, before you venture finally to decide in favour of a doctrine, which may do great mischief should it prove erroneous.

I heard of several other facts of a similar nature which tend to contradict your doctrine; but, indeed, it was added that the cow-pox had not been severe enough to extinguish the susceptibility to the small-pox.

The above-mentioned farmer thought that the disease of the cow called cow-pox spreads through a dairy in the way of other contagious diseases. The very offensive stench, which those sick cows give out from the lungs and the udder seems to indicate that the disease spreads by infection, without the interference of the milker's hands, or the *grease* of the horse's feet.

Thomas White, an eminent farrier in the neighbourhood of Calne, was of the same opinion.

By inquiring more minutely on what is asserted in page 56–57, you will, I make no doubt, find it erroneous. But I will make no farther observations, as it is far from my wish or my intention to enter into any controversy with a man of whom I have conceived a very high opinion. Let it suffice to have communicated to you, in a friendly way, a fact which may awaken your attention.

I am, with every possible sentiment of esteem,

Sir,

Your obedient humble servant,

J. Ingenhousz, Physician to the Emperor and King.

Bowood Park,
Oct. 12th. 1798.

Dr. Jenner to Dr. Ingenhousz.

Dear Sir,

I shall ever consider myself as under great obligations to you, for the very liberal manner in which you have communicated a fact to me on a subject in which at present I feel myself deeply interested; a subject of so momentous a nature that I am happy to find it has attracted the attention of some of the first medical philosophers of the present age, among whom it is no compliment in me to say that I have long classed you.

It will doubtless, in the course of time, meet with a full investigation; but as that moves on (and from the nature of the inquiry it must move slowly) I plainly foresee that many doubts will arise respecting the validity of my assertion, from causes which ought to be examined with the nicest inspection before their convictive force be fully admitted.

Truth, believe me, Sir, in this and every other physiological investigation which has occupied my attention, has ever been the object which I have endeavoured to hold in view. In the publication on the Variolæ Vaccinæ, I have given little more than a simple detail of facts which came under my own inspection, and to the public I stand pledged for its veracity. In the course of the inquiry, which occupied no inconsiderable portion of my time and attention, not a single instance occurred of a person's having the disease, either casually or from inoculation, who on subsequent exposure to variolous contagion received the infection of the small-pox, unless that inserted in page 71 may be admitted as an exception. And from the information you have given me, and from what I have obtained from others who have perused the pamphlet, I am induced to suppose that my conjecture respecting the cause of that patient's insecurity, namely, her having had the disease without any apparent affection of the system, might have been erroneous; and that the consequences might be more fairly attributable to a cause on which I shall, in my present address to you, feel it my duty to speak explicitly. Should it appear in the present instance that I have been led into error, fond as I may appear of the offspring of my labours, I had rather strangle it at once than suffer it to exist, and do a public injury. At present I have not the most distant doubt that any person, who has once felt the influence of perfect cow-pox matter, would ever be susceptible of that of the small-pox. But on the contrary, I perceive that after a disease has been excited by the matter of cow-pox in an imperfect state, the specific change of the constitution necessary to render the contagion of the small pox inert is not produced, and in this point of view, as in most others, there is a close analogy between the propagation of the cow-pox and the small-pox. Therefore I conceive it would be prudent, until further inquiry has thrown every light on the subject which it is capable of receiving, that (like those who were the objects of my experiments) all should be subjected to the test of variolous matter who have been inoculated for the cow-pox.

(The remainder of this letter is lost.)

This very mild and satisfactory letter by no means produced the effect that might have been expected on the mind of the imperial physician.

In the same proportion as Jenner was diffident and conciliatory, *he* became rude and truly imperious; resisted every explanation that was given; contended with increasing obstinacy for the accuracy of his own assertions; and, of course, became more and more convinced of what he deemed the errors of Jenner.

It will easily be believed that the reiteration of such sentiments could not fail to be disagreeable to Dr. Jenner. He soon, however, formed a proper estimate of the character of his correspondent. At first he evidently felt a degree of anxiety concerning the result of the conflict with such an opponent; but he met all the attacks with perfect good temper, till the haughtiness and self-sufficiency of his antagonist put an end to all further intercourse. He had sent another long explanatory letter to Dr. Ingenhousz, and proposed, in order to do away all appearance of shrinking from an exposure of their correspondence, that this letter should be printed, and be considered as an appendix to the Inquiry. The following letters refer to this incident :—

DR. JENNER TO MR. EDW. GARDNER.

Berkeley.

DEAR GARDNER,

We wondered at Ingenhousz's delay in answering my letters, particularly the long one that you inspected. A tempest is generally preceded by a calm. He has in some measure exemplified the remark. I know not what to do with him, and wish for your advice, after you have seen his letter. It is a matter of real moment; a matter on which perhaps much of my future peace may rest—indeed, my existence. I sometimes think that it would be most prudent to desire him to make public all he knows of the cow-pox; but would there not, in this measure, be a sort of defiance that might irritate? The grand question at present to be determined is this—shall I immediately publish an appendix, or say nothing till every bolt is flung, and then attack my adversaries?

This very man, Ingenhousz, knows no more of the real nature of the cow-pox than Master Selwyn does of Greek: yet he is among philosophers what Johnson was among the literati, and, by the way, not unlike him in figure. 'Tis no use to shoot straws at an eagle. When shall I see you?

Yours, sincerely,

E. J.

DR. JENNER TO MR. EDW. GARDNER.

DEAR GARDNER,

I fully depend upon meeting you at Eastington to-morrow to sit in council on several subjects of high import. My friends must not desert me now. Brickbats and hostile weapons of every sort are flying thick around me; but with a very little aid, a few friendly opiates seasonably administered, they will do me no injury.

Ingenhousz has declined my offer of receiving my letter in print—so that must be modelled anew. We must set off by impressing the idea that there will be no end to cavil and controversy until it be defined with precision what is, and what is not cow-pox.

The *true* has many imitations by the *false* on the cow's udder and nipples; and all is called *cow-pox*, whether on the cow or communicated to the human animal.

My experiments move on—but I have all to do single-handed. Not the least assistance from a quarter where I had the most right to expect it!!

Bodily labour I disregard, but pressures of the mind grow too heavy for me. Added to all my other cares, I am touched hard with the reigning epidemic—Impecuniosity.—Any supplies from the paper-maker?

Adieu!

Your faithful friend,

Wednesday morning. E. J.

You must be more attentive to me than you were during the last Cheltenham recess. I believe you came here only once, and then in your way from Bristol.

Dr. Jenner sent also one of his friends who was well acquainted with vaccination to endeavour by personal explanation to remove Dr. Ingenhousz's misconceptions. An amusing account of that interview is contained in this letter.

T. PAYTHERUS, ESQ. TO DR. JENNER.

December 14*th*, 1798.

DEAR JENNER,

The moment I received your letter I called on Dr. Ingenhousz; he was in the country, but expected in town the next day. Yesterday I called a second time and made an appointment for this morning, in consequence of which I have had an interview with this very interesting character.

A more determined or a more formidable opponent you need not covet or desire. Unfortunately for your hypothesis, he made his first inquiry of a Mr. Alsop, of Calne, who immediately named a person who had had the small-pox after the cow-pox. This person he was afterwards introduced to and satisfied himself of the *fact*. The second application was to Major General Hastings: he also pointed out an instance of the small-pox subsequent to the cow-pox, at Adlestrop.

Dr. Garthshore has also at Dr. Ingenhousz's request written to Dr. Pulteney, of Blandford, who in reply has assured him that the inoculators of his neighbourhood have known many instances of the small-pox happening after the cow-pox. He believes that it does in many instances produce that change in the human constitution as to render it unsusceptible of the small-pox, but not with certainty in *all* cases. He would not hear a word in defence of your opinion respecting its origin.

He is confident that a spurious small-pox cannot be produced by what you call putrescent variolous matter, and that whether the matter be kept in a wet sponge, or on cotton, either in a moist or a dry state, it will uniformly produce the small-pox. Yet he confessed in his own practice that the dried matter more generally produced a confluent small-pox. In your last letter to him you speak of the putrescent state of the cow-pox matter, and that the milk might likewise undergo a similar change. To this he objects, and says that milk will become acescent, not putrescent.

That it should render the habit unsusceptible of small-pox, and not of its own specific action, is to him incredible. You tell him in one of your letters that you have heard from Adlestrop, and that the father of the boy or girl now thinks that the small-pox preceded the cow-pox. To *think*, it seems is to *doubt*, and he says the ambiguity on the part of the father confirms the first statement instead of weakening it. His respect for your character has kept him from publishing, and he declines entering into controversy with you. Had you been a less formidable antagonist he would have flogged you long since. He spoke very handsomely of you, and desired me to assure you that nothing would have kept him from answering your letters but the desire of satisfying his mind on the subject. He desires that you will not be in haste to publish a second time on the cow-pox, but wait till you have collected a sufficient number of facts, and to secure your ground as you advance. He remarked that you would not be permitted to be judge in your own cause; that you were now before the tribunal of the public, and so long as *sub judice lis est*, you ought not to risk an opinion.

* * * * *

Instead of printing his letters to Dr. Ingenhousz in

the form of an Appendix to the *Inquiry*, as he had proposed, he determined to give them another shape, and published on the 5th of April, 1799, "Further Observations on the Variolæ Vaccinæ." The reader will see at page 3, a pleasing illustration of the spirit in which all his researches were conducted. The same sentiment will be found in his private letter to Dr. Ingenhousz.

"Ere I proceed let me be permitted to observe that truth in this, and every other physiological inquiry that has occupied my attention, has ever been the first object of my pursuit; and should it appear in the present instance that I have been led into error, fond as I may appear of the offspring of my labours, I had rather see it perish at once than exist and do a public injury."

The effect of Mr. Cline's successful vaccination in London, after Dr. Jenner had left the metropolis, was strongly manifested by an increased degree of attention on the part of the men of science. Many of the most distinguished, while they gave all due confidence to the statements of Jenner, waited with becoming prudence till additional experience should either confirm or refute the very remarkable phenomena which they had been called on to consider.

Others, with little wisdom, precipitately adventured into a field of investigation of which they had no knowledge but what they derived from Dr. Jenner himself; and on mere speculation questioned the accuracy of his details, and denied the cogency of his reasoning.

The gentlemen who distinguished themselves in this manner proceeded at the outset, at least in their intercourse with Dr. Jenner, with much apparent respect and candour. Their main object seemed to be to further the advancement of truth; and to second him in his benevolent purpose. Such unquestionably was the feeling with which he regarded their efforts. He welcomed them as valuable coadjutors in his great cause, and he continued to do so till proceedings, the character of which could not be mistaken, compelled him reluctantly to discover qualities that but ill assorted with these anticipations. Were it not that some circumstances of great consequence to himself and to the cause of vaccination were involved in the events which I am now about to mention, I would gladly pass them by without any notice.

Having concluded his correspondence with Dr. Ingenhousz he soon encountered other opponents. His opinion with respect to the origin of the Variolæ Vaccinæ in an especial manner drew forth the animadversions of his contemporaries. It was received with derision by some, and with suspicion and distrust by almost all. This is not the only instance in which he had the misfortune to shoot above the heads of his brethren; and to have doctrines treated as conjectural and ridiculous, of the truth of which he had satisfied himself by rigid and patient inquiry The whole history of his labours on the subject of vaccination, as already set forth in this work, affords pregnant illustrations of this observation.

Some grave and learned persons doubted all the assertions contained in the *Inquiry*; and, of course, set no value on the reasoning connected with them. Others thought that there was truth in the reported prophylactic virtues of cow-pox, but that Jenner had no more merit than was due to the publisher of a fact long known, though buried in provincial obscurity; that all the information respecting the laws of this singular affection remained to be discovered; and that his opinions on such high matters merited little consideration. Thus, under the head of those who refused assent to his positions altogether, or of those who questioned the accuracy of his information and the fidelity of his descriptions, may be arranged all the active detractors, who zealously bestirred themselves to overthrow Dr. Jenner's pretensions, either by direct contradiction, or by faithless and insidious efforts to reduce his merits as a discoverer, even while their mouths were filled with ardent professions of gratitude and admiration.

It will appear in the sequel that vaccination incurred much greater risk from the latter than from the former. The apparent candour of their proceedings; their eagerness and zeal in the cause; and the respectability of their professional character lent an authority to their statements to which they were by no means entitled; and moreover gave a temporary countenance to errors both in doctrine and in practice, which doubtless would have proved fatal to vaccination had not Jenner placed it on a foundation that could not be shaken. The ordeal which tried both the firmness and patience of his character, and the accuracy and faithfulness of his most minute details was brought into full and inquisitorial activity very soon after the publication of the *Inquiry*, and to that subject we must now direct our attention.

Dr. Jenner remained chiefly at Cheltenham and Berkeley from the time of his leaving London in July 1798, till the following February. During this period he was most assiduously employed in collecting additional information respecting the Variolæ Vaccinæ; and in carrying on an extensive correspondence with medical gentlemen in different parts of the kingdom: but his stock of vaccine matter having become exhausted; and being disappointed in supplies from the dairies, he could not answer all the demands that were made to him for virus.

It was not till towards the end of November that he was enabled to procure any. On the 27th of that month he inoculated two of the children of his friend Mr. Hicks, of Eastington, with matter taken the preceding day from a farm at Stonehouse. I dwell on this incident that I may, in the first place, record Mr. Hicks's confidence in the prophylactic powers of cow-pox, who had the merit of being the first gentleman that submitted his own children to the new practice; and, in the next place, to disprove an assertion subsequently made that the first vaccinations performed by Dr. Jenner, after the publication of his Inquiry, were with virus furnished by Dr. Pearson.*

The commencement of the intercourse between Dr. Pearson and Dr. Jenner augured a very different progress and termination from those which actually took place. The early letters of the former indicated not less ardour in behalf of vaccination than espect and admiration for its author. Dr. Jenner amply repaid the good-will of his correspondent by the most unreserved communication of all the knowledge which he possessed, concerned merely for the dissemination of truth, and regardless of all personal considerations.

Soon after the subject had attracted public attention Dr. Pearson rendered service to the cause of vaccination by establishing an extensive correspondence with medical men in different parts of the kingdom, by which he was enabled to prove that cow-pox was much more widely epizöotic than had been at first believed; and that all the local traditions fully confirmed Dr. Jenner's positions.

* The first personage of rank who broke through the chains of prejudice, and had her only child vaccinated, was the Lady Frances Morton (now Lady Ducie.) The Countess of Berkeley equally surmounted every prepossession in favour of the old practice, and at the commencement, as well as at every subsequent period, ardently recommended and adopted the new.

It is painful to be compelled to state that the feelings avowed to Dr. Jenner at this period ill accorded with the attempts which were afterwards made to overthrow his undoubted, and previously admitted, claims as a discoverer. On the 8th of November, 1798, and just on the eve of the publication of his pamphlet, Dr. Pearson wrote a letter to Dr. Jenner which, among other matter, contained the following expressions: “ Your name will live in the memory of mankind as long as men possess gratitude for services, and respect for benefactors; and if I can but get *matter* I am much mistaken if I do not make you live for ever.”

About the middle of November Dr. Pearson's pamphlet was published. This circumstance was thus announced to Dr. Jenner.

Dr. Pearson to Dr. Jenner.

My Dear Sir,

Unexpectedly my pamphlet made its public appearance a day or two ago. I am sorry to trouble you to say by what conveyance I can send you a copy, and to what place? If you have any commissions to execute in London you may as well have a parcel made up, and I will see it forwarded.

I observe several errors since printing, partly mine and partly those of the printer; but I know other authors discover similar errors, and that readers do not perceive them.

You cannot imagine how fastidious the people are with

regard to this business of the cow-pox. One says it is very filthy and nasty to derive it from the sore heel of horses! Another, O my God, we shall introduce the diseases of animals among us, and we have too many already of our own! A third sapient set say it is a strange odd kind of business, and they know not what to think of it! All this I hear very quietly, and recollect that a still more unfavourable reception was experienced by the inoculation for the small-pox.

I wish you could secure for me matter for inoculation, because, depend upon it, a thousand inaccurate but imposing cases will be published against the specific nature of the disease by persons who want to send their names abroad about any thing, and who will think yourself and me fair game. By way of *se defendendo* we must inoculate. I have thought it right to publish the evidence as sent to me, and also my own reasoning, because I know you are too good a philosopher to be offended at the investigation of truth, although the conclusions may be different from your own. I think, too, your principal facts will be the better established than if it had happened that I had uniformly acceded to all your doctrine.

I am, with Mrs. P.'s best compliments to Mrs. Jenner and yourself,

Your faithful Servant,

G. PEARSON.

Leicester Square, Nov. 13th, 1798.

"NOS POMA NATAMUS!!"

Dr. Pearson and other medical gentlemen went on speculating, and doubting, and deciding on Dr. Jenner's doctrines without ever having seen at the time an example of cow-pox, or having given due consideration to his arguments in favour of the opinion as to the origin of the disease. Matters continued in this state till about the 20th of January 1799, when it was discovered that the cow-pox existed in a dairy in Gray's-inn-lane, London. This occurrence was made known to Dr. Jenner by Dr. Woodville.

DR. WOODVILLE TO DR. JENNER.

Ely Place, Jan. 25, 1799.

DEAR SIR,

* * * * *

On Sunday last I was informed that the cow-pox had broke out among Mr. Harrison's cows in Gray's-inn-lane. The next day I took Mr. Tanner with me to examine them; and as he declared it to be the genuine disease, I that day inoculated six persons with the matter that he procured from a cow which appeared to be the most severely affected with this pustular complaint. On Wednesday I called again at the cow-house to make farther inquiries, when I was very much pleased to find two or three of the milkers were infected with the disease; one of whom exhibited a more beautiful specimen of the disease than that which you have represented in the first plate. From this person I charged a lancet with the matter which appeared different from that taken from the cow, as that of the former was purely lymphatic, and the latter of a purulent form. With this lymphatic matter I immediately inoculated two men at the hospital.

Finding now there could be no doubt of the disease I the same day called upon Sir Joseph Banks, Dr. Pearson, Dr. Willan, &c. to inform them of the circumstance; and these gentlemen, together with Lord Somerville, Sir William Watson, and Mr. Coleman, met me the following day at the cow-keeper's, where your book was produced; and upon comparing *your figure* with the disease it was allowed by all to be a very faithful representation, and every gentleman seemed highly gratified at seeing so good an example of the cow-pox. From this place we proceeded to the hospital where I inoculated six patients, so that the whole number inoculated by me with the cow-pox matter amounts to fourteen.

* * * * *

Dr. Jenner to Dr. Woodville.

My Dear Sir,

I am extremely obliged to you for your letter, and most sincerely wish circumstances would admit of my being at your elbow while you conduct your experiments on the interesting subject before you.

I answer your letter by return of post to suggest (what perhaps is needless) the immediate propriety of inoculating those who may resist the action of the cow-pox matter, and may have been exposed to variolous contagion at the hospital. After the description you have given there can be no doubt I think that the disease among the cows in Gray's-inn-lane is the true, and not a species of the spurious cow-pox. In the account of the appearance on the milker's hand the report of my friend Tanner merits great confidence. Whether to the cold season of the year or to what other cause it can be ascribed I know not, but out of six patients that I lately inoculated two of them only were infected. An inflammation was excited in the arms of all, and in some of those, whose constitutions would not feel it, it did not die away for more than a week, and even went on so as to leave a little scab behind.

It has not happened so, generally. However, once in the course of the last summer I was foiled in a similar way. Three or four servants at a farm were carefully inoculated with matter fresh from a cow:—they all resisted it, but in the course of the season all of them were infected by milking the cows. As every case of cow-pox is to be considered as a case of inoculation I mention these facts to you, that it may be considered whether some mode more certain of infecting the subject than that at present in use with variolous matter may not be thought of.

It would imitate the casual mode more closely were we first by scratch or puncture to create a little scab, and then, removing it, apply the virus on the abraded part.

I am shortly going to publish an appendix to my late pamphlet (which, by the way, I hope you received, as I directed it to be sent to you before I left London) to mention the precaution of destroying the pustule and the general sources of spurious cow-pox, &c &c.

I shall also point out the result of one of the cases where caustic was used soon after the symptoms of infection appeared (see page 41). This I shall concisely relate to you now. About six weeks ago I inoculated M. James (see page 40) with fresh small-pox matter, and at the same time exposed her to the effluvia of a patient. The appearances of the arm were just the same as if she never had had either small-pox or cow-pox: and on the eighth day I expected, from the appearances, she would be ill. She was a little hotter than usual during the night but slept well, and it was supposed that a rash appeared for the space of a few hours about the wrists. I inserted matter from her arm into two other subjects, a boy, and woman of fifty. The boy had about half a dozen pustules, two or three of which were fairly characterized. Their appearance was preceded by a pretty general rash. The woman, though she felt an indisposition, had not a single pustule. A person near sixty years of age, who had in the early period of her life been exposed to the contagion of the small-pox and resisted it, fully exposed herself now to this infection. She sickened in consequence and had three pustules, one of which became a perfect small-pox pustule. It would be unfair to draw positive conclusions from such scanty precedents, but yet they lead one to hope that a mild variety of the small-pox might thus be actually created.

* * * * *

The result of the trials mentioned in Dr. Woodville's letter, and of others performed under like circumstances, had well nigh proved fatal to the infant cause of vaccination. They were all instituted at the Small-pox Hospital in London; and many of the patients, besides being exposed to a variolous atmosphere, actually had small-pox matter inserted into the arm on the third and fifth days after vaccination: and strange to say, this was deemed a fair trial of the virtues of cow-pox. Every one at all acquainted with the influence of small-pox contagion might have anticipated what actually occurred. Dr. Jenner had positively asserted that pustules do not belong to the cow-pox, as he had never seen them produced by genuine vaccine matter. Dr. Woodville, on the contrary, reported that three-fifths of the patients whom he had inoculated with vaccine matter had pustules not to be distinguished from variolous ones. A statement of this kind, from such a quarter, so much at variance with what had been anticipated, excited the strongest feelings of disappointment among the principal medical men in London; and, of course, for a season threw considerable doubts on Dr. Jenner's accuracy. He from the commencement suspected the real cause of this deviation; but he was willing to give the gentlemen who began the experiments in London the benefit of every supposition that might tend to acquit them of the blunder which they had actually made. He thought it possible that there might be some peculiarity either in the constitution of the individual vaccinated, or something in the nature of the virus itself that might occasion the eruptions: but after admitting these suppositions he could not help hinting to one of the gentlemen concerned, in the earliest letter he addressed to him on the subject, that he *could not divest himself of the suspicion that the London cow-pox was somehow or other compounded with small pox.* When he became fully acquainted with the manner of conducting the practice in the Small-pox Hospital* this suspicion amounted to

* The following anecdote, illustrative of the manner of conducting vaccination at the Small-pox Hospital, has been communicated to me by a gentleman on whose authority I can rely. A medical gentleman who was particularly forward on the occasion, but who was not very well acquainted with the characters either of small-pox or of cow-pox, applied to Mr. Wachsel,

certainty; and at a later period he had the hardihood to declare to Dr. Woodville that the matter had absolutely been contaminated in the Hospital. The word *contaminated* was very distasteful to the Doctor, although it was the only one that was really expressive of the fact and was certainly not used in an offensive sense by Dr. Jenner. The disasters which arose from this contamination might well have justified a stronger expression on his part. It is impossible now to deny the fact that this *impure* matter was really disseminated over many parts of England, and also on the Continent, in place of that of the true Variolæ Vaccinæ. It was the practice of Drs. Pearson and Woodville to take matter from the patients who had eruptions, and with it to inoculate others. This curious fact was communicated to Dr. Jenner by Dr. Pearson.

the apothecary, for leave to charge some threads with vaccine virus, professedly to distribute them to his medical correspondents throughout the island. Mr. Wachsel chanced to be called out of the apartment; during his absence the doctor selected a patient, and was busily engaged in charging his threads. Mr. Wachsel observed on his return that he had fixed on a patient who had a general sprinkling of small-pox pustules, and inquired whether he intended to furnish his friends with the virus of small-pox as well as of cow-pox? He replied "with the virus of cow-pox only." "Then, Sir," said Mr. Wachsel, "you know not what you are doing, you are taking the virus of small pox." The threads thus charged, but for Mr. Wachsel's vigilance, would have been distributed as vaccine virus!!!

DR. PEARSON TO DR. JENNER.

Leicester Square, February 15th, 1799.

MY DEAR SIR,

* * * * *

Our apprehensions are now almost entirely dissipated of danger from inflammation around the inoculated parts. In my patients it has been inconsiderable indeed, not amounting to inflammation, being merely what I would call erythematous, or the same kind of affection as in scarlatina. I have inoculated two infants of fourteen days old: one of them did not take the disease the first time but did the second time, but the constitution was not apparently disordered, and there was less redness around the pustules, which were large and conical, than even in the small-pox. I therefore inoculated it yesterday in the presence of Dr. Woodville a third time, from a patient in the same room ill of the cow-pox with eruptions on the body to the amount of two or three hundred. You will be astonished at our talking of eruptions, but it now appears in Dr. Woodville's cases that as many have eruptions on the body as have them only in the parts inoculated.

I inoculated a child about fourteen months of age with matter from one of Dr. Woodville's patients in one arm, and in the other with matter from the cow. To my astonishment the disorder proceeded exactly as the inoculated small-pox. On the ninth day convulsive fits came on, and vomiting occurred with other pretty violent symptoms, which alarmed me and made me very uneasy. The eruptions appearing on the body in great number, all became tranquil. A few only suppurated; the rest died away sooner than the suppurated. I could not persuade myself but that there was some mistake, and that Dr. Woodville's lancet, which I used, had variolous matter upon it. I called him to see the patient. He assured me the lancet having been lately ground had not been used, and that the patient from whom he took the matter had but two or three, and those not like the small-pox. He had no doubt that my patient, notwithstanding the numerous eruptions, had nothing but the cow-pox. And to-day, at the hospital, a patient appears to be likely to have as many eruptions as my patient. I shall ascertain this fact by inoculating patients with the matter of the eruptions. As I am to write to-day Dr. Woodville will not send his letter for some days, in order to report the further progress. I have sent you a bit of thread with matter upon it in this letter. The stiff part will denote the part imbued with the infectious fluid. It is from Dr. Woodville.

It appears that the more severe the previous symptoms the greater the crop of eruptions: and I think the sooner the disease takes place the more eruptions appear.

In some cases the patients have been ill only *two* days, but in one case *six* days:—I mean ill with the constitutional affection. For the most part there are very few suppurated pustules. All the symptoms are relieved on the eruptions taking place.

Some of my patients made no complaint at all, nor till asked did I hear of even pain of the *armpits* or rather of the *shoulder*.

On the whole, Dr. Woodville and myself conclude that the inoculated cow-pox, as far as we have seen in at least fifty cases, is a slighter disease than the inoculated variola, and that it is not probable there is any danger to life from it.

On telling Dr. Woodville that I had been anxious about your publishing the use of the caustic, he replied "that would have damned the whole business."

Be assured that if the practice cannot be introduced without the caustic, or call it by any other name, it will never succeed with the public. I cannot yet tell whether all my patients have had sufficient affection of the constitution. There has not been time for a second inoculation and with variolous matter. Some of the patients had undergone the small-pox.

Dr. Parr's letter you shall see in town merely to satisfy you, but it contains nothing that is relative beside what I extracted from it. I must tell you that Dr. Parr has written to me to say that although he "is not yet convinced he is staggered, and begins to doubt." We shall have to experience soon a number of *gnat bites*. If the practice is likely to go forward it will excite opposition. What obligation society owes to those worthy and liberal men who favour the public with their *a priori* opinions, having never seen the disease, and not even understanding the arguments!! Tantæne animis cœlestibus iræ!

I trust we shall establish facts enough to prove whether cow-pox inoculation extinguishes that of the *variola* or not. We have got able, candid, and worthy men on our side, and proceeding as we have all done, circumspectly, I do not feel any dread from the opponents who have yet taken the field.

Mrs. P.'s best compliments to Mrs. Jenner.

Dear sir, yours,

G. Pearson.

It is somewhat inconsistent with the habits of wise men to apply such expressions as are last quoted to a series of experiments unexampled for the absence of all the precautions necessary to give them effect: and now that the danger which was threatened has been obviated we may be permitted to smile at the ability, candour, and circumspection evinced in the management of this new and delicate process. It will hereafter appear, when we come to consider the cases at Petworth, that the matter which was issued by Dr. Pearson produced pustules similar to those that were seen at the Small-pox Hospital. This, however, did not happen with all the virus so sent out. Dr. Jenner received some of it in the early part of March 1799, and used it successfully.

DR. JENNER TO DR. PEARSON.

Berkeley, March 13th, 1799

MY DEAR SIR,

I received your letter while I was writing to Dr. Woodville, and requested him to transmit to you the result of the inoculation with the London virus. I hope he did not fail to execute my wishes. Twelve patients have since been inoculated with matter produced by this virus. They all took the infection. This is the ninth day, and they appear a little ill—no eruptions yet. The character of the arm is just that of cow-pox, except that I do not see the disposition in the pustule to ulcerate as in some of the former cases. I am the more induced to believe this to be the *genuine* cow-pox from the following circumstance.

One of the boys inoculated, sickened the preceding day with the measles, which went through its course. Yet the pustule advanced with the same regularity as if the measles had not been present. Now this would not have been the case, I presume, had variolous matter been inserted into the skin under similar circumstances. No cow-pox yet in the country! Should it appear within a particular district I shall undoubtedly know it. It cannot now be long before I shall see you in town; at least I can speak with as great a certainty of being soon there as a medical man can.

I hear of a child covered over with pustules at the Small-pox Hospital. What are they?

I am glad to find that the disposition for forming eruptions among your patients does not increase, as you tell me that none of your last inoculated patients had any, and that Mr. Rolph's children went through the disease without them. Tanner I find could not succeed in giving the cow-pox to the veterinary cow in a direct way, that is, by inserting the virus into a sound part of the nipple, in the same way as all the experiments have hitherto been conducted to confute my notions with the matter of grease; but when he found a part of the nipple that was previously affected with a sore, and applied the matter there, it took effect immediately. With best respects to Mrs. Pearson,

I remain, dear sir,

Yours very truly,

E. JENNER.

Drs. Pearson and Woodville continued to exert themselves with great assiduity to diffuse the practice of vaccination. They issued a printed circular letter containing an account of their practice at the Small-pox Hospital, and inclosing a bit of thread imbued with the cow-pox virus. Sixty persons had at that time been put to the variolous test without receiving the infection. At this period, one of these gentlemen gave himself a great degree of prominence in the important question which now agitated the medical world, by placing himself in a situation that he was not entitled to fill; and which, of right, belonged to another. The developement of this design was cautiously unfolded; and had not Dr. Jenner's friends been more vigilant than himself it certainly would have succeeded to a greater extent than it did. His nephew Mr. George Jenner who was then in London, wrote an account of what was going on, and very properly warned his uncle of the injury that might be done to his fame if he did not appear there in person to vindicate his claims. Mr. Jenner's letter is so descriptive of the state of matters at this time that I deem it right to subjoin it.

G. C. Jenner to Dr. Jenner.

Norfolk Street, March 11*th*, 1799.

My dear Sir,

After what Mr. Paytherus has written to you it will be needless for me to say any thing to urge the necessity of your coming to town to wear the laurels you have gained, or to prevent their being placed on the brows of another.

I shall only state a few facts I have got possession of since I wrote to you last. Dr. Pearson is going to send circular letters to the medical gentlemen to let them know that he will supply them with cow-pox matter upon their application to him, by which means *he* will be the chief person known in the business, and consequently deprive you of that merit, or at least a great share of it, which is so justly your due. Doctor P. gave a public lecture on the cow-pox on Saturday last. Farmer Tanner was there. Doctor Pearson adopted your opinions, except with regard to the probability of the diseases originating in horses' heels. He spoke of some unsatisfactory experiments having been made by inoculating from the greasy heels; but when we consider how difficult it was to communicate the disease from one cow to another by inoculation we are not to wonder at the still greater difficulty in communicating it from the horse to the cow. The farmer says Dr. Pearson was wrong in some part of his lecture, which he *took the liberty to tell him.*

Mr. Paytherus is much disappointed not to receive any letter from you by this day's post, but hopes you may be coming up to-day and therefore did not write. All your friends agree that *now* is your time to establish your fame and fortune; but if you delay taking a personal active part any longer the opportunity will be lost for ever. If Dr. Pearson does not intend to endeavour to give the merit to himself why should he quibble about the name you gave the disease? The eruption he calls the *vaccinous* eruption.

Your affectionate nephew,

G. C. Jenner.

Mr. Paytherus has just told me that a copy of Doctor Pearson's letter was exhibited yesterday at Sir Joseph

Banks's. When I get a sight of it I will send you an account of it.

Dr. Jenner in allusion to this letter writes thus to his friend Gardner.

DR. JENNER TO MR. EDW. GARDNER.

Berkeley, Wednesday, 1799.

DEAR GARDNER,

A letter I have just received from G. Jenner informs me that Dr. Pearson on Saturday last gave a public lecture on the cow-pox, and that it was publicly exhibited at Sir J. Banks's on Sunday evening. He has also given out that he will furnish any gentlemen at a distance with the virus.

As this is probably done with the view of showing himself as the first man in the concern, should not some neatly-drawn paragraphs appear from time to time in the public prints, by no means reflecting on the conduct of P. but just to keep the idea publicly alive that P. was not the author of the discovery—I mean cow-pox inoculation.

Yours truly,

E. J.

Dr. Jenner now began seriously to feel the weight of that responsibility which rested on him from the publication of his "*Inquiry*." He had the deepest cause to lament the precipitancy of some of his professional brethren. They rejected his statements without even attempting to understand him, and permitted prejudices to stand in the way of evidence the most conclusive. He had especial reason to feel much dissatisfaction in the result of some of the trials which were made in London, and consequently could not be assured that his own character and what he certainly valued much more (the success of vaccination) might not have sustained a serious shock. He was, moreover, assailed by many adversaries in his own district, some of whom, it is to be feared, were moved by considerations not of the purest nature. The following letter, whilst it affords a pleasing illustration of his well-founded confidence in the accuracy and power of his own knowledge, nevertheless shows his lively sensibility to every attack that threatened to obstruct his favourite cause, the cause of humanity and truth.

Dr. Jenner to Mr. Edward Gardner.

Dear Gardner,

There never was a period in my existence when my situation called so loudly for the assistance of my literary friends as the present. Though my bark will, with flying colours, reach the shore at last, yet it is now in a storm.

I am beset on all sides with snarling fellows, and so ignorant withal that they know no more of the disease they write about than the animals which generate it. The last philippic that has appeared comes from Bristol, and is communicated by Dr. Sims of London. Sims gives comments on it in harsh and unjustifiable language. It is impossible for me, single-handed, to combat all my adversaries.

Standing, as I do, before so awful a tribunal, my friends will volunteer their counsel and IMMEDIATELY appear in court.

My intended pamphlet has only been looked over in a cursory way. Every sentence must be again revised and weighed in the nicest balance that human intellect can invent. The eyes of the philosophic and medical critic, prejudiced most bitterly against the hypothesis, will penetrate its inmost recesses, and discover the minutest flaw were it suffered to be present. Language I put out of the question: the matter is what I allude to.

Give me as much of your company as you can, and as speedily.

Yours, very faithfully,

E. JENNER.

Thursday, *March 7th*, 1799.

On the 21st of this month Dr. Jenner availed himself of the advice of his friends and left Berkeley for London, where he took up his residence in Norfolk-street. On the 23d he had an interview with Dr. Woodville who informed him that he had vaccinated upwards of 200 patients. To show how little the real character of cow-pox, as described by Dr. Jenner, accorded with what was said to have occurred in London, this physician, at the interview in question, actually mentioned that the cow-pox had been communicated by effluvia; and that the patient had it in *the confluent way*. Dr. Jenner's remark on this marvellous occurrence was simply, "Might not the disease have been the confluent small-pox communicated by Dr. Woodville, as *he* is always full of the infection?"

Dr. Jenner remained in London till the 14th of

June: during his stay he had frequent intercourse with most of the other medical men of eminence resident in the metropolis. As the events which occurred at the Small-pox Hospital were so different from what *he* had experienced, he was very desirous of procuring fresh cow-pox virus from the country. For this purpose he sent to Gloucestershire; and, by great exertions on the part of Mr. Robert Tanner, he procured some from North Nibley. A portion of this he gave to Mr. Knight on the 12th of April.

The "Reports" of Dr. Woodville were at this time in the press. This work was published in June, and seemed to carry with it most formidable objections to some of Dr. Jenner's main positions. It was *concluded* that the vaccine disease was not derived from the horse. It was also contended that the cow-pox was occasionally an eruptive disease of great severity; that "three or four cases out of 500 had been in considerable danger; and that one child had actually died under the effects of the disease." The other conclusions which he drew were that the matter of the vaccine disease has generally produced fewer pustules than that of small-pox, and less indisposition. Notwithstanding this admission, his "*Reports*" were eminently calculated to repress the ardent expectations that had been excited in favour of vaccination; for he thus proceeds, after mentioning the supposed fatal case of cow-pox: "Now if it be admitted that one of five hundred will die of the noculated cow-pox I confess I should not be disposed to introduce this disease into the Inoculation Hospital, because out of the last 5000 cases of variolous inoculation the number of deaths has not exceeded 1 in 600.*

Dr. Jenner had the consolation of knowing that this alarming statement, published by the authority of the physician to the Small-pox Hospital, was not confirmed by the experience of other very competent judges. With the matter which had been sent from London on the 15th of February from Dr. Woodville he inoculated his grand-nephew Stephen Jenner, and a boy of the name of Hill about four years of age. The progress of the disorder in both is described in his second pamphlet, published a few weeks before that of Dr. Woodville.

There were in each a few small spots on the arm; and, in the first, on the face; but they disappeared in a day or two, and caused no inconvenience to the patient. These children were afterwards fully exposed to small-pox contagion without effect.—With matter taken from the arm of the boy Hill Dr. Jenner next inoculated two of the children of his friend Mr. Hicks, of Eastington, and at the same time sixteen others. They all took the infection, but no pustules appeared. With matter taken from this source his nephew Henry Jenner vaccinated successfully a child of twenty hours old. The same stock supplied Dr. Marshall with virus for his inocu-

* See "Reports," p. 151.

lations, which commenced with two of his own children, on the 22d of March. Between that day and the 26th of April he employed it on 107 persons. " In only one or two of the cases," he observes, " have any other eruptions appeared than those around the spot where the matter was inserted."*

Still farther to investigate the subject, and to determine whether there was any peculiarity in the matter found in the London dairies which tended to produce eruptions, Dr. Jenner, whilst he was in town, procured some genuine cow-pox virus at Mr. Clark's farm at Kentish-town. This he immediately dispatched into Gloucestershire by his friend Mr. Tanner. With this virus Dr. Marshall resumed his inoculations. In his second letter † to Dr. Jenner

* See the letter in Dr. Jenner's " Continuation of Facts and Observations," p. 14.

† This second letter which is printed at p. 16, unfortunately had not the date attached to it: but the original, which is at this instant in my possession, bears date " Eastington, September the 8th, 1799." I am happy in being able to supply this omission, because it affords me an opportunity of vindicating Dr. Jenner from a very harsh criticism published in Dr. Woodville's observations on the cow-pox, in July 1800. By confounding the time at which the last letter was written, with the date of the first, Dr. Woodville makes it seem that Dr. Marshall could not have performed any of the vaccinations which he recorded with the matter sent by Dr. Jenner from the cow at Kentish-town; inasmuch as he did not even obtain that matter till after the date of that letter, " which is said to contain an account of its effects by inoculation."—See Woodville's " Observations on the Cow-pox," p. 10.—Now it happens, unfor-

he mentions that he had employed it in 127 cases. The following sentences, in reference to Dr. Woodville's assertions, deserve notice. "In Dr. Woodville's publication on the cow-pox I notice an extraordinary fact. He says that the generality of his patients had pustules. It certainly appears extremely extraordinary that in all my cases there never was but one pustule which appeared on a patient s elbow on the inoculated arm, and maturated. It appeared exactly like that on the incised part."

The result of Dr. Marshall's first series of inoculations was communicated by Dr. Jenner to Dr. Woodville. But this gentleman was not disposed to admit that any error had taken place in his own practice. His statements, however, of the eruptive nature of cow-pox began to be less peremptory, as will be seen from the following extract of a letter dated on the 22d August, 1799.

tunately for Dr. Woodville's position, that the second letter has no date at all; and it is as much distinguished from the first as *capital letters, inverted commas,* and *subject matter* can make it. At the time the first letter was written (namely, April 25th, 1799) Dr. M. had vaccinated 107 persons: when the second was written the number amounted to 423: of whom it is mentioned in the postscript that 127 were infected with the matter from the Kentish-town cow. With all these most palpable distinctions it is not a little remarkable that Dr. Woodville should have written as if there had been but *one* letter of Dr. Marshall's, and *that* the *first;* forgetting entirely that the *second* bore internal evidence of having been written many months after the former.

"The cow-pox inoculations certainly go on very successfully here, but we cannot, like Dr. Marshall, boast that our cases have never been attended with eruptions. Since Mr. Knight inoculated the Duke of Clarence's children he has had a case in which the pustules were sufficiently numerous to excite no small alarm for the safety of the patient, and though the disease at the hospital has not lately put on a formidable aspect it still continues *occasionally* to produce pustules."

I am sorry to be obliged to observe that Dr. Woodville, in mentioning this severe eruptive case in the practice of Mr. Knight, showed less regard to accuracy than the occasion required. Mr. Knight, who was at that time inspector general of military hospitals, was more instrumental than any other professional gentleman in diffusing the benefits of the new practice among the higher orders in London. In addition to the children of many of the nobility he was called on to vaccinate two of the children of the Duke of Clarence, at Bushy. Dr. Jenner wrote to Mr. Knight on receiving Dr. Woodville's letter. Mr. Knight's reply was quite satisfactory.

F. Knight, Esq. to Dr. Jenner.

Aug. 28*th*, 1799.

My dear Sir,

The cow-pox must have its prejudices and misrepresentations. Dr. Woodville's report does not apply to the recruit, but to a case that I had at Mortlake, where I had a country-house. You shall have the detail of it, and will then judge how much it attaches to the cow-pox. It was the small-pox, but never occasioned a momentary alarm for the safety of the patient. You shall have the detail of it, for it is not a little interesting.

With a lancet armed from one of the Duke of Clarence's family I inoculated a young girl about ten years of age. It showed no signs of early infection and on the fifth day I was going to insert fresh matter, when I felt a little elevated pimple in the part, but without the least inflammation. It continued uncommonly sluggish, and after fourteen days scarcely formed a vesication as big as a small pea, which had no indented mark in the centre, nor any surrounding inflammation.

About this period she became feverish, and had the usual symptoms of approaching small-pox, which broke out two or three days afterwards. I was at this time obliged to quit Mortlake on account of my preparations for the Continent, and taking Mrs. Knight into Gloucestershire; but anxious to know the event of this case, and how far the appearance of the arm agreed with the progress of eruption, I employed a very clever regimental assistant to go down in my absence, and report the particulars to me.

It was on the eighth day of the eruption that he saw her. The pustules were numerous, distinct, and maturated into surrounding inflammation. Those on the face were on the turn. Where the vaccine matter was inserted, near three weeks before, there was a large circular pustule *containing lymph*, with a crust of a dark blue cast inclining to black, indented in the middle, and the edges elevated above the skin. He says it had the exact appearance of the inoculated cow-pox on the *ninth day*, (but he certainly must mistake,) except the surrounding erysipelatous inflammation; as in this case there was only an inflamed line round the bottom of the pustule, in which there was one small pock elevated and maturated.

The appearance of this girl's arm was so singular from the beginning, and betrayed such marks of suspended action, that I noticed it to those about her; and, conceiving that it was mere sluggishness of habit, I ordered her to have some wine and to live better. The small-pox had prevailed greatly in the village, but she was not known to have been in the way of it. Her disease was certainly the small-pox, and as certain is it that I inoculated her with a lancet armed from one of the children at Bushy; but from whatever source the small-pox sprang, it kept no pace with the pustule on the arm; which lingering as it was in its local action, never betrayed the least similitude to the small-pox pustule. Pray let me have your sentiments on the case.

* * * * *

Notwithstanding the prejudices which were excited against cow-pox by the report of its being an eruptive disease, it continued to gain ground. Many distinguished individuals, in different parts of the kingdom, eagerly availed themselves of the promised benefit of the practice, and exerted themselves to diffuse it as widely as possible. It is a gratifying thing to be able to remark that the ladies of England were conspicuous in this work. Lady Peyton, sister of Lord Rous, must be ranked among the first who, by personal efforts, stimulated the professional gentlemen in her neighbourhood to adopt vaccination; and she herself subsequently became one of the most energetic and successful of vaccinators.

This is not the only instance in which Dr. Jenner received most efficient and valuable assistance from individuals who did not belong to the profession. Indeed it will presently appear that the cause of vaccination would have been most materially injured if the good sense, and fair-dealing, and resolution of those to whom I refer, had not enabled him to withstand, and finally to overthrow the opposition which he encountered in other quarters. I have already mentioned Henry Hicks as his friend and counsellor: I have also mentioned how sedulously he promoted vaccination by first submitting his own children to it, and then diffusing it in his neighbourhood: I have now to attempt to commemorate his services in another line. He made himself perfectly acquainted with all the details of cowpox inoculation; and about this time he brought this knowledge into practice. He commenced a series of inoculations; and evinced an accuracy and fidelity which would have done honour to the most enlightened physician. To show that this is no exaggerated praise I will mention that in a letter to Dr. Jenner, dated the second of August, he communicated an observation which at a subsequent period was also noticed by Mr. Bryce, of Edinburgh, and which led that gentleman to propose the only thing like an improvement in the practice of vaccination that was not suggested by Dr. Jenner himself. In two of the persons whom Mr. Hicks had vaccinated he found that "the arms came on so slowly that on the sixth day the pustules were not more forward than they generally are on the third or fourth: added to which, there appeared so much irregularity in them that I thought it better to show their arms to Mr. Darke." Two days afterwards Mr. Hicks determined to re-vaccinate them. The result he communicated in the following words: "The second inoculation seemed to make immense strides to overtake the first; and, what is wonderful, the first pustules began to change their character and put on the true vaccine appearance." He also made another remark which proves how quickly he detected every peculiarity in the progress of the affection. He adds, "I must not forget to tell you that Chamberlain's child, before the inoculation, had some eruptions on its arm and back, and as the vaccine pustule came on, these pustules assumed the exact vaccine character, and became perfect cow-pox pustules." It is due to Mr. Hicks to declare that there is more original and satisfactory information to be derived from his remarks than from those of all the individuals who had written on the subject, from the time of the publication of Dr. Jenner's Inquiry.

Another gentleman, who had become acquainted with Dr. Jenner at Cheltenham and had obtained in conversation with him a knowledge of the subject of cow-pox, felt so deep an interest in its success that he not only occupied himself in investigating the history of the casual disease, but likewise in propagating it by inoculation. The gentleman to whom I allude is William Fermor, Esq. of Tusmore, in Oxfordshire. By diligent inquiry he discovered that the cow-pox was occasionally epizöotic in his own neighbourhood; and in a little work which he published in May 1800, he detailed several cases of individuals who had caught the infection from the cow; and who had subsequently resisted small-pox by inoculation and exposure in every way. Not satisfied with this very valuable information he resolved to have a series of cow-pox inoculations instituted under his own eye. As he was not himself a medical man he was assisted on this occasion by the Rev. George Jenner, the nephew of the Doctor. He carried the matter with him from Berkeley, and the experiments were conducted with the concurrence and under the inspection of many of the most learned professors of Oxford, and the most distinguished medical men in the neighbourhood. Among these may be particularly mentioned Sir Christopher Pegge, Drs. Wall, and Williams, and Mr. Grosvenor. In a short time 326 persons were vaccinated, of all ages from two months to sixty-nine years. Of these 173 were subsequently inoculated with small-pox, but never felt its influence.*

* A very interesting correspondence was carried on, for some time, between Mr. Fermor and Dr. Jenner on the subject of

Whilst things were thus proceeding in England the knowledge of Dr. Jenner's surprising discovery was exciting the deepest interest wherever it was heard of on the Continent. The unhappy war which then raged prevented direct intercourse with France and many other parts. The "*Inquiry*" nevertheless found its way, in the course of this year, (1799) to Geneva, Hanover, and Vienna. In the first-mentioned place Drs. Odier and Peschier collected all the information that could be obtained on the subject, and communicated it to the scientific world through the medium of the Bibliothèque Britannique. In Vienna, the cause of vaccination was taken up by Dr. De Carro with a zeal commensurate to its importance, and fostered and disseminated with a degree of wisdom and energy which has not been exceeded on the part of any of the eminent individuals who have advocated or advanced it. Another opportunity will soon occur for recording his great and most beneficial efforts in diffusing vaccination over a great part of Europe, and finally, through Turkey into our possessions in Asia.

vaccination, a subject to which (as has been said) Mr. F. paid close and judicious attention. In one of his early letters inviting Dr. Jenner to his seat in Oxfordshire he playfully says of one of the chief antivaccinists, first premising that he had had, during a late tour, " various conversations with medical men on the subject of the cow-pox," " I told Dr. Moseley that in his assertion against it he had acted the part of the devil's advocate at a canonization, who was to say all the harm he could against the saint in order that his life might be thoroughly scrutinized, and his merits appear the more conspicuous."

At this time it is more proper to dwell for a season on his own personal relations to Dr. Jenner. They never had the happiness of meeting, but I believe that I do not hazard an unfounded statement when I express my conviction that few or none of Dr. Jenner's friends had formed a more correct estimate of his character, more truly loved the purity and benevolence of his moral feeling, or more deeply venerated his surpassing genius, than he of whom I now speak. His first letter to Dr. Jenner is dated at Vienna, on the 14th of September, 1799. It is altogether so interesting, and describes in so satisfactory a manner the very first inoculations for cow-pox on the Continent, that I think it proper to subjoin it here.

Dr. De Carro to Dr. Jenner.

Sir, *Vienna,* 14*th Sept.* 1799.

Although I have not the honour to be known to you I hope that you will excuse the liberty I take of addressing myself to you, convinced that you are sensible of the utility that men, pursuing the same inquiries, should as much as possible communicate together.

Before I proceed I must assure you that among the innumerable admirers of your discovery, none feels more the importance of it, and is ready to do more to propagate its benefits, than I am. As soon as I heard of it I wrote to London to have your work, and desired a medical friend of mine to procure me some vaccine matter from a good source. He sent me two threads that came from Dr. Pearson, with some accounts of that gentleman's successes in the brilliant road which you have opened. The first subject I found for inoculation was the son of a physician of this town, in whom it produced two pustules, of which I need give you no other description than to tell you that they were so perfectly similar to the second and third plate of your work that one might have thought that that child's arms had been the model from which your engraver had drawn them. Encouraged by the mildness of the disease I did not hesitate to inoculate from the fresh matter of that child my eldest boy, and ten days afterwards my second boy from the arm of his brother. In my two children the disease and the appearance of the pustules were precisely similar to your description and engravings.

Three months afterwards these three children have been inoculated with variolous matter, without any effect whatever, except, in one of them, a pustule on the inoculated part, but without further effect on the system.

I have kept a quantity of matter with which the sleeves of their shirts were impregnated, sufficient for any number of other inoculations, paying the greatest attention not to collect the matter but when the pustules were in their state of greatest activity, viz. before they became ulcerated. as it happened by the scratching of my children, which I never could prevent. Though your discovery had produced a great sensation among our medical people your work was not known at Vienna before I took the pleasure of spreading it; therefore I did not find any subject for inoculation till I had inoculated my children again with variolic matter. In the month of July I inoculated two children with threads taken from these shirts, and in one of them alone the virus took effect; but the distance at which that child lived from town, and some neglect on the part of his parents, prevented me from gathering fresh matter from his inoculation. In the month of September I diluted with some warm water some vaccine matter which was in great plenty upon the shirts; and inoculated, in the usual way, the daughter of a man whose example might have been followed by many, and it produced no effect whatever. On the 9th instant I have inoculated two twins of three years old, and yesterday, according to appearances, four punctures were so dry as to make me hope for success. However, as I had not contented myself with punctures on each child I have likewise inserted threads in the epidermis, and they looked yesterday still inflamed and red; at least the lips of the wound were not closed nor crusty; therefore I do not give up all hope of success. If these two children fail I am determined not to inoculate any more till I get fresh matter from England, for fear of discouraging such parents who are inclined to lay themselves under obligations to you. I am in daily expectation of that matter which comes from Dr. Pearson and Mr. Coleman of the veterinary school in London. However, as I wish to proceed with all the precaution which the introduction of a new method requires, I should take it as an infinite favour if you would be so good as to send me some of your genuine vaccine matter (from cows, if possible) which would give a great deal of confidence to the public of Vienna, as coming from the father of the discovery. If my request does not appear indiscreet to you you have nothing else to do than to send it to London to any person who would take the trouble of remitting it to Lord Grenville's office, at the direction of *Mr. Stratton, Secretary of Legation, at the Right Hon. Lord Minto's, Vienna;* to be sent by the first messenger with a note for me. It is perhaps necessary that I should tell you who is the person who takes this liberty with you. I am from Geneva; I have studied and taken my degrees at Edinburgh, and practise medicine at Vienna, since six years.

Though you have given some directions about the manner of preserving vaccine matter be so good as to write me a few words upon the method which your experience has taught you to be the best. This is so much more necessary as it does not appear that cows in any part of the great Austrian Monarchy are subject to that disease; and even if they were I should not be bold enough to inoculate with their matter, as it appears that more veterinary knowledge than I have is necessary to distinguish the various diseases of cows. I have sent threads to Geneva, where inoculations have been immediately tried, but I do not know yet their result.

I have the honour to be with the highest respect,

Sir,

Your most obedient

and much obliged humble Servant,

J. De Carro, M. D.

Rauherstein, No. 983.

A Vienne, en Autriche.

P. S.—This morning the two pustules of the twins above mentioned were so much elevated that no doubt could be entertained of the success. I continue, however, in my request.*

After the inoculation of his own children, and other individuals, Dr. De Carro transmitted threads embued with the vaccine matter to Geneva, but they did not succeed. He then sent others, which were impregnated from the arm of Count Mottet. This gentle-

* In perusing this, and other letters of Dr. De Carro, it will be remembered that though he writes English, in general, accurately, yet it is not to be expected that his style shall be, at all times, free from foreign idiom.

man had previously had small pox, and he submitted to vaccination merely as an experiment, in order to settle the question whether it was possible to have cow-pox after small-pox. The matter from this source was conveyed from Vienna to Geneva by Dr. Peschier, who had been a witness of Dr. De Carro's previous vaccinations. This matter was immediately employed both at Geneva and at Colombier. Unfortunately all the persons subsequently had small-pox either in the natural way, or by inoculation. Of the former several died. This calamity, so inauspicious to the progress of vaccination, could only have occurred in the commencement of the practice when the true character of the vaccine pustule was imperfectly understood, and the knowledge necessary to conduct so delicate a process had not been acquired. The matter, in fact, taken from Count Mottet was not the vaccine, and the disease which it excited in the arms of the children at Geneva had certainly none of the properties of the genuine disease. Had the gentlemen who conducted these inoculations been fully aware of Dr. Jenner's instructions they would not have exposed those children to variolous infection, after having observed the irregularity of the progress and character of the pustules on their arms, which, from first to last, were spurious in their nature.

These untoward events did not materially obstruct the progress of vaccination in Geneva. Before the end of January 1801, fifteen hundred persons were successfully vaccinated. Dr. Odier much distinguished himself by his exertions on this occasion. He drew up a memoir on the inoculation of the Variolæ Vaccinæ, by order of the Minister of the Interior in France, which was sent to all the officers of health in the department.

With this memoir was also distributed a paper of "advice to fathers and mothers," eloquently and affectionately recommending vaccination, and offering it gratuitously to all who were not in a condition to pay. This paper was delivered to the parents, by the clergymen, when they brought their children to receive the rite of baptism. It was signed by the physicians Vieusseux, Odier, Vignier, Manget, Veillard, Coindet, De La Rive, Peschier, and the surgeons Jurine, Fine, Maunoir.

Dr. De Carro's vaccinations commenced in the month of May 1799. A short time afterwards, the practice was introduced at Hanover by Dr. Ballhorn and M. Stromeyer. The former gentleman translated Dr. Jenner's first work into German. This was communicated to Dr. Jenner in the Latin language by the learned Hanoverian physician in a letter dated on the 5th of November, 1799.

The eruptions, which attended many of the early cases of vaccination in London, were unfortunately also propagated in different parts of the country, where the *contaminated* matter had been distributed by Dr. Pearson.

It will now be necessary to return to the proceedings of this gentleman, both as they regarded the nature of the vaccine disease itself and the measures which he contrived for its dissemination. Among other professional gentlemen who used this contaminated matter was Mr. Andrè, surgeon at Petworth. Fourteen persons were inoculated with it; and all had variolous-like eruptions, some being loaded with heavy burdens.

Lord Egremont, who had always felt a peculiar zeal for the success of vaccination, was surprised and dismayed by such untoward circumstances occurring in his neighbourhood. Had his lordship acted as some professional men did the prejudices which had already been awakened both against Dr. Jenner and cow-pox would have been much strengthened. With the dignity and prudence becoming a nobleman of such exalted rank he did not permit his mind to be swayed till he had taken proper measures to obtain information respecting these anomalies, from Dr. Jenner himself. His Lordship wrote him a long letter on this occasion, giving an ample detail of the occurrences at Petworth. Dr. Jenner* replied in this very interesting manner.

* He also received an account from his friend the Reverend Mr. Ferryman, who was then at Petworth. This gentleman on seeing the eruptions did not hesitate to say that the disease was not the cow-pox, but the small-pox. In consequence of this Lord Egremont most humanely had *all* the patients conveyed to his own house to prevent the disorder from spreading. Notwithstanding this precaution, two persons caught the disease; of whom one died. See Mr. Ferryman's letter in Ring's Treatise on Cow-pox, page 90.

Dr. Jenner to Lord Egremont.

My Lord,

I am extremely obliged to your Lordship for your kindness in giving me so fully the account of the late inoculation at Petworth; a subject which, before, I did not clearly understand; and which, of course, had given me much vexation. I will just briefly lay before you part of the history of the cow-pox inoculation since my experiments were first publicly made known; which may tend in some measure to explain in what manner pustules may be produced.

About a twelvemonth ago Dr. Woodville, physician to the Small-pox Hospital, procured some virus from a cow at one of the London milk farms, and inoculated with it several patients at the Small-pox Hospital. Fearful that the infection was not advancing properly in some of their arms he inoculated them (some on the 3d, others on the 5th day afterwards) with small-pox matter. Both inoculations took effect; and thus, *in my opinion*, a foundation was laid for much subsequent error and confusion; for the virus thus generated became the source of future inoculations, not only in the hospital but in London, and many parts of the country.

Hearing a murmur among medical people that the cow-pox was not the simple disease I had described, but that in many instances it produced as many eruptions and was attended with as much severity as the small-pox, I went to town with the view of inquiring into the cause of this deviation. Dr. Woodville at once invited me to the Small-pox Hospital, and very ingenuously told me the whole of his proceedings. The inoculated patients were shown to me, and though some were without eruptions and exhibited the appearance of the true cow-pox, others were very full of them, and I could not discern any difference between them and the perfect small-pox. I therefore did not hesitate to tell the Doctor that it clearly appeared to me that the small-pox had crept into the constitution with the cow-pox; that I did not consider them as two distinct diseases, but as *varieties* only of the same disease; and therefore they might co-exist in the same constitution, and that thus a mixed disease had been produced. I communicated also the same sentiment to Dr. Pearson who was then, and had been, busily employed not only in inoculating from this source, but in dispersing threads embued in the virus to various places in our own country, and to many parts of the Continent. Foreseeing what was likely to ensue from these hasty measures I remonstrated against them, but was not listened to. In many places where the threads were sent a disease like a mild small-pox frequently appeared; yet, curious to relate, the matter, after it had been used six or seven months, gave up the variolous character entirely, and assumed the vaccine; the pustules declined more and more, and at length became extinct. I made some experiments myself with this matter, and saw a few pustules on my first patients; but in my subsequent inoculations there were none. From what I once saw at the hospital I had reason to think that some of these threads sent out were not only stained with small-pox matter from the contamination spoken of, but that they had sometimes a dip in a real small-pox arm; as the patients were all mingled together at the hospital, and stood with their arms bare, ready to afford matter one among another. Without making any further trials with matter from the cow managed in another way Dr. W. published a volume containing the result of his practice, which certainly damped the spirits of many who had from my representation taken up a high opinion of the cow-pock inoculation. A thought now struck me that, if possible, it would be proper to procure matter from a London cow, and compare its effects with that generated in the country.

Unwilling to determine in a hurry, I procured matter from a London cow, conceiving it possible that the animal in this situation might generate matter possessing qualities differing in some measure from that which is more in a state of nature in our meadows here: but the result convinced me that the virus was the same, as 500 people were inoculated from this source without the appearance of any pustules. But this history, my lord, does not tell you by what means the pustules appeared at Petworth:—but it informs you how errors may arise, and how they may be persisted in. There is another source which I fear will be too common. Lancets are often carried in the pocket of a surgeon with small-pox dried upon them, for the purpose of inoculation. A gentleman some time ago sent a lancet here to have it charged, as it is called, with cow-pox matter: perceiving it stained at the point with some dried fluid, it was sent back; when he immediately recollected that his lancet was prepared with the matter of the small-pox. What confusion might have happened from this; and how narrowly we escaped it! For it was but an equal chance probably, that, had the lancet been used, a direct small-pox might have been produced; for the chance was equal whether it produced one disease or the other.

It may be necessary to observe, it is improbable that a mixture of the two matters used in this way would have produced a mixed disease, as two different diseased actions cannot go forward in one and the same part at the same time, so that the disease would have been either the perfect cow-pox or the perfect small-pox.

The matter which was made use of, I hear, came from Dr. Pearson; and doubtless Mr. Keate will have candour, and, I hope, industry enough to trace the error to its source. That there was an error somewhere, of which Mr. Keate became the innocent cause, is a fact that I think will not admit of controversy. I have sometimes seen, perhaps in one case in a hundred, a few scattered pimples about the body, and sometimes rashes: but these have arisen from the inflammation and irritation of the arm, for it is very well known that many acrid substances applied to the skin, so as to produce local inflammation, will frequently occasion a similar appearance.

The cases of the boys your lordship describes infected by the matter I sent, were exactly those of the true cow-pox. Many hundreds of people of all ages were lately inoculated in a part of this county where the clothing manufactory prevailed, but not one of them lost a day's work, nor had they any pustules.

The inclosed virus is secured in a way that I imagine it cannot fail to infect, if Mr. Andrè will reduce it to a fluid state by moistening it with water on the point of his lancet previous to its insertion. When he has once infected a patient, he may afterwards go on to a certainty, as the fresh fluid matter is more efficacious than that which has been dried. The inclosed is just taken: it has borne a journey across the Atlantic, and been found perfectly good on its arrival.

* * * * *

To prove how anxiously he viewed the events which were passing at Petworth, he had scarcely dispatched the preceding letter, before he was induced to transmit another.

DR. JENNER TO LORD EGREMONT.

MY LORD,

I am almost ashamed to trouble your lordship again so soon on the subject of the cow-pox, but conceiving myself the general guardian of its inoculation, and my own reputation as being in some degree involved in the misconduct of others, the more I reflect on the late untoward event at Petworth, the more my anxiety increases. A thought has occurred to me, which, if your lordship will allow me to put it into practice, may clear up every doubt which has arisen in the minds of all parties respecting the late disaster. A comparison might be made on the spot between the inoculation of Mr. Keate and that of a person more perfectly conversant with the disease. My nephew, Mr. H. Jenner, of whom I have spoken in my publication as assisting me in the pursuit, would, if such a thing should meet with your lordship's approbation, set off at once for Petworth, and inoculate any number of people that might be willing to submit to it.

To add to my anxieties, I have heard from Lady C—— that the small-pox has been given to a domestic of her ladyship's for the cow-pox. But the matter, I find, was derived from the same source as that which was used in your lordship's family.

* * * * *

It was not thought necessary by his Lordship to avail himself of Dr. Jenner's offer, to send his nephew to Petworth. The gentleman who conducted the first inoculations with the matter furnished from London, also made use of that transmitted by Dr.

Jenner. The ensuing letter to Mr. Gardner announces the result of this comparative trial.

Dr. Jenner to Mr. Edward Gardner.

December, 1791.

Dear Gardner,

I am sorry the night was so unpleasant, and that your catarrh was increased by your journey home. I fear you fall too frequently into the common custom of seeking warmth after exposure to cold, which I am sure is a common means of both giving catarrh, and increasing the malady when present in a slight degree.

Great news from St. James's. The King has sent me a very civil message, so you will produce a *page* to wait upon his majesty and express the obligation.

From Lord Egremont I have also had a pleasant account. The matter I sent his lordship has dissipated all doubts and prejudices, he says, from the minds of the people around him. Forty have been inoculated with the virus, and all had the disease as I describe it. You must observe that fourteen had previously been inoculated with matter from Pearson, and ALL had variolous-like eruptions. Some were loaded with heavy burdens. But yet we must again review our observations on the pustular subject, for if I should produce them on any one patient, Pearson would triumph; and I must confess I don't see why cow-pock, though not small-pock pustules should not appear. I have seen one on a woman's forehead like the pustule on the arm, but smaller, and one on a girl's wrist. But more of this when we meet; and if you please, as your cold is not well, it shall be put off to Monday. Of this I will write by your messenger and apprize H. H.

I think the surgeon at Petworth who inoculated both sets of patients, mine, if I may call them so, and Pearson's, should publish immediately the result of the two inoculations. A letter in the newspapers would have a good effect, recommending the country to follow the example of London and Bath in forming general institutions for the gratuitous inoculation of the cow-pox. By this means the small-pox must soon hide its diminished head, and I could come forward to P—— with a good grace. What a triumph it would be if Gloucester would show a disposition to come forward! Trye would be at the head of the medical department. Do you know any of the leading people there who are not in —— shackles?

I believe you will now receive the most material of the journals, &c. I will put any thing else in my pocket on Monday that I can find.

Yours truly,

E. Jenner.

The preceding letters are not dated; but I am enabled to prove by entries in his journals that they must have been sent off some time between the first of December and the twenty-fourth of that month.

Dr. Jenner was at the time living at Berkeley, having returned to that place from Cheltenham, with his family, on the fourteenth of October. While he remained, whether at Cheltenham or Berkeley, he was partly occupied in his professional affairs; but his correspondence, connected with the subject of vaccination, had so much increased as to leave him little leisure for other employment. He received many applications for cow-pox matter: and, as has been already seen, had a great weight of labour thrown upon him in attempting to rectify mistakes which had been committed by others in the practice of vaccination. He was very careful in preserving the history of the virus that he sent out, and for the most part, likewise, transcribed with his own hand the letters of moment which he wrote at this period.

In the early part of December this year (1799,) he sent vaccine matter both to Berlin and Vienna; the first in consequence of an application from the Princess Louisa, of Prussia, communicated to him by Mrs. Colonel Walker. The second in compliance with the special request of Dr. De Carro.

The application for the Princess Louisa of Russia was replied to in the following letter.

Dr. Jenner to Mrs. Walker.

Dear Madam,

It would be highly gratifying to me to see the vaccine disease introduced at Berlin in the very respectable way you mention. I should be very happy to consign the matter to Her Royal Highness the Princess Louisa's Physician with every direction that could be conveyed through the medium of a letter, but yet on so important an occasion, it would be still more gratifying were I to be allowed to appoint a person to Berlin, whom I could with confidence recommend as perfectly conversant with the cow-pox in all its stages. My nephew, Mr. Jenner, who has assisted in conducting my experiments on the subject, and who has inoculated considerable numbers, would be very ready to accept the embassy. But on mentioning this, perhaps I presume too much. My motive is the possibility of making a mistake.

In three or four days I shall be able to take some matter from a pustule that is ripening on the arm of a healthy boy. Sad confusion has been made in London, and in some other places, by the use of imperfect matter. Since my pamphlets were written, a vast deal of information has flowed in upon me, which, I am happy to say, all tends to establish the grand point, that the cow-pox perfectly secures the constitution from contagion of the small-pox. Five thousand persons have now been inoculated; the greater part of whom have been exposed to the infection of the small-pox by inoculation, and in every other way, without the least effect.

I shall very shortly republish what I have written on the subject, with an appendix. Whatever may be new or interesting in the appendix, I will transcribe for the satisfaction of the Princess, fearing it will not be printed so soon as you would wish to send it.

With respectful compliments to Colonel Walker,

I remain,

Dear Madam,

Yours, &c.

E. Jenner.

To prove how eagerly persons exerted themselves at this period to communicate to their neighbourhood the advantages of vaccination, I cannot refrain from mentioning the Rev. Mr. Holt, Rector of Finmere, near Buckingham. He had in the summer of this year been strongly impressed with the benefits of vaccination from a conversation with Mr. Abernethy. This clergyman found no opposition to the introduction of the practice, because the people were not unacquainted with the virtues of cow-pox when caught in the natural way. The results of his experiments were communicated in a letter to Mr. Abernethy, dated Finmere, November 6th, 1799. The last named gentleman forwarded it to the Medical and Physical Journal, and it was printed in the number for December. Dr. Jenner was much gratified with the satisfactory information contained in it, as well as with the skill and caution evinced by the reverend gentleman, and expressed himself to that effect.

Dr. Jenner to Mr. Abernethy.

Dear Sir,

I thought of being in London long before this time, and to have called in Bedford-Row to have thanked you for your valuable publication and the very polite note that accompanied it, which I received on the eve of my departure from town in the spring.

I am happy in seeing that through you the cow-pock inoculation has already moved in a channel, which from being adopted, may expedite its general diffusion through this country. I allude to the paper of the Rev. Mr. Holt in the Medical and Physical Journal of the present month.

I should be extremely obliged to you to inform me from what source Mr. H. derived his matter. He speaks of pustules similar to those excited by inoculation, except that they were smaller. I have sometimes (though very rarely) seen a few pimples scattered over the body, which for the most part have disappeared quickly; but I have sometimes known them remain long enough to show a little fluid at their apex. I have attributed them to the inflammation and irritation of the arm, as many acrid substances applied to the skin, so as to excite these effects, will produce similar appearances. They occurred so rarely that I have had no opportunity of investigating the nature of them; however, if their effluvia (provided in the first instance they have arisen from pure vaccine virus) should produce similar appearances, we may call the disease the small-pox at once. Indeed I have never considered the cow-pox and small-pox as distinct diseases, but the latter as a variety of the former.*

Excuse the trouble I give you, and believe me,

Dear Sir,

December 19, 1799. E. J.

* In a former chapter while illustrating the natural history of the Variolæ Vaccinæ I mentioned, that they had not appeared recently in a fatal form among the cattle, as they did in the days of Layard, and Ramazzini, and Lancisi. Since that portion of this work was sent to the press, I have received a communication which proves that even now it is sometimes fatal, and occasionally shows itself with some of its most malignant features. I could have wished to have inserted the information which I subjoin when endeavouring to sustain my general argument in Chapter Seventh, and several of those which precede. As however that cannot now be done I must place it in its present situation, conscious that it is well worthy of consideration on its own account; but especially so when it is remembered how strongly it corroborates the testimony of former writers, and sustains the doctrine which I have endeavoured to unfold.

Mr. Williams, of Dursley, was kind enough to procure for me the following statement; it was drawn up by Mr. Tiley, veterinary surgeon, who is well acquainted with the disease. It is

The conduct of one of the earliest and one of the most inveterate of the antivaccinists brought into the

supposed that the infection was conveyed to the cow that died, by a servant of a farmer named Stratford, who daily dressed a greasy-heeled horse.

"The beginning of May 1825, I was sent for to Mr. William Long's of Clinger, to see a cow that was very ill. It being, as I considered it at first sight, such an extraordinary case that I was at a loss to say what the real complaint was, as I had never seen one so affected before; but on examination I found it to be an aggravated case of cow-pox, and progressively arrived to the state it was then at. On my examining the teats I found several blackish scabs peculiar to the disease; and the whole skin, with the exception of no part of it from the base of the horn to the end of the tail, and to the hoofs, was one continued disease, not of vesicles nor scabs, but a discharge similar to that produced by a blister. Even the nose, and to the very edges of the lips, were affected the same as the other parts of the skin. Every symptom of violent fever was present; no attention having been paid to that previous to my seeing her: she lingered, and died the tenth day after I saw her. In the mean time I was employed to attend the other cows which were affected at the same time, then ten in number; and the same as I had seen the disease before; but every day brought newly-affected subjects till the whole pack, between forty and fifty, had the disease. The youngest of the cows, from two years old to five, were, more or less, affected about the skin (partially so) the same as the one that died. Till now I considered this disease in the cow to be topical. Constitutional remedies were employed; and every subject, except the first, recovered. One of Mr. Long's servants being asked by a neighbour of his to milk a cow for him that was rather obstreperous, which he did, and communicated the disease to the pack, which affected them in a similar manner. The whole of them did well. All the servants of Mr. L. had sore fingers through milking the cows, except one."

field an honest and indefatigable champion, who continued faithful to the last, and promoted the cause of vaccination as much by his personal efforts, as by his zeal and ability in overthrowing the arguments of its opponents. Dr. Moseley, physician to Chelsea Hospital, thought fit in a treatise on sugar to bring forward an irrelevant attack on cow-pox, and then gave his professional brethren a specimen of that elegant and classical phraseology which so peculiarly distinguished his subsequent lucubrations. His station, much more than the force of his argument, gave weight to his observations. He wrote in great ignorance of the subject; and with all the bitterness and prejudice that generally attend on ignorance.

Mr. Ring very successfully exposed and refuted his remarks. The doctor had seen in distant prospect an awful aggravation of human ills, from an admixture of bestial humours, which the cow mania, as he elegantly termed it, threatened to inflict upon our race. He even predicted an alteration in "the human form divine," and that another brood of minotaurs would overspread the land—

"Semibovemque virum, semivirumque bovem."

Mr. Ring's efforts to remove these apprehensions from the mind of the doctor, led to his forming an acquaintance with Dr. Jenner. Mr. Ring's own opinion on vaccination had at this period been somewhat modified by the publications of Dr. Pearson and Woodville. Some time elapsed before these impressions were completely removed. When this took place, and when the full force of Dr. Jenner's accurate, perspicuous, and satisfactory details were confirmed and illustrated by his own experience, no one more truly appreciated the genius of the author of vaccination, or entertained a deeper feeling of respect and admiration for the great services which he had rendered to mankind.

From this time Mr. Ring devoted a great part of his professional life to the cause of vaccination. He investigated every adverse case that he heard of in London: he offered gratuitous vaccination to all who would accept it; and he marshalled the chief medical men in London who had satisfied themselves of its efficacy, by procuring their signatures to a testimony which I subjoin.

" Many unfounded reports having been circulated, which have a tendency to prejudice the public against the inoculation for cow-pox: We the undersigned physicians and surgeons think it our duty to declare our opinion that those persons who have had the cow-pox are perfectly secure from the future infection of the small-pox. We also declare that the inoculated cow-pox is a much milder and safer disease than the inoculated small-pox."

The above document bears the signatures of thirty-three of the most eminent physicians, and of forty distinguished surgeons of the metropolis, amongst whom are the well-known names of Baillie,

Vaughan, now Sir Henry Halford, Cline, Cooper, Abernethy, Lettsom, Willan, Garthshore, Maton, Lynn, Blair, Dundas, Good, John Pearson, James Moore, Saunders, Croft, Garnett, &c. &c.

The correspondence between Mr. Ring and Dr. Jenner commenced at a time when the eruptions generated at the Small-pox Hospital were the subject of much attention. A few of the letters which passed on this occasion may not be uninteresting here, although Dr. Jenner's sentiments on the origin of the eruptions have already been disclosed in his letters to Lord Egremont.

DR. JENNER TO MR. RING.

Cheltenham, 16*th August*, 1799.

DEAR SIR,

The very candid and satisfactory manner in which you have delivered your sentiments on the Variolæ Vaccinæ, cannot but be gratifying to the public in general, and to my feelings in particular. I write to express my thanks to you. At the same time allow me to make a few observations on the origin of the pustules which have appeared under vaccine inoculation, as this occurrence seems to have led you into an erroneous inference.

You observe that eruptions have appeared among those who have been inoculated in the country as well as in the metropolis, and also that the infection has been communicated by effluvia even from the inoculated pustule.

Let me call your attention to the source of the infection you allude to. It was that which was generated at the Small-pox Hospital. From the time I first heard that pustules similar to the variolous had appeared among the patients inoculated there with the vaccine virus I strongly suspected, from a coincidence of circumstances, that by some imperceptible avenue the variolous virus might have crept into the constitution at the same time. Subsequent occurrences tend strongly to confirm this supposition. My last publication (Further Observations, &c.) was sent out before I could so fully decide on this important point as I can at present. Conceiving the London cow to be more out of a state of nature than the animal fed in the country meadows I could not say positively whether the virus generated by one or the other might not in some measure differ, and therefore was unwilling to decide until this had been ascertained by experiment. Some time in April the cow-pox appeared at one of the great milk farms in the neighbourhood of town. With this virus several patients in the country were immediately inoculated. The result was just the same as in my former experiments; that is, it produced the true cow-pox pustule on the part where it was inserted, but no secondary pustules; nor has a single pustule appeared in any one instance wherein the matter was taken from this source for the purpose of inoculation, and the cases now amount to more than seventy. From variolated pustules one cannot be surprised to hear that a disease has been communicated by effluvia. By no means that I could devise have I been able to infect a person by the effluvia of the simple cow-pock pustule, although I have tried several.

Among others, I have suffered children two or three times in a day to inhale by the mouth and nostrils the effluvia of pustules on the arms of others, when the matter has been in its most active state, and the pustules punctured in several places to give the matter its fullest effect.

There is another fact that strengthens the supposition

of the matter's being contaminated at the Small-pox Hospital. The variolous appearance among the patients is more and more retiring. Out of the last 110 cases Dr. Woodville remarks only seven had pustules.

The cow-pox then maintains its ground having nearly destroyed the co-operating effects of the small-pox.

And this event gives strength to what I have from the commencement of my experiments imagined that the latter is a malignant variety of the former; the parental root being the cow-pox. It is a little vexatious to find that so many should take up the subject, and give their decisions to the public without understanding it in the least; but after the castigation that one of these gentlemen has experienced from your hands I hope they will in future be more cautious.

I remain, dear sir,
with great respect,
Your much obliged and
obedient humble servant,
EDWARD JENNER.

MR. JOHN RING TO DR. JENNER.

DEAR SIR,

After your very polite attention to my former request I am really quite ashamed to be troublesome again so soon, but from fear of losing the matter I spoke so favourably of in my last, from want of a regular succession of patients—I had inoculated only two the last time, in both of whom the arm failed to inflame, so that my matter is now extinct.

I have inclosed a couple of lancets which I should esteem myself much obliged to you if you will arm with matter. I shall in future observe other precautions, and preserve dry matter, not trusting so implicitly to the chance of inoculation taking place.

I am extremely indebted to you for your different communications. On this subject you will do me the favour to observe a profound silence. I sincerely thank you for your good wishes, but cannot indulge much hope of success, not having the least idea of writing on the subject till within three weeks; and scarce a leisure hour at any one time since.

If you will allow the Medical Society the honour of enrolling you among its corresponding members I shall be happy to propose you, when the fate of my dissertation is determined. Before that period it would create a suspicion of the author.

It must be delivered on or before the first of November.

I am, with much respect,
Dear Sir,
Yours most sincerely,
JOHN RING.

New-Street, Hanover-Square,
Oct. 25th, 1799.

DR. JENNER TO MR. RING.

DEAR SIR,

When I had the pleasure of receiving your letter there was no cow-pox matter here in a fit state to send you. That which is inclosed was taken about four days ago and, if soon made use of, will doubtless prove efficacious. This matter is from the source mentioned at the conclusion of my second pamphlet. It has been passing from one patient to another for upwards of six months, and except in the single instance I have mentioned, I have seen no pustules produced by it: indeed, in that instance, they did not maturate.

It was my intention to have filled this sheet with some further observations on this singular disease, but I am called away, and do not like to detain the threads for another post.

I remain, &c.

E. J.

Sept. 18*th*, 1799.

Mr. Ring, Surgeon, New Street,
Hanover-Square, London.

Towards the end of this year attempts were made in different parts of the kingdom to diffuse the benefits of cow-pox by forming institutions for gratuitous inoculation, and for affording supplies of lymph to all who might apply for it. The city of Bath, under the suggestion of Mr. Creaser, was among the first to forward such a plan: but in London measures of a very peculiar nature had been adopted to set on foot a similar establishment. They were such as to place Dr. Jenner in a very embarrassing situation, and to give just cause of offence to all who considered what was due to him and to the public. Not long after he had left London (on the 14th of June) the gentleman who had promised to confer upon him a never-dying fame began to carry his purpose into effect. His first step was to form a vaccine board, of which the chief place was allotted to himself,—the inferior departments being also filled up in conformity with his wishes. The board, thus constructed, received a degree of countenance worthy of a metropolitan charity: His Royal Highness the Duke of York having permitted himself to be named as patron. Lord Egremont also consented to hold an office in it.

Matters having been thus adjusted an account of the proceedings was forwarded to Dr. Jenner, from London, by the most conspicuous agent in this business. The terms in which the communication was made afford a very curious specimen of the writer's notion of *justice and liberality*. They also show that he set no small value upon his own labours, and the "*new lights*" which led him astray.

DR. PEARSON TO DR. JENNER.

London, Dec. 10*th*, 1799.

MY DEAR SIR,

I wish ever to be governed in life by the rule of doing justice; and, if I can, acting liberally to my fellow-labourers. I trust I have acted consistently to you; and, if I have differed in opinion on some points, it was because new lights broke in upon me; but I trust in such instances I, too, acted consistently and was more anxious to bestow commendation than to be studious to point out faults. Agreeably to my principle I now address you to say that we have made some progress in the institution of a charity for inoculating the vaccine pock. I do not know that I can confer any honour on you by proposing you (if I am able) to the directors of our establishment, nor do I well know what to propose to you. It occurs to me that it might not be disagreeable to you to be an extra corresponding physician, and I can see no objection to this proposal at our meeting. The medical establishment consists of two physicians of the college, two consulting physicians, two surgeons, and three visiting apothecaries. We have got very high patronage, but the institution is not yet completely organized, nor will be so for some time. Rush, Keate, and his nephew, Gunning, Brande, Devaynes, belong to the medical departments. Exactly what Woodville will or can be, on account of his connexion with the Small-pox Hospital, I cannot tell: but he authorizes me to say in a letter from him, "that he wishes to give his assistance and promote the undertaking."

No expense is to be attached to your situation except a guinea a-year as a subscriber, and indeed I think you ought to be exempt from that, as you cannot send any patients: but you may depute some proxy in town. I confess I was surprised that you neither called nor sent to me for the last *two* months you were in town. However, if it was because you were so much occupied, I certainly excuse you. I hope you will excuse haste in this letter, but it will serve to assure you that I remain, with great consideration,

Yours truly,

G. Pearson.

Compliments of Mrs. P. and myself to Mrs. Jenner.

It will very easily be conceived that this letter, notwithstanding certain suspicions which had been excited concerning the conduct of the author, would occasion considerable surprise in the breast of Dr. Jenner. Mild, and generous, and humble as he was, he could not but express what he felt on the occasion.

He conveyed a just and dignified reproof to the writer in the following words:—

DR. JENNER TO DR. PEARSON.

Berkeley, Dec. 17, 1799.

SIR,

I received your letter of the 10th instant, and confess I felt surprised at the information it conveys.

It appears to me somewhat extraordinary that an institution formed upon so large a scale, and that has for its object the inoculation of the cow-pox, should have been set on foot and almost completely organized without my receiving the most distant intimation of it. The institution itself cannot, of course, but be highly flattering to me, as I am thereby convinced that the importance of the fact I imparted is acknowledged by men of the first abilities. But at the same time allow me to observe that if the vaccine inoculation, from unguarded conduct, should sink into disrepute (and you must admit, Sir, that in more than one instance has its reputation suffered) I alone must bear the odium. To you, or any other of the gentlemen whose names you mention as filling up the medical departments, it cannot possibly attach.

At the present crisis I feel so sensibly the importance of the business that I shall certainly take an early opportunity of being in London. For the present I must beg leave to decline the *honour* intended me.

I remain, Sir,

Your obedient Servant,

E. JENNER.

The reply to these remarks was, if possible, still

more illustrative of the writer's mode of conferring honour and immortality. As Dr. Jenner did not, I believe, deem it worthy of an answer I shall not present it to my reader.

While these transactions were going on Dr. Jenner was in constant correspondence with Lord Egremont. In one of his letters he alludes to the successful inoculations at Petworth with the vaccine matter he himself had furnished; and likewise to the new institutions which were forming in Bath and London. In reference to the latter he observes:—

"The new institution in London for vaccine inoculation, considered abstractedly, cannot but be flattering to my feelings; but many will scarcely believe that not the least intimation of the business was given to me until all was organized, and then Dr. Pearson in a letter burst the whole open to my view at once, and kindly offered me the post of extra corresponding physician!"

Dr. Jenner left Berkeley for London on the twenty-eighth of January, 1800, with his friend Dr. Hicks, by the way of Bath. His object was to forward the steps which were in progress in that place for diffusing vaccination; but more especially to give his personal attention to the strange proceedings which were carrying on in London and its vicinity. Immediately before he left the country he had occasion to reply to some of the statements which had appeared in the periodical journals respecting vaccine eruptions. He published a short letter dated from Berkeley, January 13th, 1800. The last sentence of this letter was to this effect. "Time will develope the mystery before us: at present I very much suspect that where variolous pustules have appeared variolous matter has occasioned them." Not long after this period he published a continuation of "Facts and Observations relative to the Variolæ Vaccinæ," in which, with admirable precision and moderation, he entered upon the questions at issue between himself and Drs. Pearson and Woodville.

In this work he observes that he cannot "feel disposed to imagine that eruptions similar to those described by Dr. Woodville have ever been produced by the pure uncontaminated cow-pock virus; on the contrary, I do suppose that those which the Doctor speaks of originated in the action of variolous matter which crept into the constitution." Although observations of this kind had been made by Dr. Jenner to Dr. Woodville long before he published his "Reports," and although they were repeatedly brought before him subsequently, he did not express (whatever he might have felt,) any of the indignation which he permitted himself to disclose when the same facts appeared in print. He did not notice them in his "Reports," and continued apparently in friendly intercourse with Dr. Jenner till the publication of the third pamphlet, which contained the observations above quoted. In July

1800, Dr. Woodville sent forth his Observations on the cow-pox, which he inscribed in a harsh and angry address to Dr. Jenner. I will not perpetuate the remembrance of his unjustifiable severity on this occasion, by recording his language. It excited no hostile feeling in the breast of Jenner; and it was freely forgiven when the first opportunity of reconciliation presented itself.

Dr. Jenner arrived at Adam Street, Adelphi, on the thirty-first of January, 1800. One of the great objects of his journey to London having been to deliberate with Lord Egremont and his other friends respecting the establishment of a Vaccine Institution he wrote the following letter to his lordship soon after his arrival.

Dr. Jenner to Lord Egremont.

My Lord,

I am just favoured with your lordship's letter, and feel extremely obliged to you for the friendship with which you honour me. I should be happy to profit by your lordship's advice in an affair wherein not only my own future happiness, but that of society, is much involved. Dr. P—— has certainly dealt very unhandsomely by me, and has convinced me by his conduct that I can never more have any private concerns with him: but before a new arrangement is made for vaccine inoculation, I hope to have the satisfaction of seeing your lordship at Petworth. Mr. L. visits your lordship on Sunday, and I hope to accompany him. I have looked forward to the honour of an interview which, I trust, will produce a plan from which this country may derive the advantages of the new antidote for the small-pox.

* * * * *

The new institution was a subject of considerable interest with many individuals of the highest rank, who had acquired a knowledge of the benefits of vaccination, and who, with the most benevolent and praiseworthy feeling, wished to lend the sanction of their name and station in diffusing them as widely as possible. In doing this they were certainly not aware that the disinterested author of the practice was likely to suffer any wrong; or that the doctrines which he established, and on which the safety and success of the practice rested, incurred any risk of being misunderstood or perverted either from ignorance or design. In fact, the establishment, according to its original plan, could not have been carried into execution without impugning these doctrines in a manner the most objectionable in itself and peculiarly offensive to Dr. Jenner. It tended immediately to place in the most conspicuous and commanding situation those who questioned the truth of his statements in the most material points, and who made their own fatal mistakes the ground of difference from him. These things were certainly not at first understood either by the patron or the president of the institution. His Royal Highness the Duke of York, with his characteristic attention to those under his command, had recommended the practice of vaccination in the army. Being, therefore, well aware of its advantages he did not hesitate, on application being made, to give the high authority of his name to an institution ostensibly designed to promote that practice. Other members of the royal family, the Duke of Clarence more especially who had more time to devote to such matters, at a very early period saw through the injustice of the proceedings and very plainly delivered his sentiments on the subject to some of the individuals chiefly concerned. He likewise desired that Dr. Jenner might be introduced to him on his arrival in London.

This introduction took place early in February. On this occasion Dr. Jenner had the honour of a long conference with his royal highness. That conference related to the subject of vaccination in general, but was more particularly directed to the means of diffusing the benefits of cow-pox inoculation as widely as possible. Dr. Jenner's own views on that subject were very clearly conveyed to his Royal Highness; and they were subsequently communicated to Lord Egremont.

PROPOSALS BY DR. JENNER FOR A PUBLIC INSTITUTION FOR VACCINE INOCULATION.

(FOR LORD EGREMONT.)

Having now pursued the inquiry into the nature of the cow-pox to so great an extent as to be able positively to declare that those who have gone through this mild disease are rendered perfectly secure from the contagion of the small-pox; and being convinced from numberless instances that the occupations of the mechanic or the labourer will meet with no interruption during its progress, and the infected and uninfected may mingle together in the most perfect safety, I conceive that an institution for the gratuitous inoculation of the lower classes of society in the metropolis would be attended with the most beneficial consequences, and that it might be so constituted as to diffuse its benefits throughout every part of the British Empire.

EDW. JENNER.

London, March 16th, 1800.

In order to diffuse the advantages of the institution for promoting the inoculation of the cow-pox as widely as possible, it is proposed:

1st. That communications be made to the principal medical gentlemen in London and throughout the British Empire, acquainting them with the nature of the Institution and soliciting their associating as honorary members.

2dly. That a Physician be appointed who shall superintend the medical department.

3dly. That a house be appropriated in some convenient part of this metropolis containing the necessary apartments for a medical attendant, a secretary, porter, &c. Apartments also for the reception of the patients sent for inoculation, and for the occasional reception of those who may choose to aid the charity.

4thly. That virus for inoculating the cow-pox be sent to all such honorary members as may make a proper application for it at the apartments of the Institution, and that none be sent forth without the signature of the superintending Physician as a test of its being genuine.

5thly. That the virus be accompanied with directions for its use, and (to guard against error) with some general observations on the nature of the disease.

6thly. That the Institution be supported by voluntary contribution.

7thly. That an annual subscriber of
be a Governor.

8thly. That the Governors meet at the apartments the first day of every month for the inspection of the reports relative to the general progress of the inoculation, &c. &c.

9thly. That an abstract of the reports be published as often as it may be deemed proper.

On the 15th of February Dr. Jenner had an opportunity of discussing the subject more fully with his lordship, he having on that day gone to Petworth with his friend Mr. Ladbroke. He remained at this splendid mansion of his noble host till the 24th of the same month. During his stay he was much gratified with the personal attentions he received, but still more with the ability and zeal displayed by its distinguished owner in making himself acquainted with all the details of vaccination, and by investigating the origin of those disastrous occurrences which, for a season, so much obscured its character.

The trials made under his lordship's eye with the matter furnished by Dr. Jenner happily removed every doubt as to the accuracy of his description of the disease, nearly two hundred persons having been inoculated before he left Petworth, without *one* deviation from the ordinary course having taken place.

On the 25th of this month he received a message from His Royal Highness the Duke of York desiring an interview. That interview took place on the 1st of March. Mr. Knight was present. On this occasion the subject of the vaccine institution was very fully discussed, and received the greatest consideration from His Royal Highness. From this period up to the 17th a great deal of correspondence took place between the officers of the new institution and the eminent individuals who had promised to patronize it. The main object of these negotiations was to vindicate the conduct adopted towards Dr. Jenner; to hold out additional offers of conciliation to him; and to prevent, if possible, that secession from the institution which was anticipated by them on the part of the royal Patron, and of the noble President.

Some of Dr. Jenner's friends were, at one time, inclined to think that he still might connect himself with the institution on certain arrangements being made. His own temper and feelings always inclined him to make concessions, and he certainly would have done so on this occasion had his personal feelings been alone concerned; but the conduct of the individuals who framed the institution proved that the cause of vaccination could not be safely committed to their hands; and that an establishment, which had been organized as this was, could not receive his sanction without his appearing to abandon those truths which he had advanced respecting the nature of Variolæ Vaccinæ. No success could attend a coalition of this kind. The misrepresentations which had been published and the blunders which had been committed had already too much endangered the safety of the new practice; and even a slight consideration of what had been done in these respects could not fail to show the impropriety of requiring Dr. Jenner to give them any countenance by co-operating with persons who still continued to uphold sentiments directly opposite to his own. These reasons, which satisfied his own mind, proved in the end equally influential with his friends; and Dr. Jenner was informed by Lord Egremont at an interview on the 17th that both his lordship and his Royal Highness the Duke of York had resolved to withdraw from the institution altogether.

Dr. Jenner's firmness and prudence in this affair gave unqualified satisfaction to his friends; and supported as he was by the handsome and efficient interposition of Lord Egremont, he was enabled to defeat the ambitious designs of those who sought for high patronage in proceedings of a very questionable nature. It is not my wish to dwell longer on this unpleasant topic: and I have abstained from printing many of the documents from which the preceding facts have been drawn: indeed I would gladly have passed them by altogether had not the character of Dr. Jenner, and still more the character of vaccination, been materially affected by them.

It certainly was the feeling of all those elevated personages who wished to assist in forming a vaccine institution that none could be established which did not assign to the author of the discovery that situation of dignity and influence which was due to his merit, and which would enable him to direct the practice with vigour and effect. This feeling was strongly evinced when the Royal Jennerian Society was formed; and it was also very characteristically expressed at this time by his friend the late munificent Mr. Angerstein. He said "*that he would not mind a subscription of one hundred or two hundred pounds in an institution organized by the man who was best competent to set about it, but that he would have nothing to do with one grafted on the present blunders.*"

All Dr. Jenner's medical friends of the first consequence in London cordially recommended and approved of the steps which he took, and although they could not subscribe so largely as Mr. Angerstein, this test of their sincerity was not wanting on their part.

Dr. Jenner had also the gratification of knowing that those who were most distinguished in the profession had no part in these blunders, and that all whose good opinion he most valued really *did* appreciate both the scientific character of his inquiries, and the splendid practical results to which they led. A letter from a British philosopher, alike eminent for the extent and accuracy of his knowledge and the originality and depth of his genius, embodying these sentiments forms a striking contrast to the vague and unsatisfactory correspondence which he had with some other of his medical brethren: I therefore gladly introduce it here.

Dr. Wollaston to Dr. Jenner.

1800.

Sir,

I return you many thanks for your observations, and with them beg leave to propose one question for your consideration.

You have proved to the satisfaction of every candid person that there is a disease of the very mildest kind communicated by inoculation, which perfectly secures the constitution from the small-pox.

You have ascertained that unless great precaution is taken in procuring the fluid for inoculation a disease of a more violent kind, and in no degree beneficial, may be produced.

You have described the appearance which the fluid for inoculation ought to have, and have named a period (earlier than in small-pox) beyond which it cannot be depended upon, on account of the changes which it may have undergone; but you have left it for future experience to determine on what that change depends.

Query. Does not this change depend on that species of erysipelas which produces the blushing areola?

May not this erysipelas supersede the action of the vaccine virus although it be incapable of subduing, but is only superadded to, that of the variolous?

The disease produced by degenerated matter appears, conformably to this hypothesis, erysipelatous.

You may possibly have seen reason to form a different opinion in the course of your practice, or from that of Mr. D. (page 30 of your "Farther Observations,") who took matter from an arm in this state.

If, on the contrary, this conjecture be well founded, how soon can this cause be supposed to operate?

Does the areola ever take place earlier than the tenth day, as appears to have been the case, page 26 of your "Continuation," &c.?

Has a transparent fluid ever produced a wrong disease if taken before the blushing areola began to appear?

In submitting these queries for your consideration believe me influenced by unfeigned respect for the author of the most valuable communication ever made to the public, and permit me Sir to subscribe myself,

Your very humble Servant,

W. H. Wollaston.

In a former letter Dr. Jenner alluded playfully to "great news from St. James's." This expression referred to the King's gracious permission to him to dedicate the second edition of the "Inquiry" to his Majesty; and to his reception at the palace.

On the 7th of March he went to St. James's with Lord Berkeley, and presented his treatises on the cow-pock to the King: his Majesty received him very graciously.

In allusion to this intended ceremony he writes thus to his friend Mr. Shrapnell, who was then at Cranford in attendance on Lady Berkeley. "Pray acquaint Lord Berkeley I shall be ready to accept his kind offer of accompanying me to St. James's any day he may appoint in the course of the week after this. The work will then be finished, and clad in crimson. What will you give for a sight of me all in velvet, girt with a sword too? What a queer creature is a human being!"

Towards the end of this month he also had the honour of a private interview, at Carlton House, with his Royal Highness the Prince of Wales. He was introduced by the secretary the late Sir John M'Mahon, Bart. His Royal Highness on this, as on every other occasion, received him with marked respect; and at future periods showed the interest he felt for the cause of vaccination, by the personal efforts which he was pleased to make for its advancement.

Dr. Jenner's time in London was very fully occupied in visiting his professional friends, and in attending to the daily increasing importance of vaccination. To one of his correspondents, Mr. Shrapnell, at this period he says, "I have not made half my calls yet in town, although I fag from eleven till four." To the same gentleman he thus expresses himself in reference to the progress of vaccination. "Pray write without delay to Tierney, and tell him how rapidly the cow-pox is marching over the metropolis and, indeed, through the whole island. The death of the three children under inoculation with the small-pox will probably give that practice the Brutus-stab here, and sink for ever the tyrant small-pox. Would Tierney like to have a little virus, that the cow-pox inoculation may be set going under his own eye at Edinburgh? I should be happy to furnish him.

Let him know that my new edition mentioning his name, with the appendix, is published. A very little attention would place the practice in its proper light in Edinburgh, a thing devoutly to be wished."

Soon after this Dr. Jenner received two letters from Mr., now Sir Matthew Tierney, Bart., which delineates the state of professional feeling at Edinburgh at that time.

M. J. TIERNEY, ESQ. TO DR. JENNER.

No. 2, Fisher's Drummond Street,
Edinburgh, 21st March, 1800.

DEAR SIR,

In a letter I lately received from Mr. Shrapnell I with much pleasure find the attention paid to your (I must say valuable) *discovery* of the vaccine matter I beg leave to congratulate you on the advantages it is likely to bring to society, and the honour, so justly merited, you received yourself. I presume he informed you of the state in which it is here. Not knowing much of its effects its real value is not yet attended to. Dr. Gregory the Professor of Physic here knew very little about it, and of course did not encourage it. I gave him the sum of my experience on it, and he now seems to entertain more favourable opinions of it. Indeed, he did me the unwished-for honour of reading my accounts to his class. Since then, the students here seem anxious to see and know the disease better. A Mr. Anderson, a surgeon at Leith, is the only person *here* who has tried it, and his accounts are strongly favourable.

From the conviction I have of its advantages I think it a duty I owe to society at large to extend it as much as in my power: and on this account many of my friends here have earnestly solicited me to get some of the matter, not that I expect to have an opportunity of using it myself here, but conceive it may be a *second* focus from which it may extend itself more and more rapidly.

As you know how fond of variety all young medical men are, with this intent and as Mr. Shrapnell tells me, you would be so good as to send me some of the matter and, as coming through you, every suspicion of error would be done away, I am the more anxious to have it from you and shall feel much obliged if you send me some as soon as possible. A friend of mine proposes giving in a paper on this disease to the Medical Society, the greater part of which I shall contribute to, not having an opportunity of writing myself. This too may be a further means of extending it, as it wants no more but to be known and received by every medical man. I regret very much that during my absence from the regiment many of the men were inoculated, and I have not had an opportunity of attending to them. It is surprising it is not universally adopted in the army. The advantages from it are self-evident.

M. J. Tierney, Esq. to Dr. Jenner.

No. 2, Fisher's Drummond Street,
Edinburgh, 15*th April,* 1800.

Dear Sir,

I can with much satisfaction address you now, and am sure the success of the first introduction and the advantages likely in a very short time to result from the vaccine virus *here* will be the pleasantest acknowledgement for your last kind favor.

In my former you were informed that it was very little attended to here: but on receiving the matter from you I mentioned it to Dr. Gregory, (Professor of Practice of Physic) and with his usual liberality of mind and to show his confidence in my former statement, he wished me to inoculate *his* youngest child who is ten months old, and even teething. I did so, and have now the satisfaction to say the disease has gone through its stages even milder than any I saw before. This is the thirteenth day since inoculation. The inflammation is much reduced, and I have no doubt the puncture will get well without further trouble. Nothing was done either in regimen or application to the part. So mild was the constitutional irritation that the Doctor *himself* could not say whether it may not be occasioned by teething. This you will most readily see is the most effectual mode of spreading the disease here, and in fact it has already had that effect; many persons applying to have their children inoculated with it. Mr. Anderson of Leith, whom I mentioned in my last, is the only person who attended to the disease. He inoculated, since May last, 150 persons in all of whom its progress was much *milder* than it has been observed to be in England; but he has observed some curious phenomena in its progress, having had three children at different times under its influence, whose cuticle was abraded in different parts of the body; in one by a prior eruption on the back. He observed pustules to appear on those parts, from which he took matter and produced the disease in others with it. In *no* case where the cuticle was sound did he observe pustules or eruptions. I asked him, *naturally*, if it was not possible the matter from the puncture might have been applied to these parts in dressing or undressing the child; but he seems to think this could not be the case. He never had occasion to apply the mercurial ointment, and on the whole, his accounts are even *more favourable* than any others I have heard. He further says (which by the by is a considerable advantage) that the prejudice of the people against the vaccine disease is *much less* than against the inoculation with small-pox. In this country religious opinions direct the people a great deal.

Its being received by the Professors here will certainly be a means of spreading it more rapidly, and I flatter myself this is now established.

* * * * *

At the formation of the Bath Vaccine Institution Dr. Jenner wrote to his friend Lord Somerville to request his acceptance of its presidentship. His Lordship, who was then in Lisbon, gave his assent in a letter dated 4th March.

On the 27th of the same month Dr. Jenner went to St. James's Palace, to the Queen's drawing-room, when he had the honour of being presented to her Majesty by her chamberlain the Earl of Morton. Her Majesty, on this occasion, asked him many questions relative to the progress of cow-pox; and received him with marked attention.

On the 12th of April Dr. Jenner received, whilst in London, some matter which had been generated on the cow by inoculation with the virus of *grease* by Mr. T. Tanner. Some part of this matter he transmitted to Mr. Wachsel, of the Small-pox Hospital.

On the 15th of the same month his Royal Highness the Duke of York sent a message to Dr. Jenner to request him to go to Colchester to vaccinate the 85th regiment. He could not himself obey this request, but he dispatched his nephew George Jenner, who set off on the 25th. He could scarcely have entered upon an undertaking more annoying to himself or more inauspicious to the character of vaccination than he encountered in his attempts to vaccinate the 85th regiment. He was under the necessity of waiting a considerable time before he could communicate the disease to any one. He found the whole of the regiment, together with the women and children, labouring under the itch. He commenced his inoculations, but they all proved unsuccessful. On the 2nd of May he found that one child had taken the cow-pox, and he adds that "all the men are cured of the itch, and in two days will be washed and fit for inoculation." These anticipations proved rather premature, for he observed on the 13th of May that he could not succeed in communicating the true cow-pox to those whose constitutions had been under the influence of the itch.

It appears to me of moment to allude to these occurrences, because they afford very convincing proofs of the truth of a doctrine which Jenner subsequently adopted and invariably maintained to the last hour of his life, namely, that any cutaneous disease, however slight in appearance, was capable of interfering with the regular course of the cow-pox and of preventing it from exercising its full protecting influence.

The occurrences at Colchester ought to have great weight, and to induce medical men to pay more attention to his directions on this subject than they have hitherto done.

Though Dr. Jenner could not himself undertake to vaccinate the regiment at Colchester he went twice to that place to inspect the progress of the practice.

Although the practice of vaccination encountered no active popular hostility at its commencement, there were, nevertheless, some instances in which prejudice and ignorance led the lower class to manifest their feelings in a violent and intemperate manner. Mr. Gooch communicated some curious facts of this kind, and gave evidence, at the same time, of his own zeal, as well as that of his lady, in forwarding the practice.

T. S. Gooch, Esq. to Dr. Jenner.

Holbecks, Hadleigh, Suffolk,
April 24th, 1800.

Sir,

Having understood from Lady Peyton that you are always pleased to hear any communication on the cow-pox I trouble you with a remark or two on what has happened to our patients under that disorder. Mrs. Gooch and myself have inoculated 611 persons with cow-pox virus, and have not had one patient whose arm has been at all sore, so as to require any application to it.

I see by your last publication you suppose it impossible

for a person inoculated with the pure *uncontaminated* cow-pox virus to have pustules; I beg leave to mention on that subject that we have had six people with evident pustules, from which we might have inoculated. Two of them had pustules on the eye, and four on the inoculated arm near the elbow. We have had a proof of the possibility of one person inoculating himself accidentally by rubbing his eye on the arm of another under the disorder; therefore the pustules on the eyes may have been owing to this. We inoculated five persons in one family; two others of this family took it in an extraordinary manner; viz. one by sleeping in the same bed, and rubbing its eye on an infected arm; and the other was inoculated by its brother, a child of five years old, who did it by a scratch of a pin on which he had put matter.

We had our virus from Mrs. Gooch's sister, Lady Rous, who had it immediately from you.

I am happy to inform you that in spite of all ignorant prejudice, and wilful misrepresentations, this wonderful discovery is spreading far and wide in this county. The first people we inoculated in Hadleigh were absolutely pelted, and drove into their houses, if they appeared out; we have now persuaded our apothecary to inoculate the whole town (7 or 800 persons) and our hundred-house is now under inoculation (about 350 persons.) A physician at Ipswich, Dr. Hamilton, has taken it up in a very liberal manner and is extending it very much. I beg pardon for troubling you so much, and am, Sir,

Your obedient humble servant,

T. S. GOOCH.

From this time till the period when Dr. Jenner left London he continued actively engaged in promoting the cause of vaccination, by conferences with his medical brethren; by discussions at the medical societies; and by attending those public meetings where it is so usual to forward measures of general interest and utility. His presence on such occasions always afforded an opportunity of introducing the subject of his discovery. The effect of proceedings of this nature is very considerable in this country, where they are generally reported in the newspapers and thus are speedily diffused over the kingdom. He had interviews with many noblemen on the subject, and received from them strong marks of attention and respect. Amongst those who took a more especial interest in this matter may be mentioned Lord Hervey and the Earls of Aylesbury and Ossory. He was also consulted by many of them professionally, and was likewise employed in vaccinating their children.

Dr. Jenner left London with his nephew George, and his man Richard, on the twenty-third of June. They slept at Buckingham and arrived the next day at Tusmore, the seat of his friend William Fermor, Esq. where he dined. In the evening he went to Oxford. The following day he was introduced to Dr. Marlow, the Vice-Chancellor of the University. He also had intercourse with Dr. Wall, Chemical Professor; Dr. Williams, Regius Professor of Botany; Sir Christopher Pegge, Reader in Anatomy; and Mr. Grosvenor, Surgeon to the Radcliffe Infirmary. These gentlemen on this occasion signed the following testimonial drawn up by Sir C. Pegge :—

" We, whose names are undersigned, are fully satisfied, upon the conviction of our own observation, that the cow-pox is not only an infinitely milder disease than the small-pox but has the advantage of not being contagious, and is an effectual remedy against the small-pox."

He quitted Oxford on the twenty-seventh, and arrived at Berkeley on the same evening.

CHAPTER X.

INTRODUCTION OF VACCINATION INTO AMERICA, FRANCE, SPAIN, MEDITERRANEAN, CONSTANTINOPLE, BAGDAD, BOMBAY, &c.

WHILE Dr. Jenner was thus employed his discovery was making rapid progress throughout the world. Early in the year 1799 his first work reached the shores of North America.

Dr. Lettsom transmitted a copy of it to Dr. Waterhouse, Professor of the theory and practice of Physic in the University of Cambridge, Massachusets. The tidings of Dr. Jenner's discovery were received in America very much in the same manner as they had been in other countries. A judicious few at once felt and acknowledged the strength of his facts and their important consequences. Some doubted, others abstained from expressing any opinion; whilst the greater number treated the whole subject with ridicule. Dr. Waterhouse was not slow to estimate the advantages of the discovery. It is my duty, therefore, not merely to record his services, as the first who made known the blessings of vaccination to his countrymen, but likewise to speak of him as the firm, consistent, and ardent admirer and friend of Jenner. In the year 1800 a correspondence commenced between them which was kept up with increasing interest and attachment till nearly the close of Dr. Jenner's life.

As in the New World the ordinary method of making known discoveries, even in medicine, was through the medium of the newspapers, Dr. Waterhouse published in the Columbian Sentinel of March 12th, 1799, a short account of the cow-pox. The article was headed "SOMETHING CURIOUS IN THE MEDICAL LINE." Not long afterwards he brought the subject before the American Academy of Arts and Sciences. The illustrious President of the United States, John Adams, who was likewise President of the Academy, was at the meeting and received the communication in a manner worthy of an individual who had proved himself alike capable of directing the resources of a great and free people, and in promoting the advancement of every useful art and science.

After several unsuccessful attempts to obtain cow-pox matter from England Dr. Waterhouse at length succeeded in getting some from Dr. Haygarth, of Bath, who forwarded it from Bristol. It was procured from Dr. Jenner's stock by Mr. Creaser. With this matter he inoculated seven of his own children, six of whom went through the disease in the usual manner. In order to confirm the doctrine of its prophylactic powers he resolved to have them inoculated with small-pox matter in the most public manner. With this intention he wrote to Dr. Aspinwall, physician to the Small-pox Hospital in the neighbourhood of Boston, requesting him to perform the experiments. This gentleman assented to the proposal. Three of the children were sent to the Small-pox Hospital. One of them, twelve years old, was selected for the trial. Active small-pox matter was inserted by two punctures: an infected thread was likewise drawn through the skin, and the patient then left in the Hospital. On the fourth day there was some slight appearance of infection; but it died away, and left no traces of its action.

The successful vaccinations in Dr. Waterhouse's family soon turned the tide of popular feeling in favour of cow-pox. The zeal of the medical men was excited to an unparalleled degree; but, unfortunately, their discretion did not keep pace with it. They disregarded the cautions of Dr. Waterhouse, and paid no attention either to the state of the matter with which they inoculated or to the progress of the pustule. It appears, likewise, that the cupidity of persons not of the medical profession was stimulated on this occasion, and the manner in which they carried on their traffic was alike indicative of their avarice and their ignorance. The followers of this trade obtained the shirt-sleeves of patients which had been stiffened by the purulent discharge from an ulcer consequent on vaccination. These they cut into strips, and sold about the country as impregnated with the true vaccine virus. Several hundred persons were actually inoculated with this poison which, in several cases, produced great disturbance in the constitution. These blunders, it is to be feared, were not confined to vagrant quacks, inasmuch as several medical men were not quite blameless in this respect.

Soon after these doings the character of vaccination was much injured from another cause.

A vessel arrived from London at Marblehead. A common sailor on board was supposed to have cow-pox: matter was accordingly taken from him, and was used extensively. It was soon discovered that small-pox matter had been employed, and *that* disease spread rapidly through the neighbourhood.

The occurrences at Marblehead led Dr. Waterhouse to believe that the vaccine virus had degenerated; he, therefore, sent a very urgent request to Dr. Lettsom, begging him to apply to Dr. Jenner for a fresh supply. He stated that he had gained some credit by following Dr. Jenner's footsteps; but that having lost his track he turned to him again for directions. He then adds, "a letter from him, should he allow me to publish it or any part of it, might set this benevolent business a-going again next spring. Could I likewise say to the American public that I had received matter from Dr. Jenner himself, it would have a very good effect indeed."

Dr. Jenner complied with both these requests, and the matter which he sent out arrived early in the spring of 1801. The letter which accompanied it contained a long and satisfactory explanation of the deviations from the regular course of the disease, together with some rules for the successful conducting of the practice. The letter will be found at length at page 110 of Dr. Waterhouse's work on Cow-pox, published in 1802. He forwarded some of this matter to the President Jefferson, in whose hands it completely succeeded. This distinguished individual did not think it beneath him to set an example to his fellow-citizens. In the course of July and August he, with his sons-in-law, vaccinated in their own families and in those of their neighbours, nearly two hundred persons.

Dr. Jenner, while diffusing the Variolæ Vaccinæ to other parts of the world, did not forget our colony of Newfoundland. He had sent matter, through his nephew George, to his friend Clinch at Trinity. This gentleman used it successfully himself and carried it to St. John's, where it was extensively employed by Dr. Macurdy. He, in a letter to Admiral Pole dated December 19th, 1800, mentioned that the practice, notwithstanding some untoward circumstances which had occurred among those who were first vaccinated at Portugal Cove, was followed up with the greatest success. He sent matter to Ferryland, Placentia, and Halifax.

When the Inquiry was published the intercourse between this country and France was almost entirely suspended in consequence of the war. The blessings of the discovery were therefore not so soon experienced at Paris as they probably would have been had better times prevailed.

Vaccination had been pretty extensively practised in remoter cities, and had actually been wafted to the shores of America before it had reached the French capital. I believe the first notice of the subject that appeared in any French writer was contained in the learned and elaborate work of MM. Valentin and Desoteux on the History and Practice of Variolous Inoculation. The discovery of Jenner is mentioned in a note at page 301, where an abstract of some of his opinions is delivered, and the reader is referred to Nos. 69, 70, 71 and 72 of the Bibliothèque Britannique for farther extracts from the Inquiry.

Although the work of MM. Valentin and Desoteux occupied in a particular manner the attention of the "*Ecole de Medicine de Paris,*" it does not appear from their report which was presented to the Minister of the Interior, and which was signed by their president Thouret, that the note which alluded to cow-pox inoculation and the great benefits which were promised from that practice, had excited any degree of curiosity. But this indifference did not continue long; for in the following year (1800) a joint committee was named by the National Institute and the Ecole de Medicine to obtain information on this most interesting subject. About this time Dr. Colladon of Geneva returned from England to the Continent, and carried with him virus to Paris. With this virus trials were made at the Salpêtriere under the superintendence of M. Pinel. But they unluckily did not prove successful.

The committee, however, appointed by the Institute and the School of Medicine had happily anticipated the effects of this disappointment, and provided against them. They had previously dispatched Dr. Aubert to England, with a series of questions drawn up under their authority, in order to elicit precise and accurate intelligence. Before his return the enthusiasm which warmed every philanthropic bosom in this country found its way to the French capital, and roused a spirit of energy and benevolence which neither the horrors of war nor the agitations connected with a half-extinguished revolution could quench.

M. Larochefoucault Liancourt, who had himself witnessed in England the happy effects of vaccination, laboured with zeal and perseverance to carry them to his countrymen. He commenced a subscription for establishing an institution for vaccine inoculation in Paris. And among the first in the list of names for this honourable purpose were found those of the Minister of the Interior Lucien Bonaparte, and of M. Froshot the Prefect of the Department of the Seine.

The subscription was successful; and the house of Dr. Colon, which he had generously offered for the purpose of a vaccinating station, was occupied on the 5th of April, 1800. The committee appointed to superintend this establishment forwarded an official communication to London, to the Vaccine Institution, for virus with directions for its use; all that had been formerly sent from Geneva and England having failed.

During the time that Dr. Aubert was in London he had frequent intercourse with Dr. Jenner, but he sought his information respecting cow-pox chiefly in the Small-pox Hospital. There he imbibed the prejudices which those who carried on vaccination in that place had adopted; and on his return to Paris he published a report which seemed to sanction the opinion that the Variolæ Vaccinæ were really an eruptive disease. After a short time a true knowledge of the affection was attained in France, and a few successful vaccinations dissipated all doubt and excited a spirit which soon spread a knowledge of the practice to the remotest parts of the kingdom.

The Prefect of the Department of the Seine founded a central hospital for the practice of vaccination. At Rheims, Rouen, Amiens, Brussels, Marseilles, Bourdeaux, &c. &c. associations were formed under the auspices of the Minister of the Interior to promote the same object.

In the month of January 1800, the Count De la Roque, who then resided in London, translated Dr.

Jenner's "*Inquiry*" into French. This translation was forwarded to Paris, where it was published. It was received with so much avidity that three editions were sold in less than seven months. These facts were made known to Dr. Jenner by the Count in a letter dated August 5th, who at the same time mentioned that he had likewise translated the second memoir entitled "Further Observations on the Cow-pock, &c." and was about to send it to France.

From Paris the practice spread into Spain during the latter part of this year (1800). Don Francesco Piguilem, a physician of Puigerde, obtained some virus which was employed with perfect success in the month of December. An announcement of this fact appeared in the Gazetta Real of Madrid of the 6th of January, 1801. The Spanish Government from the first evinced a degree of energy in promoting the practice of vaccination which did not usually mark its other proceedings. Senor de Condado, who represented his Catholic Majesty at the Cisalpine Republic, had previously sent an account of vaccination to his court; and this court, degraded and corrupted though it was, had at a subsequent period the wisdom to plan and the virtue to execute one of the most benevolent designs that ever shed a lustre around the proceedings of any state or government.

On the 3rd of May, 1801, an announcement appeared in the Madrid Gazette of a translation of Dr. Jenner's "Inquiry," and by a letter from Don Lope de Mazarredo to Dr. Jenner dated Bilboa

October 3rd, 1801, I perceive that the new practice had become general throughout the whole country. The feelings of respect for the discoverer were strongly manifested in Madrid at this time, he having been elected an honorary member of the Royal Economical Society. The diploma announcing this distinction was not received by him till some years afterwards.

The rapidly increasing interest attached to the subject of vaccination brought with it a daily addition to Dr. Jenner's labours. The service to which he had devoted himself was that of mankind, and it was chiefly to him, in this stage of his proceedings, that foreigners as well as his countrymen looked for information and guidance. He was not, however, without zealous friends. Among these no one proved himself a more judicious and enlightened champion of vaccination than Mr. Dunning of Plymouth-dock. He at this time published his "Observations on the Inoculated Cow-pox." While preparing this treatise he had occasion to write to Dr. Jenner. The correspondence which was thus begun continued with little intermission; and I am certain that few of those with whom he held intercourse enjoyed a greater share of his esteem and attachment than did Mr. Dunning.

During the following year Mr. Dunning, at the instance of the Medical Society of Plymouth, applied to Dr. Jenner to sit for his portrait to Northcote. With this flattering request he was induced to comply; and it now adorns their hall. A mezzotinto engraving was subsequently made from this painting: it wants the peculiar expression of Jenner's countenance, and does not faithfully display his manner; but, on the whole, it is a better portrait than some which have appeared since.

Early in July this year a mission of a peculiarly interesting nature took its departure from England, in order to carry the Variolæ Vaccinæ to Gibraltar and Malta, and from thence to all the nations bordering on the Mediterranean. Dr. Marshall, who had so much distinguished himself by his successful vaccinations at Stonehouse, and Dr. John Walker undertook this office. They sailed from Portsmouth on the 1st of July under the sanction of the British Government, and with special letters of recommendation from his Royal Highness the Duke of York to our military governors on foreign stations. On their arrival at Gibraltar they vaccinated eleven seamen on board his majesty's ship Endymion; and all the soldiers of the garrison who had not had the small-pox. A certificate of the latter occurrence was transmitted by the surgeon-major to the commander-in-chief, the Duke of York.

The following letter from Dr. Marshall gives so interesting an account of their progress, and is at the same time connected with so many public events of a highly momentous nature, that I cannot forbear presenting it to my readers.

DR. MARSHALL TO DR. JENNER.

MY DEAR SIR,

Since my last letter to you from hence the progress of the cow-pox inoculation has been rapid, and is now generally adopted, I may say without exception, in this Island; the governor has also patronized an institution for the cow-pox or Jennerian inoculation, the rules of which I shall transcribe and send you with this. At Gibraltar, where we made our first essay after leaving England, and where we were received with the greatest attention by the governor General O'Hara, we were gratified with observing the cow-pox proceed in the usual mild, and easy progress to its termination as in England; nor did we perceive that the unusual heat of the climate (in the month of August) in the smallest degree aggravated the symptoms, though the soldiers of the garrison continued their fatiguing duties as customary previous to their inoculation, nor was any alteration made either in their diet or allowance of wine. The children of the inhabitants also experienced its mild and gentle progress, nor in any one instance were its symptoms in the least aggravated.

The morning after my arrival, Lord Keith issued the following general memorandum to the fleet.

"*H. M. Ship Foudroyant,*
Gibraltar Bay, Oct. 19th, 1800.

"GENERAL MEMORANDUM.

"Any soldiers, seamen, or marines in the Fleet, who may not have had the small-pox and wish to avoid that dreadful malady, may by application to Dr. Marshall on board the flag ship, be inoculated with the cow-pox, which, without pain or illness, or requiring particular diet or state of body, or leaving any marks, effectually excludes all possibility of the patient's ever being affected with the small-pox.

"By command of the Vice-Admiral,

(Signed,)

PHILIP BEAVER.

To the respective Captains of the Fleet."

Immediately after the issuing of this order its effects were almost rendered nugatory by the dispersion of the fleet to several different rendezvous: of course the practice was confined to a few ships in Gibraltar and Teteran Bays; however, upon arriving at Minorca, it was introduced into several other ships of the fleet, and I found the inhabitants eager to avail themselves of it.

The morning after I arrived at Mahon I inoculated several children, and so anxious was I to give a proof of its efficacy that, on the fourth day after the insertion of the matter of the cow-pox, I inoculated the patient with the variolous matter (taking him into the room, and to the bedside of a patient in the small-pox at its height) in the presence of the physicians, surgeons, and principal inhabitants of Mahon. This trial so publicly made, and from which the little vaccinian came off triumphant, firmly established its character in Minorca; and as the small-pox at the time was proceeding with rapidity, patients daily falling victims to its horrid ravages, every one became anxious to participate in this most happy discovery, calling down blessings upon the head of its promulger to the world.

At Malta I again joined my friend Dr. Walker, and as the small-pox had made its appearance on board the Alexander and other ships lying in the harbour when the fleet arrived, several of the seamen of which ships had already died in the disease, the admiral became alarmed in the probability that so dreadful a malady would spread through the whole fleet which, in its then crowded state, would be attended with great hazard to the lives of a number of brave seamen, and ultimately the exertions and services of the fleet in the expedition upon which it is now employed: he therefore immediately issued the following order:—

[This order is exactly similar to the foregoing.]

The army under the command of Sir Ralph Abercromby being in general landed, he became anxious for their safety upon their returning on board again, as, should the small-pox contagion break out in the army, it was probable a great number of brave men might be snatched by its dreadful effects from the service of their country. He therefore gave orders for the general inoculation of the army with the cow-pox; this in some measure was carried into effect, but as their stay here was too limited to permit the inoculation of the whole, either of the army or navy, Dr. Walker proceeded up with the expedition.

Since Dr. Walker has left this I have been fully employed inoculating the inhabitants of this island, who through the laudable exertions of the governor Captain Ball have universally adopted the practice; several of our cow-pox patients have been subjected to the test of the small-pox but without effect; and as the small-pox is now in La Valette, the anxiety of the inhabitants to participate in the benefit of the Jennerian inoculation is great indeed.

The last part of your treatise has been translated into Italian, and is now printing here. The two former parts will also be translated in a few days, as also a short address written by me at the request of the Governor and distributed through the island. These I shall do myself the pleasure of sending to you by the first conveyance. The establishment of an institution here for the inoculation, and the name given to it of the Institution for the Jennerian Inoculation, will in a small degree serve to show you the high respect and gratitude they feel to you for the benefit they are enjoying from your discovery. The world will speedily follow the example of this little island, and do justice to the man who has conferred the greatest possible benefit upon society.

From hence, as soon as I have inoculated the troops now here under the orders of General Pigot and with which I am now employed, I intend to proceed to Palermo and Naples, at which places the introduction of the Jennerian inoculation is anxiously expected; from thence I intend doing myself the pleasure of again writing to you, and in the interim I beg to subscribe myself,

My dear sir,

Your obliged friend,

J. H. Marshall.

La Valette, January 7th, 1801.

Dr. Marshall continued at Malta from December till the ensuing March, during which time (to use the words of Sir Alexander Ball in his certificate to Lord Hawkesbury) he rendered the most essential service to the inhabitants by the introduction of the vaccine inoculation, by which the ravages of the small-pox, so dreadful in this climate, were prevented. Sir Alexander added " I further certify that he has performed this service, without receiving any pecuniary reward from me, as I conceive that the British Government know best how to appreciate and remunerate his services."

Before they left the island Drs. Marshall and Walker, with Drs. Caraccini and Cassar, had the satisfaction of laying the foundation for a vaccine establishment, under the patronage of his excellency the governor.

From Malta Dr. Marshall went to Naples and Palermo, where he resided several months. When he was about to return to England the King of the two Sicilies directed General Acton to deliver him an introductory letter to the Prince Castelcicala, then Ambassador at the British court, expressive of his majesty's satisfaction with his successful labours and conduct; and desiring that those sentiments might be communicated to the British Government; and that the prince himself should do all the best offices in his power to Dr. Marshall.

Copie d'une Lettre du Prince Castelcicala, à Milord Hawkesbury, du 25 Fevrier, 1802.

J'ai l'honneur de remettre à votre Excellence copie d'une depêche que Monsieur le Chevalier Acton m'a écrite par ordre de Sa Majesté Sicilienne.

Les deux Siciles ont une grande obligation au Docteur Marshall pour y avoir introduit et propagé avec le plus grand succès l'inoculation de la vaccine. Votre Excellence sera certainement très aise que l'on doit cette obligation à une sujet Britannique; j'ose recommander à sa puissant protection une personne aussi digne, et à qui le Roi mon maître a témoigné d'une manière non équivoque toute sa satisfaction.

J'ai l'honneur d'être, &c.

(Signé) CASTELCICALA.

Dr. John Walker proceeded with the expedition under Sir Ralph Abercrombie to Egypt; he vaccinated all the seamen and soldiers of the expedition, and received both from the admiral, and from the commander-in-chief, very strong testimonies of his zeal and success in the new practice. When we consider how many great results hung upon the issue of this expedition it is not too much to assert that a practice, which so effectually protected our gallant troops from that dire enemy the small-pox, must have materially contributed to the success of this campaign at once so glorious and so important to England.

Dr. Marshall returned from his interesting expedition in January 1802. Some extracts from an account of his proceedings, transmitted from Paris on his way home, will be read with satisfaction.

DR. MARSHALL TO DR. JENNER.

Paris, January 26th, 1802.

MY DEAR SIR,

Having finished my vaccine tour I am at length arrived at Paris on my way home. You have doubtless received my letters from Gibraltar, Minorca, Malta, Sicily, and

Naples, in which I informed you of the progress made in the extension of the vaccine inoculation.

I will now just give you, knowing it must be pleasing to its discoverer, an account or rather sketch (for I shall not trouble you with detail) of what I have been doing since my departure from England in his Majesty's ship the Endymion, in July 1800.

The first trial was made upon a black, a seaman on board the vessel, with matter sent me by my friend Mr. Ring. It perfectly succeeded, and from his arm several more of the crew were inoculated; amongst the rest a marine, who on the eighth day of the disease got drunk with spirits, and fell asleep upon the muzzle of one of the guns on the outside of the vessel; in this state, as is supposed, a sudden heel of the ship threw him into the sea, where he must have perished had it not been for the very active exertions of Mr. Valentine the first lieutenant, who, to his honour, regardless of every thing but the preservation of the life of a fellow-creature, immediately leaped overboard from the mizen chains, not even taking time to strip off his coat, and had the happiness, after a considerable struggle, to bring him on board apparently lifeless, in which state he continued for some time.

So singular a circumstance occurring at this particular period of the disease naturally excited my attention, and I attentively watched the progress of the vaccine pustule, not knowing what would be the result of such an accident in this stage of the complaint, and I found that for about twenty-four hours, during which time he experienced a slight degree of fever occasioned by the quantity of liquor he had drunk and the means used to recover him, the cow-pox appeared to be stationary; but afterwards went through its regular course as in the others.

The cow-pox was introduced at Palermo, in the

island of Sicily, where the ravages of the small-pox had always been experienced with unusual violence, and in which city eight thousand persons had perished the preceding year from that destructive malady alone.

Here it was also adopted with enthusiastic ardour, and from the very gracious reception with which His Majesty was pleased to receive me, added to the very laudable exertions used in its favour by the Government, its practice soon became general; though not before it had undergone every possible test of its preventive powers in resisting the infection of the small-pox both by inoculation with variolous matter, and by exposure of the cow-pox patients to patients in the confluent small-pox.

It was not unusual to see in the mornings of the public inoculation at the Hospital a procession of men, women, and children, conducted through the streets by a priest carrying a cross, come to be inoculated. By these popular means it met not with opposition, and the common people expressed themselves certain that it was a blessing sent from Heaven, though discovered by one heretic and practised by another.

At Naples I found the inclinations of the inhabitants, from the accounts they had received from Palermo, favourable to its practice.

An Hospital for the inoculation of the Jennerian disease was immediately established, and every endeavour used to extend its benefits through the kingdom.

* * * * *

Towards the end of this year (1800) Sir Gilbert Blane, who was one of the Commissioners of the Sick and Hurt, became desirous of introducing the practice into the navy. Earl Spencer, the First Lord of the Admiralty, acquiesced in this design; and an order was issued by his lordship, and instructions were given to the navy-surgeons accordingly. This measure, and the corresponding one previously enforced in the army by the orders of the Commander-in-chief, were certainly the most important that had hitherto been adopted in this country for the propagation of cow-pox by public authority. They were important as they regarded the encouragement thus given to the practice of vaccination: but they were still more so as they regarded the efficiency of our naval and military force, the safety of the community, and the public expenditure; the breaking out of small-pox in our fleets and armies having on too many occasions crippled their exertions, defeated the measures of Government, and destroyed many valuable men.

Dr. Trotter, who was then physician to the fleet, seconded the recommendation of the Admiralty with all that zeal and energy by which he was so much distinguished. He had, indeed, earnestly recommended the measure when vaccination first came before the public; and for this most praiseworthy and patriotic exertion he drew down upon himself no small measure of censure from some of his ignorant and inconsiderate contemporaries. He observes, in a letter dated Dec. 9th, 1800, "that the Jennerian inoculation will be deservedly recorded as one of the greatest blessings to the navy of Great

Britain that was ever extended to it." This venerable physician had, at a future period, an opportunity of evincing the sincerity of these sentiments by presenting to Dr. Jenner, in conjunction with the other medical officers of the navy, a splendid token of their admiration for the discoverer of the vaccine inoculation, and a proof of their confidence in its prophylactic power. These gentlemen voted a gold medal, which was presented to Dr. Jenner in February 1801. The medal represents Apollo as the god of physic introducing a young seaman, recovered from the vaccine inoculation, to Britannia; who in return extends a civic crown, on which is written JENNER. Above, "Alba nautis Stella refulsit." Below, "1081."

On the *reverse* an anchor; over it "Georgio Tertio rege;" and under it, "Spencer duce."

This medal was forwarded to Dr. Jenner by Dr Trotter with a letter written with great warmth and eloquence, and not less honourable to the feelings of the writer than just and gratifying to him to whom it was addressed.

DR. TROTTER TO DR. JENNER.

SIR,

You are, perhaps, no stranger to the information of the new inoculation being directed throughout the navy by Admiralty authority. The inquiries which had been instituted in the Channel for the last seven years had called the attention of the surgeons to guard against the introduction of the small-pox among seamen, which in more than a hundred instances during that time had been imported by ships; twenty of these have occurred within the last six months in this fleet only. Amidst subjects so ill prepared for its reception more than the common proportion of deaths has been the consequence. Such was the tenour of our researches, when Dr. Jenner announced to the world the vaccine inoculation as *a preservative* against variolous infection. " *Tandem veneris augur Apollo.*"

As far as the new practice has extended among us it has been followed with the usual success, and so mild that the subjects of it have not been considered in the number of sick on the list.

But the value of conducting the vaccine inoculation with spirit and perseverance throughout the navy may be best estimated by calculating the seamen at 10,000, who are unconscious of having had the small-pox. In this proportion I am justified by the experience of musters in infected ships. How dignified the councils of any nation that by timely precaution shall ward off so much probable misery!

The medical officers have not been passive spectators of an event so singular in the history of animated nature; an event which the philosopher will contemplate with wonder, and the friend of his species view with exultation.

Although secluded by their office from the earliest communication with the progress of medical science, what relates to the vaccine disease has been earnestly sought after; and the whole of your opinions and practice have excited uncommon attention amongst us. I am therefore requested to present you, in the name of those gentlemen, with a gold medal and suitable devices; at once expressive *of their sentiments in favour of the new inoculation, and to commemorate its introduction into this department of public service.* With the more pleasure I comply with the wishes of my worthy associates, as I am confident that no token of respect bestowed on a benefactor of the human race was ever conferred from more honourable or disinterested motives. It will not be the less acceptable to Dr. Jenner that it comes from a body of officers connected by the exercise of their profession with the most brilliant period of our naval annals.

As far as their authority has influence they thus offer their warmest support to the cause. The progress of truth is sometimes slow, but always certain. It is not in the nature of medical investigation long to resist the evidence of facts; and it is far less the province of medicine to check the current of charitable feelings, or to circumscribe the duties of benevolence. We must therefore hope that, while the liberal discussion it has undergone shall secure the suffrages of the enlightened mind, the love of offspring will confirm its favourable reception throughout domestic life.

Accept, Sir, in the name of my naval friends, my hearty congratulations on the honours that await your professional exertions. May the present age have the justice and public spirit to remunerate what posterity will be glad to appreciate. May the medical faculty have virtue and candour sufficient to acknowledge the value of your labours. May your example be a model to the rising members of that profession which you adorn; and may you be blessed with length of days to see your discoveries the means of abridging and preventing disease, pain, and deformity throughout the habitable globe.

Inclosed I have the honour to transmit a list of the medical gentlemen and their stations in his Majesty's naval service. I beg, with all personal esteem and regard, to subscribe myself,

Sir, Yours, &c.

T. TROTTER

Plymouth Dock,
February 20th, 1801.

To Dr. Trotter, from Dr. Jenner.

Sir,

I beg you to accept my sincere acknowledgements for the distinguished honour conferred upon me by your presenting me with a gold medal in the name of the medical officers of his Majesty's navy.

Since there is no situation occupied by medical men where the value of the discovery of vaccine inoculation could be more justly appreciated, from no quarter could such a mark of attention for the endeavours I have exerted in pointing out the means of annihilating the small-pox have been received by me with greater pride, or warmer emotions of gratitude.

If any thing could enhance the estimation in which I shall ever hold such a mark of distinction, conferred by such respectable characters, it is that it is presented to me by a man who has cultivated the most useful science with so much success: and introduced so many valuable improvements into the navy of Great Britain.

* * * * *

Whilst efforts were making to transmit the Variolæ Vaccinæ to the eastern parts of Europe, as well as into Asia and Africa through the medium of the Mediterranean, exertions of a similar nature were directed by Dr. Jenner to diffuse the antidote to other parts of the globe. He especially wished to impart it to our distant possessions in the East.

He sent out his different publications, and large supplies of virus by the Queen East Indiaman, but she never reached her destination, having been lost at sea. After this disastrous event he continued to renew his attempts by almost every ship that left our shores, but they all failed. In the mean time most urgent demands were made for vaccine virus, in consequence of the devastations that small-pox was committing, particularly in the island of Ceylon.

He was twice sent for by Lord Hobart, to the Secretary of State's office, to deliberate on this emergency. He represented in the strongest terms the necessity of employing means more effectual than sending out dried matter, and pointed them out in the most satisfactory manner. He proposed that on board some ship going to India twenty recruits, or men of any description who had not had the small-pox, should be selected; and that he should be allowed to appoint a surgeon to attend them perfectly conversant with the vaccine inoculation. Thus he engaged that the disease should be carried in its most perfect state to any of our settlements. After some deliberation these reasonable and practicable proposals were rejected. Dr. Jenner resolved himself to endeavour to effect what the Government was unwilling to attempt. A vessel properly equipped was all that was necessary to insure the success of his benevolent wishes. He knew that there were persons in this great country who would not be backward in co-operating with him for such a purpose. His design was to raise as speedily as possible, by subscription, sufficient to defray the expenses of the voyage. He wrote to Dr. Lettsom on the subject; and to prove how zealously he was affected in this matter, he desired that his own name might be put down for one thousand guineas.

Before this design could be carried into execution tidings arrived from the East which happily rendered it unnecessary, Dr. De Carro having succeeded in forwarding vaccine matter from Vienna to Constantinople, and from thence to Bombay, in a manner which will be hereafter described.

Two letters, one from Dr. Underwood of Madras, and the other from Dr. Scott of Bombay, refer to these transactions.

Dr. Underwood to Dr. Jenner.

Madras, February 28th, 1801.

Dear Sir,

Permit me to thank you most sincerely for having kindly sent per Queen your publications on the cow-pox, together with matter. I have been made acquainted with your friendly attention by my friends in London, who mentioned that you intended to send some matter to me overland, in the event of the loss of that sent per Queen. This unfortunately proves the case, and I lament it most sincerely, and anxiously look for recovering some from you overland. I have read with very great pleasure your remarks on that disease, and feel most particularly anxious to introduce and extend it in the country, under the greatest confidence that it would save many lives. It has been with extreme pleasure that I have hitherto embraced every opportunity of inoculating with the variolous matter, but the loss of a beautiful little patient has humbled me, and I confess I never now take up a lancet for that purpose but with fear and trembling.

We generally inoculate here in the month of January, and I trust before that time arrives to be in possession of vaccine matter.

Mr. White of the Royal Establishment, who arrived here the other day, favoured me with a perusal of your book, and a little vaccine which he had taken a few days before he left London: with this I inoculated my son and nephew; but unfortunately both failed, in my opinion from the matter being kept too long. The moment I had made use of the matter I made Dr. Anderson our Physician General acquainted with it, and forwarded him the books. He is much pleased with your exertions, and I am sure will do every thing in his power to promote its success. Lamenting extremely this failure I am in careful pursuit of the disease in this part. The other day a cow was brought to me said to have the small-pox, and the natives assure me it is common amongst them, but that this is not the season of the year. On the teat of this cow there appeared one large pustule of a white colour, and several others dry. I took the matter from this, and inoculated four cows on the teats. One of the four had several lumps like glandular obstructions: on the seventh day and on the ninth one pustule, but which was very unfortunately broke before I was sent for. I shall continue the search, and shall have great pleasure in acquainting you of my success. If I am so fortunate however, I would not rely on it, but entreat you will have the goodness to forward me, through Mr. James Curtis, No. 1, Ludgate Hill, vaccine by every opportunity until we are so fortunate as to succeed. Dr. Anderson this day sent me a horse of his which had greasy heels. I inoculated two cows from it, and Mr. White has taken some of the grease round to Calcutta to make the same experiment. He is to sail for Calcutta in a day or two. Should it please the Almighty to bless this undertaking, and enable us to inoculate for the cow-pock, the natives will very easily be persuaded to make use of it very generally: but as for the small-pox they are decidedly against it with very few exceptions, and it is with the utmost difficulty they will allow matter to be taken from them. May I take the liberty of requesting the matter to be sent out in different ways—betwixt two pieces of polished glass, carefully covered at the edges with wet skin or cotton carefully sealed down in a phial, and probably the glasses that are used for phosphoric matches would do well:—however you will be much the best judge how to forward it. The Court of Directors would most assuredly forward matter by their overland dispatches if proper application was made to them. Sincerely thanking you for your very polite attention, and most ardently wishing you every success in a disease you have introduced with so much honour to yourself and real benefit to mankind,

I have the honour to be, with sincere regard,

Your most obedient servant,

John Underwood,
Surgeon to His Majesty's Naval
Hospital, Madras.

Dr. Helenus Scott, Bombay, to Dr. Jenner.

Sir,

I received the letter with the vaccine matter which you were so good as to send me by Colonel Oakes. Before his arrival we were fortunate enough to produce the disease here, and from this place it has spread to almost every part of Hindostan. We have already inoculated between

2 and 3000 children in Bombay. It gives me great pleasure to observe that the natives begin to acquire confidence in this practice, as well as knowledge of the true appearances of the disease. I have no doubt but that every day will increase its reputation among them, and spread to a wider extent the benefit of your happy discovery. Among so many others I have the strongest motives for being thankful to you, for I have three children, born in three successive years, who were the first in India to enjoy the protection of the cow-pox. I set an example in my own family which has been followed by every European family here, and has or will be followed by every other in India.

I beg to enclose you a number of our newspapers on this subject. From their dates you will be able to trace the progress of the vaccine disease from Constantinople to its present state in India. We are, I suppose, in the first instance indebted to Dr. De Carro of Vienna for sending it to Constantinople. You will see by the accompanying papers that our intentions have always been right, our opinions sometimes wrong. Upon the whole, the practice is established; and much praise is due to the Government of this country for their liberal views, as well as to the zeal of the faculty in India.

If I can do any thing for you in this country I beg of you to employ me.

It gives me pleasure to reflect that my situation, as a member of the Medical Board of this Presidency, has enabled me to give some assistance in spreading the reputation of your discovery. Who must not envy your reflections on this subject!

I beg of you to believe me, with much regard and esteem,

Your very obedient servant,

H. SCOTT.

By the unwearied labours of Dr. De Carro vaccination had already penetrated a great part of Germany and Poland. From him, likewise, it emanated to Venice, Lombardy, and other parts of Italy. He was now engaged in an enterprise embracing a wider field. During the summer of the year 1800, an English gentleman and lady, Mr. and Mrs. Nesbit, were passing through Germany on their way to the capital of the Ottoman empire. They were going to visit their daughter Lady Elgin, whose husband Lord Elgin was British Ambassador to the Porte. They remained a short time at Vienna, and became acquainted with De Carro then in the height of his vaccine inquiries. He spoke with enthusiasm upon his favourite topic Vaccination. They had left England imperfectly acquainted with its importance, but his conversation excited so lively an interest that, shortly after their arrival in Constantinople, Lord Elgin sent a request to Dr. De Carro for some vaccine virus. The first subject upon whom it was tried was the infant son of that nobleman. The inoculation was performed on the 6th of September by Dr. White, who afterwards fell a victim to his imprudent inoculation of himself with the plague.

His lordship communicated the result of the trial to Dr. De Carro in the following letter :—

Constantinople, Dec. 23d, 1800.

SIR,

Though much hurried at this moment I am happy in being able to acknowledge my obligation to you for the vaccine matter. It took effect on the third trial, and indeed fully answered the favourable account you had given of it. My child was so little ill that I should be at a loss to say how he was affected; nor could I determine that the infection had taken place, had I not been prepared, by your letter, for the almost imperceptible appearances which were observed. I mean this, in regard to the constitutional complaint: for the affections on the incisions were perfectly clear.

My former failures in the operation having induced Dr. Whyte (who officiated) to touch the skin with the lancet in three places, in each of which there was a pustule. The matter from his arm has been applied to several children here with equal success.

The matter you last sent, conveyed in a quill and a glass phial, has been given to the captain of an American frigate now here. It has taken most favourably in one of the instances to which he applied it; and he proposes inoculating further from the first patient.

I shall be happy to write you further on the subject by an early opportunity. Meanwhile I was anxious to say, in this hasty manner, how much I feel indebted for so great a benefit as I have thus received from you; and am, by your means, the instrument of introducing into this country.

I have the honour, &c.

ELGIN.

To Dr. De Carro,
Vienna.

Doctors Hesse, Pezzone, and Auban, then practising at Constantinople, adopted vaccination with ardour. It was for a while confined to a few European families; by degrees, however, it vanquished Mahommedan prejudices, and was, at last, admitted within the walls of the Seraglio, having found advocates in Dr. Roini, and Lorenzo Noccioti — the physician and surgeon to the Grand Seignior—Roini presented him with an extract from Dr. De Carro's observations and experiments, translated into the Turkish language. That prince, who had suffered severely from the small-pox, expressed his regret that a discovery, which would have saved him so much pain, had not been made during his early days, and he added a desire that it should be adopted throughout his dominions. Roini vaccinated a child belonging to a servant of the Seraglio, and after a time the Hekim Bacchi, or chief physician, expressed a wish that three of his own children should receive it. Its progress was, nevertheless, not very rapid. When Dr. Hesse left Constantinople in 1802 the virus was lost, but it was speedily renewed by Dr. De Carro; and in 1803 it was calculated that in the different quarters and suburbs of Constantinople between five and six thousand had been vaccinated.

Lord Elgin and his lady continued most sedulously to promote the practice. During a tour that they made in 1802 through the islands of the Archipelago they, with the aid of Dr. Scott a physician in their suite, carried it to Athens, Argos,

Corinth, and other cities of Greece. The English consul at Salonichi Mr. Charneu, and Dr. La Font, a French physician long resident there, interested themselves heartily in the cause; and the latter even entertained a hope that it might prove an antidote to the plague. Salonichi was the birth-place of the aged inoculator Thessala, who had, about a century before, attracted so much attention at Constantinople by her inoculations. The inhabitants of Salonichi, Turks, Greeks, Armenians, and Franks, received the new practice with so much eagerness that 1130 were vaccinated in the course of eight months.

Larissa in Thessaly, and Macedonia were equally anxious to partake of the blessing. Dr. Moreschi who had published two works on vaccination, and had introduced it into the Venetian states and a great part of Italy, next directed his efforts to the western side of Greece, and by an opposite geographical track carried the practice to the same point. In a spirit of honest exultation, and with feelings enlivened by classical associations, he thus triumphantly recounts his success. "My glasses and threads, impregnated with vaccine, after passing through most of the provinces situated on the Adriatic, having been sent to Drs. Marochia and Mirowish physicians of Prau, Spalatro and Salona, places renowned in the history of Rome, have gone almost to Athens; and have reached Theachi, the famous Ithaca of Ulysses, and Patras in Peloponnesus. During the last winter I gave vaccine to several persons of distinction and merchants of this city, (Venice) for their friends and correspondents of Cephalonia, Cerigo, and Zante. Persons worthy of credit have assured me that vaccination is practised with success in the islands of the Ionian Sea; and others, who have returned from Corfu, that the vaccine of that island has passed to Butronto in Macedonia,

——" celsam Buthroti ascendimus arcem."

No sooner had the account of Dr. Jenner's discovery reached India than the greatest anxiety was evinced by every professional man there to become possessed of an agent capable of securing the inhabitants from a scourge more dreadful in its effects, in that climate, than in almost any other. Dr. Jenner himself felt this so strongly that, after sending virus by almost every conveyance without success, he at last proposed the noble scheme for its certain transport already mentioned. In the mean time the governments at the different Presidencies were not inattentive to this great object. The Honourable Jonathan Duncan, governor of Bombay, early felt the importance of securing the benefit for India as speedily as possible. He therefore wrote on the 21st March, 1801, to Lord Elgin at Constantinople requesting his co-operation in forwarding virus to Bombay, by the way of Bagdad and Bussora. He also sent at the same time to Mr. Harford Jones, (now Sir H. Jones Brydges) resident at Bagdad, to the same purport. On the 8th of September, 1801, Lord Elgin forwarded one quill containing vaccine matter to Governor Duncan, and likewise sent supplies to Mr. Manesty, at Bussora, with directions to that gentleman to try to reproduce it there, and then to forward it to Bombay. His Lordship had in the preceding year sent virus to him with which his own son was inoculated, but in neither case did it succeed.

The Medical Board in Bombay likewise evinced their zeal by addressing the Governor on the 4th of August, 1801, and suggested the measures for the conveyance of the virus which were afterwards successful. The Governor sent the letter of the Medical Board to the Court of Directors, earnestly soliciting their assistance.

After receiving many urgent requests for matter Lord Elgin wrote in January 1802 to the Honourable Arthur Paget, to this effect,—" I have so many applications for vaccine virus from Bussora, the East Indies, and Ceylon, that I beg you will immediately apply to Dr. De Carro, and request him to send some by every courier." Mr. Harford Jones, too, requested Dr. De Carro to send him the virus, with necessary instructions, direct from Vienna. Many disappointments having arisen from the difficulty of transmitting it in an active state, he employed every expedient that ingenuity or experience could suggest to obviate them.

Various methods had been tried. Impregnated lint, or threads enclosed between plates, or in bottles and in tubes closed up with wax. The practice of imbuing the points of common steel lancets was soon abandoned. To these succeeded lancets of silver, silver gilt, gold and ivory. After a series of trials he gave the preference to ivory, which he considered to be, in all respects, the most secure vehicle for transporting the virus. On lancets of this material it was sent from Breslau to Moscow, where, under the patronage and actual inspection of the Russian Empress, it completely succeeded.

As that which was destined for Bagdad would be exposed to the accidents of a long journey, in a climate heated by a scorching sun, he took special care to protect it as much as possible from external influence. He sent some on lancets of silver, silver-gilt, and ivory: he also impregnated some English lint with the vaccine fluid, and enclosed it between glasses; and when he had properly secured them he dipped them at a wax-chandler's till they formed a solid ball, which he enclosed in a box filled with shreds of paper. In this state the packet was safely conveyed across the Bosphorus, and passed over the whole line of deserts; and he had the satisfaction of hearing that, on its arrival on the banks of the Tigris, its contents were still liquid, and succeeded on the first trial. It was received on the 31st of March, 1802. Mr. Harford Jones immediately delivered it into the hands of Dr. Short, who employed it on the 5th of April. It was forwarded by Dr. Short to Mr. Milne, surgeon to the British Consul at Bussora. This gentleman before the 17th of the following June had vaccinated forty persons; among whom were the crews of some vessels departing for Bombay. He also sent it to Bushire, in the Persian Gulf; and to Muscat. a port in the Arabian sea. Before the end of June it reached our establishment at Bombay.

In this manner was this precious antidote first transmitted to the continent of Asia. While recording the principal events in this interesting history it ought not to be omitted that the vaccine virus, which served to protect the Eastern part of the world, came not from the English stock, but was derived from the cows of Lombardy. It had been sent from Milan by Dr. Sacco to Vienna; and Dr. De Carro was making his first trials with it at the time he received the application from Mr. Jones at Bagdad. These trials having completely succeeded he was enabled to comply with the request in the manner just described.

The Recovery, the vessel which carried the virus from Bussora to Bombay, left the former place late in May. She was three weeks on her voyage.

On her arrival between twenty and thirty subjects were inoculated, but only one instance of success occurred, and that was in the hands of Dr. Helenus Scott. On the 14th of June he inoculated Anna Dusthall, a healthy child about three years old, the daughter of a female servant belonging to Captain Hardie. The progress of her affection was watched with the utmost anxiety; and the medical gentlemen had the happiness to find that it accorded so completely with the description of Dr. Jenner that they felt quite assured that they had, at last, gained possession of their long-wished object, the genuine Variolæ Vaccinæ. On the eighth day from her inoculation five children were vaccinated, and successfully. A relation of these proceedings was published in the Bombay Courier by Drs. Moir and Scott. This communication is dated on the second of July, 1802. It gives a very interesting account of the momentous event, and makes honourable mention of all the individuals who had assisted in bringing it about. Not the least important part of this report was that both the Hindoos and Parsees at Bombay and Surat evinced the utmost desire to have their children inoculated with the "vaccine disease."

The Medical Board, anxious to diffuse the acquisition they had made throughout India, forwarded virus to Bengal, Madras, Ceylon, &c. It very soon took effect at Hyderabad and Trincomalee, and was spread with great rapidity from place to place.

The Government most cordially seconded every effort of the Medical Board. Dr. Desborah, surgeon to the Residency at Poonah, having failed in producing the infection with virus sent to him at different periods, Government humanely directed that subjects affected with the cow-pox might be sent to secure the introduction of the disease into the capital of the Marhatta empire. A Brahmin was despatched with two children under the infection. He, with one of the children, reached Poonah in six days.

The exertions of Dr. Anderson at Fort St. George were unceasing. He established an extensive correspondence with the medical gentlemen at the different stations, and he circulated the reports which he received from them through the medium of the Madras Gazette. He likewise succeeded in transmitting the virus in an active state to Bengal. On this event Dr. Fleming, the first member of the Medical Board, addressed a letter to the Governor-General in Council. His Excellency the Marquess Wellesley was pleased to direct that that letter, with its enclosure, should be published for general information. This letter announced the important fact that a boy who had been vaccinated by Captain Anderson, commander of the ship Hunter, on the twelfth of November, on its passage from Madras to Calcutta, had arrived at that place on the seventeenth with the disease in an active state. Three children were immediately inoculated, and, on the following day, eight others, in all of whom it completely succeeded.

Dr. Fleming then observed "The settlement being now, as I conceive, in complete possession of the benefit derived to mankind from Dr. Jenner's celebrated discovery I take the liberty of submitting to your Excellency's consideration my opinion on the best mode of preserving the continuance of so great a blessing, and spreading it as rapidly as possible throughout the provinces."

He, therefore, recommended that a surgeon of approved skill and assiduity should be appointed to the charge of preserving a constant supply of recent genuine matter; that it should be a part of his duty to vaccinate the children of the Natives; and to instruct the Hindoo and Mohammedan physicians in the proper mode of performing the operation. In order to induce the Natives to adopt vaccination he likewise proposed that an address should be published in the Persian, Hindoo, Bengalese, and Sanscrit languages, giving a succinct account of the discovery in which the curious and, to the Hindoo, very interesting circumstance that this *wonderful preventive was originally procured from the body of the cow, should be emphatically marked.* Next, an explanation of the essential advantages which vaccination possesses over the small-pox inoculation: and, lastly, an ardent exhortation to the natives to lose no time in availing themselves of this inestimable benefit. This letter bore date November 29th, 1802.

Shortly afterwards the Governor-General in Council was pleased to order "That the high approbation of his Excellency in Council be signified to Dr. James Anderson, physician, &c. &c. on the establishment of Fort St. George.

"To Captain Anderson, commander of the ship Hunter; to John Fleming, esq.; and to Messrs. Russells, Hare, and Shoolbred and the other medical gentlemen employed on this important occasion, for their diligence and ability in promoting at this Presidency the successful introduction of Dr. Jenner's discovery.

"That Mr. William Russel be appointed to superintend the further promotion of the benefits of Dr. Jenner's discovery throughout the provinces subject to the immediate government of this presidency. And that a notification be prepared and published in the Persian, Hindoo, Bengalese, and Sanscrit languages, according to the suggestion of Mr. Fleming."

These different measures proved very successful. The prejudices of multitudes were in favour of the practice; and the superstitious veneration in which the Hindoos regard the cow contributed greatly to this result. Although the Brahmins for the most part were friendly, they were not uniformly so. In order to gain their approbation two different attempts were made to prove that their own records contained descriptions of the benign preventive which they were required to sanction. The facts connected with these well-intended devices will be found mentioned in another part of this work. There is great reason to believe that they have recently been misunderstood, and have given rise to the assertion that the Hindoos were really acquainted with vaccine inoculation in early times, when in fact the manuscripts which seemed to support that opinion were modern fabrications; certainly intended to beguile by their apparent antiquity, but never designed by their authors to be treated as real documents on which claims, adverse to those of Dr. Jenner, could be founded.

In Ceylon the inhabitants at first manifested some reluctance to receive vaccination. This reluctance soon gave way, and by the excellent measures adopted by Governor North, under the direction of the Medical Superintendent General, Thomas Christie, Esq. it was so speedily diffused that he was enabled to state that two thousand had been vaccinated in the district of Columbo alone, during one month. This gentleman's account of the introduction, progress, and success of vaccination at Ceylon, published after his return to England in the year 1811, is an extremely valuable document. It shows how wisely and energetically he carried into effect the arrangements for general vaccination, and proves beyond all doubt that, when it is so used, variolous contagion may be certainly and completely controlled. In this respect Dr. Jenner always set a high value on the facts which were furnished from Ceylon, and he himself urged Dr. Christie to publish the account I have referred to, believing that it might render essential service to his great object of exterminating small-pox.

To Sweden and to Ceylon Dr. Jenner was in the habit of pointing when he wished to prove what his discovery might accomplish; or when he lamented that fatal obstinacy of his fellow-creatures which, with such examples before them, could induce them to reject blessings within their reach. He had an opportunity during the latter years of his life of enjoying much friendly intercourse with Dr. Christie, in Cheltenham; and I am certain that intimate knowledge of that gentleman strengthened the feelings of respect and regard which his previous conduct had given rise to.

Should the reader desire further information on these points I would refer him to Dr. Keir's account of the introduction of the cow-pox into India, published at Bombay in 1803; and to the History of Vaccination in Turkey, Greece, and the East Indies, by Dr. De Carro, published in French at Vienna in the year 1804.

This excellent physician, while labouring with indefatigable zeal to promote the practice of vaccination in every part of the world, kept up a constant correspondence with its distinguished author. The warmth of his attachment to Dr. Jenner, the energy and ability displayed in collecting information from all quarters, and the eloquent manner with which he communicates information, are so conspicuous that I should feel strongly tempted to lay all his letters before the public, did I not fear that I should thereby be prevented from recording as I ought the efforts of other zealous friends of Jenner. But the following extracts so satisfactorily elucidate the preceding history that I think they will be more acceptable than any narrative that could be constructed from them. That these may be the more fully understood I prefix part of a letter from Dr. Jenner to this learned physician.

DR. JENNER TO DR. DE CARRO.

March 28th, 1803.

DEAR SIR,

Since the commencement of our correspondence, great as my satisfaction has been in the perusal of your letters I do not recollect when you have favoured me with one that has afforded me pleasure equal to the last. The regret I have experienced, at finding that every endeavour to send the vaccine virus to India in perfection again and again failed, is scarcely to be described to you; judge, then, what pleasure you convey in assuring me that my wishes are accomplished. I am confident that had not the opponents, in this country, to my ideas of the origin of the disease been so absurdly clamorous (particularly the *par nobile fratrum*) the Asiatics would long since have enjoyed the blessings of vaccination, and many a victim been rescued from an untimely grave. The decisive experiments of Dr. Loy on this subject have silenced the tongues of these gentlemen for ever.

I am happy in seeing this interesting work translated by you, and hope it will travel the world over.

It is very extraordinary, but certainly a fact, that the plate which I gave in my first publication of the equine pustule (although its origin was detailed) was by almost every reader considered as the *vaccine.* There are probably some varieties in the pustules which arise among horses. You will observe, by a reference to my publication, that the virus in the instance I now allude to was so very active that it infected every person who dressed the horse.

I am happy to find an opinion taken up by me, and mentioned in my first publication, has so able a supporter as yourself. I thought it highly probable that the small-pox might be a *malignant variety* of the cow-pox. But this idea was scouted by my countrymen, particularly P. and W. * * * * *

To Dr. De Carro, Vienna.

Vienna, April 22, 1803.

Dear Sir,

Nothing more was wanting to make me enjoy fully the satisfaction of having introduced the vaccination in India than to learn it had long been one of your wishes, and that the accomplishment of it by me has given you so much pleasure. I hope to show you soon the astonishing progress that it is making in that part of the world. I receive daily new and interesting documents from the East, which induce me to delay still for some time the publication of the work that I spoke to you of in my last letter.

If you have felt so much pleasure in hearing that your discovery is known and practised in India, I hope that my late intelligence of the true cow-pox, produced at Milan with the *giardoni* on Dr. Sacco s own horse and that of one of his neighbours, has not been less agreeable to you. The first does honour to your heart, the second to your head ; but indeed it is long since I am quite at a loss to determine which of the two is best in you.

You have lately given a strong proof of your moderation, by the silence you have kept with your antagonists.

———'s conduct (I am less acquainted with that of W.) borders on insanity. I am extremely glad that you have treated it with the contempt it deserves. On the other side I am happy to see that it has not been left unnoticed, and that your friends have taken upon themselves to show his malevolent and ridiculous designs.

I do not know whether you are well informed of the great improvement which MM. Ballhorn and Stromeyer have made to the glasses invented by you. They have taught us a simple and easy manner to preserve the vaccine lymph fluid during an indefinite term. The Hanoverian vaccinators take a small bit of English *charpie*, which you call, I believe, *dry lint*. The quantity must be, of course, equal to the concavity of the glass. The pustule then is punctured by a circular or half-circular incision with the lancet, so as to open a greater number of the cells forming the vaccine pustule covered with the same pellicle. The lint is applied upon the pustule on the most woolly side, so as to act better as a syphon. It pumps in a very short time a sufficient quantity of vaccine fluid to saturate it as completely as if it had been dipped in a glass of water, particularly if the lint is now and then gently pressed with the point or the back of the lancet. When it is quite full you take it with the lancet, and place it carefully in the cavity of the glass; you put a drop of oil, or a little mucilage upon the internal surface of the glasses; you make the flat bit of glass slide upon the *charpie*, so as to exclude the air as much as possible; you tie the two bits with thread, and seal the edges. To prevent the access of light I commonly fold it in a black paper, and when I was desired to send it to Bagdad, I took the precaution of going to a wax-chandler's, and surrounded the sealed-up glasses with so much wax as to make balls. With this careful manner it arrived *still fluid* on the banks of the Tigris.

* * * * *

Dr. De Carro to Dr. Jenner.

Vienna, 21st June, 1803.

My dear Sir,

My friend Dr. Marcet wrote to me lately that the account I have sent to you of Dr. Sacco's experiments have afforded you great satisfaction. The motive which induces me to write to you to-day is another confirmation of your theory which has taken place in a country where you scarcely expect it from, the more so that it is accompanied with veterinary observations which appear to me very nice and curious.

Monsieur La Font, a French physician established at Salonica in Macedonia, has been one of the most active vaccinators I know on the continent; his last letter, of the third of June, mentions that he has since last Autumn vaccinated 1130 persons. He first heard of your discovery on the occasion of Lord Elgin travelling in Greece with Dr. Scott; during which journey his lordship and the doctor took a particular care of propagating vaccination. The English consul at Salonica went to Athens to meet Lord Elgin; there he saw a great number of young Athenians with vaccine pustules. Not a word had yet been heard, at Salonica, of your discovery, and he desired Dr. Scott to give him vaccine matter to put into the hands of Dr. La Font, and Lady Elgin was so kind as to give to the consul a copy of my work, for the instruction of his physician. The first Athenian matter did not succeed, but seeing its failure, Dr. La Font applied directly to me, and my ivory lancets produced their effects at the first trial. Since that time I have been in regular correspondence with that physician, who appears to me to be possessed of much learning, prudence and activity.

Some time afterwards I sent him a translation of Dr. Loy's experiments, and desired him to make as many veterinary observations and experiments as he could. He has some reasons to suppose that the cow-pox reigns in that country, according to the report of several Albanese peasants. As to the *grease* (which he calls javart,) he says that the farriers at Salonica know it very well. Dr. La Font began his experiments with the kind of grease which the Macedonian farriers call the *variolous.* He found a horse which had been attacked with feverish symptoms, that ceased as soon as the eruption appeared. The fore legs were much swelled; the left had four ulcers, one upon the heel, a second some inches higher, a third on the articulation, a fourth near the breast. The eruption on the legs was, he says, very like the small-pox, but none was to be seen on the other parts of the body. He took matter from the upper ulcer which was of twelve davs' standing. The matter was limpid, but a little yellowish and *filamentous,* (thready;) first a cow was submitted to this inoculation, but without success; secondly, a girl twelve years old, without effect; but this girl had been vaccinated some months before without success, and was suspected to have had the small-pox; thirdly, two boys, one six, the other five years old, were inoculated with the same *equine* matter; and in both a pustule appeared, which followed the regular course of a vaccine pustule. The colour was less white, and more purple than usual. Those two children had a pretty strong fever, for which some cooling medicines were administered. Those inoculated with matter from them underwent the disease in its usual mild way.

These particulars, I hope, will silence all those who still doubt of the truth of your doctrine. These observations enhance the merit of your discovery. The *means* of making it were every where; yet nobody before you had the least idea of that singular connexion between the grease, the cow-pox, and the small-pox.

* * * * *

In pursuing the progress of vaccination into the East we have for a time lost sight of Dr. Jenner's personal history, and to that subject we now return.

He went to Cheltenham on the 13th of July (1800.) Mrs. Jenner and the children did not arrive there till the 29th. He remained till the 6th of November, on which day he departed with his family for London. He arrived in Bond-street on the 8th.

During his stay at Cheltenham he had some little relaxation from the incessant efforts which we have seen him compelled to make. Some of the mistakes in the practice of vaccination had, in part, been obviated, and he had the satisfaction of knowing that it was diffusing itself in every direction.

His own personal endeavours were, as heretofore, laborious. He offered gratuitous inoculation to all the poor who thought fit to apply at stated periods. His benevolent invitations were, in the main, very generally accepted, parents bringing their children in great numbers both from the town and the adjoining parishes.

I believe it was at this time that the incident I am about to mention occurred. Notwithstanding his repeated notices of gratuitous aid one parish had hitherto obstinately held back. This year, however, he found the people bringing their children in great numbers. Of course he wished to know by what means they had become converts to the new inoculation. He found that arguments of a very authoritative nature had brought about the change. The small-pox, in the course of the preceding year, had been introduced into the parish, and proved extremely fatal; but it was not this circumstance, nor yet the security of those who had been vaccinated in the adjoining parishes, that brought cow-pox inoculation into favour. The cost of coffins for those who were cut off by small-pox proved burdensome to the parish; the churchwardens, therefore, moved by this argument effectually exerted their authority and compelled the people to avail themselves of Dr. Jenner's kind offer.

His exertions, as usual, in London were incessant in prosecuting his favourite object. They were, however, interrupted by a considerable attack of illness which, for a short time, confined him to his bed. The disease at its commencement had somewhat of a typhoïd character. His former sufferings from a similar disease were never forgotten, and he continued to the end of his days to entertain great apprehensions of any such attacks.

He was soon enabled to resume his active duties. His vaccinations among the higher ranks were very numerous; and many of the nobility were desirous of having from his own mouth information on the great question which then engrossed so much of the public attention. A party of this kind assembled at the house of Earl Spencer on the twentieth of December. Besides his Lordship and the Countess there were present Lord Lucan, Lord Camden, Lord Macartney, Mr. Grenville, &c. &c. Jenner, whether in conversation or in writing, never failed to treat this or any other subject with great eloquence and perspicuity: his knowledge of it was complete, and his feelings being at the same time deeply engaged, his language assumed a degree of animation and precision highly impressive.

While engaged in these public acts (for so they may be called) he seized every opportunity of cultivating his domestic affections, and in promoting or participating in the little amusements and enjoyments of his children. He took them to see the different sights which London afforded. Among others, the illuminations in honour of the Queen's birth-day delighted them exceedingly, and his parental feelings in sympathizing with the pleasures of his children are expressively recorded by the insertion of this trifling incident in his memoranda.—Next day, January the twentieth, he went with his son Edward to place him at school at Mr. Evans's, in Islington.

CHAPTER XI.

PUBLICATION OF THE ACCOUNT OF THE ORIGIN OF VACCINE INOCULATION — INTRODUCTION OF VACCINATION INTO DENMARK, SWEDEN, RUSSIA, &C. &C—DISCOVERY OF THE VARIOLÆ VACCINÆ IN LOMBARDY, &C.

About this time Dr. Davids, of Rotterdam, went to Paris for the express purpose of seeing the practice of vaccination, and of bringing it into his own country. He was not successful in this attempt; the virus which he carried from Paris having failed he subsequently received some from Boulogne, which proved effectual. These facts he communicated to Dr. Jenner in a letter dated on the first of January, 1801. This gentleman also translated and published Dr. Jenner's *Inquiry*.

As Dr. Davids had gone to Paris for instruction, so in like manner another foreign physician came from the Continent in order to derive information from the great author of vaccination himself.

Dr. Reumont of Aix-la-Chapelle is the person to whom I allude. He had frequent intercourse with Dr. Jenner, from whom he received great civilities and every necessary information connected with his object.

Early in this year Dr. Jenner published his account of the origin of the vaccine inoculation. Among other public bodies to whom he sent this little work he did not forget the National Institute of France. He presented a copy to that learned body accompanied by the letter here inserted.

DR. JENNER TO THE PRESIDENT OF THE NATIONAL INSTITUTE OF FRANCE.

SIR,

It is impossible for those who delight in science, on whatever part of the globe fortune may have chanced to place them, to contemplate without admiration the ardour with which her paths are cultivated by those illustrious characters who form the National Institute of France.

Conscious that it is the wish of that society to receive as well as to diffuse knowledge, permit me, Sir, to present you with a few pages containing the History of the Origin of Vaccine Inoculation.

The speedy adoption at Paris of the plan I had the happiness to announce for the annihilation of the small-pox, and the strenuous efforts of the French in making it known throughout their dominions, have filled me with sentiments of the highest respect for their talents in appreciating those discoveries which tend to meliorate the condition of human nature.

I have the honour to be,

Sir,

with high consideration,

Your most obedient humble servant,

EDWARD JENNER.

This civility was acknowledged on the part of the Institute in an official note.

FROM THE NATIONAL INSTITUTE OF SCIENCES AND ARTS.

Paris, the 16 *Thermidor,*
9th Year of the French Republic.

The Office of the National Institute to M. Edward Jenner, M. D. F. R. S. Lond.

The National Institute in exerting itself to propagate the inoculation of the Vaccine has only fulfilled one of the most important functions which the Constitutional Law of France imposes.

To gather new truths from whatever quarter they may come, to spread them when they are useful, this is its first duty. It is sufficient to let you know how far your discovery, which appears to unite those two qualities in the most eminent degree, ought to excite the attention of this society, and the profound esteem it ought to inspire for its author.

The Institute charges us to make you acquainted with its sentiments, thanking you for the dissertation which you have addressed to it.

We beg of you at the same time to persuade yourself with the assurance of our particular consideration.

COULOMB, P^t.
G. CUVIER, S^e.
DE LAMBRE.

I have already mentioned that the true Vaccine virus had been lost in America, and that Dr. Jenner had used every means to send a fresh supply to Dr.

Waterhouse, which that physician acknowledged in a letter worthy of publication here. It presents a good specimen of the nature of the epistolary correspondence already alluded to as having long subsisted between two friends, whose views were so peculiarly directed to one common object, the promotion of the vaccine practice throughout the world.

DR. WATERHOUSE TO DR. JENNER.

Cambridge, April 24th, 1801.

DEAR SIR,

Being just informed of a ship's sailing to-morrow for London I have only time to acknowledge the receipt of your most excellent letter: the answer to it must be postponed a week or two longer. With it I received a supply of vaccine matter, which came to hand thirty-eight days after the date of your letter, for which you have my most cordial thanks. I have inoculated with it, and found it good, and here send you the first crop from it. When my good friend Dr. Lettsom has sent me curious melon seeds I have sent him as soon as possible some seed raised from them, that he might see whether our soil, atmosphere, and mode of culture effected any alteration from the original stock. The same I have now done with your Vaccine virus. It was taken on the 9th day, for the pustule afforded none on the 8th. I took the patient into my own house that I might watch the progress of the local affection, which I did with the microscope. It is now the tenth day in the morning, and I expect the efflorescence will, in ten or twelve hours more, put on the appearance of your *tenth day representation* in the coloured engravings.

It is impossible for me to express the great satisfaction your letter gave me. The subject was before involved in a mist: your letter was a ray of light, which ray must be reflected for the benefit of the western world. Oh! that it were possible for this ray to become still more brilliant and even generative at the point of repercussion.

I entirely agree with you as to the cause of our late failures in inoculation. The case at Geneva, under Dr. Odier, was ours exactly. One inch and a half of infected thread from Dr. Haygarth was the whole stock from whence perhaps 3,000 persons have been inoculated, but I fear the greatest part of them have been spurious. I here enclose a newspaper containing a communication written in the *clouds* last December. I will allow you to smile at my mercurial and antimonial process, and likewise at my sextuple quantity of *deteriorated* virus! You know not what it is to be perplexed in this business. That prince of physiologists, John Hunter, once told me that "he loved to be puzzled, for then he was sure he should learn something valuable." Burthensome as it was at the time I do not now regret my perplexity. When I had lost my way, and wandered into the wilds of conjecture, I stood still. I gave out that the winter was an unfavourable season for this new inoculation, and by that means I suspended the practice throughout the country from that period until the arrival of fresh matter and your letter. Now we are going on again, but not with the faith and spirit of the last season. Some unlucky cases have damped the ardour of a people who received this new inoculation with a candour, liberality, and even generosity, much to their credit. The first political and literary characters in our nation are still warm advocates for the practice. I have lately received a request from head-quarters to supply the matter, and give the instructions to the regimental surgeons for inoculating the corps of artillerists and engineers stationed at different places on our coast.

Accept my thanks for the coloured plate. It is indeed a happy expedient, and honours the graphic art. It is thought here to be so important that I am anxious to know if I can with propriety procure more of them. I should wish to possess a couple of dozen to be deposited in the hands of some of our leading practitioners, or clergymen, in different parts of the United States by way of standards. If this could be done I would propose that your artist or bookseller should send them to my bookseller, Mr. Mawman in the Poultry, who will pay for them and transmit them to me. Could I procure two or three, delineating the appearances on the skin of the negro, I would send them into such of our southern states as are blackened by these degraded beings. I have lately had letters from Virginia, respecting matter, instructions, &c.

I have been informed from a quarter not likely to be deceived, that cows (contrary to my assertion in page 22 of my pamphlet) have been known to have the small-pox.

The account is this. At one of our periodical inoculations, which occurs in New England once in eight or nine years, several persons drove their cows to an hospital near a populous village, in order that their families might have the daily benefit of their milk. These cows were milked by persons in all stages of the small-pox: the consequence was, the cows had an eruptive disorder on their teats and udders, so like the small-pox pustule, that every one in the hospital, as well as the physician who told me, declared the cows had the small-pox. Since the cow-pox has been talked of this account has been revived and credited. Have you found any thing like this in England?

I inoculated one of my cows with the Vaccine virus,

and obtained from her a crop of matter on the ninth day, which produced the disease in the human subject to perfection. Is this experiment known among you? As I operated myself there was no avenue opened for deception in the whole experiment.

I have invariably found that weakly children have been benefited by the vaccine inoculation, and some it has cured of the hooping-cough.

Could you believe that not a single case of the cow-pox inoculation has yet occurred in Philadelphia? A young physician applied to me a few days since from that for the infection. It seems that the leading physician there pronounces it too beastly and indelicate for polished society! It is impossible to think of this without calling to mind Mr. Ring's solemn appeal to Dr. Moseley respecting *cows' milk, beef steaks,* and *mutton chops*. Please to present my best compliments to that gentleman, and tell him that this single stroke of wit, so much in the spirit of our *Franklin* who always decorated philosophy with a smile, has done me more service than half the publications I have read on the philosophy of vaccination.

You very politely express a wish for more of my letters on the Vaccine or any other subject. In order to damp this desire and surfeit you at once, I have directed my bookseller to send you a whole volume of them, which the partiality of Dr. Lettsom has brought into light.

I here send you the Massachusets Register for the present year. It may possibly afford you some information as to our literary societies, &c. and may give you some new ideas respecting the present state of a country which was characteristically denominated by the English a century ago "THE WILDERNESS."

I need not, I think, say how highly I should prize the correspondence of Dr. Jenner on any subject; but more especially on that for which he is so deservedly celebrated; and who, according to my understanding, is the only *clear, consistent*, UNCONFUSED writer on the cow-pox that has yet appeared.

I reiterate my thanks for your kindness, and beg you to accept the assurances of high consideration and esteem!

BENJAMIN WATERHOUSE.

P. S. As the library of this University is by far the largest in the United States, and is the grand deposit of rare and valuable books in this quarter of the world and will long continue so, I cannot resist expressing my wish that a copy of your invaluable work may be deposited there by its author. I presume my motives for wishing this, and hinting it, stand in no need of an apology. By a law of the Commonwealth, an author to secure his copy-right must deposit a copy of his work in this library; and books sent to it come free from duty. This library, museum, and other public rooms are constantly visited by strangers as among the curiosities of the country. When I had the honour of waiting on the Duke of Kent through them, he expressed his surprise at such a collection of books and natural productions in about thirty years, *for the small-pox destroyed the chief of what had been collected since* 1638: that is to say, it raged in Boston, and the legislature on that account occupied one of the public rooms in the hall, which contained the library; when it by some accident took fire and was, one alcove excepted, totally destroyed. Thanks be to Dr. Jenner, such an accident from *such a cause* can never happen again.

The ever-active mind of Dr. Jenner was now investigating the history of some of the other disorders of the inferior animals. Among these was the dog-distemper. He found that animal to be very susceptible of the Variolæ Vaccinæ; and he believed that after having undergone their influence, it was rendered unsusceptible of *the distemper*. Several of our great fox-hunters eagerly availed themselves of this hint, and had their hounds vaccinated. This practice is, I believe, not now in use.

A paper on the dog-distemper was some years afterwards presented to the Medico-Chirurgical Society of London, by Dr. Jenner. This paper, together with an account of two cases of small-pox communicated to the *fœtus in utero*, was printed in the first volume of their transactions in the year 1809.

It has been seen that whilst the practice of vaccination was extending rapidly over the world Dr. Jenner had the mortification to learn that a perfect knowledge of the manner in which it ought to be conducted did not keep pace with it. The consequence was a very natural degree of disquiet in his mind and great alarm and agitation on the part of the public from cases of alleged failure, which were assiduously circulated in newspapers and magazines, accompanied with all the exaggerations and misstatements that ignorance and prejudice could invent.

Dr. Jenner's method of dealing with rumours of that kind may be gathered from a letter written on such an occasion.

DR. JENNER TO MR. BODDINGTON.

London, April 21st, 1801.

DEAR SIR,

I am extremely obliged to you for your letter which I long waited for with great anxiety; not an anxiety arising from any fear as to the result of my inquiry, but from a wish to satisfy the minds of those who were thrown into consternation from the reports which prevailed respecting Master Boddington's case. It was a case, Sir, on which I would willingly have rested all the merits of the vaccine discovery. Your letter and that of the medical gentleman who inserted the variolous matter fully justify me in saying I should have been right, for I never wish for a more perfect specimen of the non-effective powers of the small-pox upon the constitution after the cow-pox, than that you have laid before me. How a gentleman, following a profession the guardian angel of which is fame, should have so committed himself as to have called this a case of small-pox after cow-pox, is not only astonishing to me, but must be so to all who know any thing of the animal economy. He should have known that upon the skin of every human being that possesses a more than ordinary share of irritability the insertion of the variolous virus (whether the person has previously had the cow-pox or small-pox) will produce either a pustule or a vesicle capable of communicating the small-pox, and frequently attended with extensive inflammation. He should have known, too, that an inflammation, by whatever artificial means excited upon a detached part of the skin, is capable not only of exciting some degree of indisposition but also eruptions. Within this fortnight I have seen these appearances from a small blister on the child of a nobleman here. From the irritation of the blister the eruption became almost universal.

I remain, dear Sir,

Your obliged and obedient Servant,

E. Jenner.

Dr. Jenner has often been blamed, but I really believe inconsiderately, for encouraging unprofessional persons to practise vaccination. Every individual in this situation, with whom he had to do, actually studied the subject, and implicitly followed his directions. Not so, many of his professional brethren. They would not condescend to be instructed, and they openly disregarded his rules. Of these one of the most important was violated by the first public body that was formed in London for the purpose of vaccination. They actually printed and distributed a paper stating that the virus may safely be taken from the pustule so late as the thirteenth day!!!

It certainly would have been desirable to have confined the practice from the beginning to the hands of medical men. It is, nevertheless, true that from the causes just mentioned *they* fell into greater error, and did thereby much more endanger the character of vaccination than those who, from their general ignorance of medical subjects, might have been considered as more liable to go astray.

With these exceptions, every thing connected with vaccination was going on prosperously. His communications from remote parts, as well as from different places in our own country, all conveyed the pleasing intelligence of a rapid and satisfactory extension of the practice. It continued to excite great interest among the public at large.

I have mentioned several of the nobility and gentry who distinguished themselves by their zeal. I have now to add to that number the Earl of Carnarvon, who about this time gave Dr. Jenner an account of his vaccinations, which had been very extensive and successful.

This was not always the case with those of medical men. An incident proving this was communicated to Dr. Jenner, and he made it known to Lord Egremont, because from its connexion with two public bodies it might materially affect the character of vaccination. The Sick and Hurt Office had sent to the public Institution for Vaccine virus. They were furnished with matter issuing from an ulcer on a child's arm. The child had been vaccinated sixteen or eighteen days. Dr. Jenner invariably exclaimed against proceedings of this kind. He was even conscious that he himself had, in the outset, been less scrupulous as to the time of taking matter than he ought to have been. His sentiments on these points are thus expressed by himself:—

DR. JENNER TO MR. RING.

Bond Street, July 1st, 1801.

MY DEAR SIR,

You have often, I believe, heard me say (and I now repeat it) that it would not in the least surprise me to hear that some of those who became my vaccine patients in the early part of my inoculation were infected with the small-pox. I had not then learned to discriminate between the efficacy of virus taken at an early, and at a late period of the pustule. Accordingly I have long since requested my nephew, whenever an opportunity offered, to reinoculate those people. I am confident that the cases you mention will be productive of good consequences, although they must create some temporary uneasiness, and give a little local check to vaccination. They will tend to stamp perfection on what I called "the golden rule" which, backed by your persuasive pen, will tend to make practitioners more careful than ever in their conduct. The precepts given out by P—— allow an unlimited time for taking the Vaccine virus from the pustule. This I have often exclaimed against. It was not long ago that an apothecary wanted to take it from a pustule on the arm of a patient of mine, after the incrustation had begun to form three or four days.

Your case would certainly have raised a clamour a year or two ago; but now the phenomena of cow-pox have been so fully examined, and are so well understood, none but the ignorant and illiberal will lay any stress on it for a moment.

Yours very truly,

Edward Jenner.

The golden rule alluded to in this letter regarded the time of taking the Vaccine fluid for the purpose of inoculation. This Dr. Jenner maintained ought to be done at an early period of its formation, and before the appearance of the areola. The next rule which he insisted on was that the pustule, when excited should be permitted to go through all its stages in an uninterrupted manner. If any deviation appeared in its progress he always forbade the employment of virus from such a pustule for farther inoculations.

These rules, together with the injunctions which he subsequently delivered on the effect of certain diseases of the skin in interfering with the progress and character of the Vaccine pustule, may be considered as comprehending the main points to be attended to by those who undertake to conduct the practice of Vaccine inoculation.

On the 22nd of April he attended a meeting of the Lyceum Medicum Londinense, when he was admitted a fellow. This fact is worthy of being mentioned, inasmuch as this society was founded by his friend John Hunter, and Dr. Fordyce.

As it becomes me to seize every opportunity to prove the accuracy of his observation, and the truth of his doctrines, I may mention some of the remarks which he made at this time touching the influence of cutaneous diseases in modifying the progress of the vaccine pustule. One of the entries in his journal is to the following effect: "Inoculated Lady C. F. a second time. It is very evident that *that* affection of the skin called red gum, deadens the effect of the Vaccine virus. This infant was covered with it when inoculated four days ago. The same thing happened to Mrs. D.'s infant."

Dr. Jenner's conjecture respecting the power of

vaccination in preventing the dog-distemper has been already mentioned. Further to put this opinion to the test he, together with his nephew Mr. George Jenner, vaccinated in the month of June about twenty of his Majesty's stag-hounds. On the 12th of the same month he vaccinated a child from its own arm. This fact, together with that recorded by Mr. Hicks, proves how completely Dr. Jenner's mind was possessed of that knowledge which led Mr. Bryce, at a subsequent period, to propose *his* test of perfect vaccination.

Dr. Jenner remained in London till the 22d July, on which day he left it with his family and arrived at Cheltenham on the evening of the 23d. He remained there till towards the end of November, when he went for a short time to Berkeley.

In the summer of this year Dr. Jenner gave some cow-pox matter to Dr. Marcet to be sent to Copenhagen. About two months after, this physician received an interesting account of its success, which he published in the London Medical and Physical Journal. His Majesty the King of Denmark evinced his paternal solicitude for the welfare of his subjects by appointing a committee to collect information concerning vaccine inoculation, and to propose such regulations as might be needful for diffusing it in his dominions. This committee agreed on their report on the 5th of December 1801. It was then laid before his Majesty; when he was pleased to give his approbation to all the proposals of the committee, and commanded the same to be duly enforced.

Professor Winslow particularly distinguished himself both as a member of this committee and as a sedulous vaccinator. The report was drawn up with great judgment, and laid the foundation for all the judicious and wholesome regulations which for a period of nearly twenty years gave Denmark a total immunity from small-pox.

From Denmark the practice was disseminated into Sweden, where means not less efficient were adopted for its propagation, and, as has already appeared, with a corresponding degree of success.

In the course of this year also the practice of vaccination was introduced into the West India islands with like benefit.

For the most part this practice was received with greater eagerness on the Continent than in England. Both France and Germany had, nevertheless, their Moseleys, their Birches and their Rowleys. Ehrmann, of Frankfort, in an especial manner rendered himself conspicuous by the violence and extravagance of his opposition to the new practice. A learned divine of the church of England (Massey) who preached a sermon against small-pox inoculation, in London, 1722, announced it as no new art, inasmuch as Job, he asserted, had been inoculated by the devil. Erhmann took rather a bolder flight, and attempted to prove from quotations of the prophetical parts of Scripture, and the writings of the fathers of the church, that the Vaccine was nothing less than Antichrist. The harsh and unsparing tone of his writings obtained for him the appellation of the *Marat* of the Vaccine.

The progress of vaccination was, however, more impeded by the misjudging zeal and mistakes of its professed supporters than by all the efforts of its most determined opposers. Had the former made themselves well acquainted with the subject, and conducted the process with care and attention, all the attempts of the latter would have been powerless. Still, as it was, nothing could effectually stay the course of this wonderful discovery.

The Cisalpine Republic during the summer of this year gave an authoritative sanction to the practice. Instructions drawn up by Dr. Sacco were disseminated throughout the whole territory; and that gentleman was appointed Director of Vaccination. I give his first letter to Dr. Jenner, announcing these facts, and his discovery of the indigenous Vaccine virus in Lombardy: together with Jenner's answer.

Dr. Sacco to Dr. Jenner.

Sir,

It is to the Genius of Medicine, to the favourite child of nature that I have the honour to write. The name of Jenner will be always beloved by all posterity. All sensible minds will feel pleased with you. Finally, population, by your incalculable and very interesting discovery, will realize more than another tenth of its increase. We must be very grateful to you, Sir; you have given the thread by which all others have guided their experiences. I also am one of those who, on the line which you have marked out, have endeavoured to render myself useful to humanity. After long researches I have at last found the virus indigenous in Lombardy; and with this virus there have already been more than eight thousand inoculations performed with the most happy success.

Several hundreds of these have since been subjected to the variolous inoculation, and have resisted it. The Cisalpine Government has distinguished itself, and wished to render it general throughout the whole territory of the Republic. It has done me the honour to appoint me Director of the Vaccination, which goes on very prosperously.

I should have wished, Sir, immediately on the publication of my work to have paid a little tribute of gratitude and esteem to a man such as you are, but the circumstances of the war have hindered me. Receive, my very estimable Sir, the homage of one who at a distance follows your steps, and who expects always from his master useful instructions. On this subject there is no English work known in Italy except your first publication and late observations. Some other books of old date are known only by extracts.

I would therefore, Sir, beg of you if you have any thing new to do me the honour to prepare for me a feast: I shall be under the greatest obligations to you for it. You will also do me the greatest pleasure in communicating to me the state of vaccination at present in England. And as I must give a report to Government, which will be printed, I shall then have the satisfaction to announce new and interesting things from the native country of the Vaccine. How happy should I be to make a journey to England that I might become acquainted with my preceptor, and to see all the establishments of that country. I pray to Heaven to inspire my country with the idea of the necessity of sending some one who may know the great men produced in that happy soil. Circumstances become more favourable every day: let us therefore hope. Honour me, Sir, I beg, with your correspondence, and be sure of my highest esteem.

You may, Sir, address your answer and any later observations to M. Woodville, whom I have requested to send me every thing. I hope you will not be displeased with me if I have not been completely convinced that the grease in the horse has been the cause of the origin of the cow-pox. I have several experiments to the contrary. If you still have some which confirm it I beg of you to communicate them to me, in order that I may do you justice.

I believe I have explained what inoculators seek, by which, making four or six punctures, the pustules do not all arise in every place. When the absorption in the lymphatics does not take place, they fail: and, on the contrary, when the puncture happens in a small lymphatic vessel, then it is sure that the Vaccine will appear in the skin, although every-where covered with small vessels; there are, however, small interstices in which they are not.

If the puncture be made in this interstice it fails. One may explain this by almost all the phenomena which have been accounted for before. You see that the glands in general are the most affected. I request you to tell me your opinion on this theory.

May you live, my dear Sir, a long while for the good of humanity, and for the sake of all those who love you.

P. S. I have sent to Mr. Woodville vaccine matter from Lombardy. I have requested him to return me some from

London. Sir, you may also send me some from the country. You will oblige me infinitely—the comparison may produce some effect.

Your very humble servant,

LOUIS SACCO.

Milan, Oct. 16*th*, 1801.

DR. JENNER TO DR. SACCO, OF MILAN.

DEAR SIR,

Accept my best acknowledgements for your very kind attention. I am extremely gratified by your goodness in sending me your pamphlet on Vaccine Inoculation, your obliging letter, and above all the virus from the plains of Lombardy. I am confident that wherever the horse and the cow are domesticated together, and the same human being that attends the one, under a peculiar malady of the foot, milks the cow also, that there the disease called the cow-pox may arise.

Until your communication reached me I was not aware that the new practice had been known in Milan. How much do I rejoice to observe that it is not only known there, but that it has already been conducted so extensively and so ably

* * * * *

Measures not less influential were adopted in the Prussian dominions. The virus which had been sent by Dr. Jenner to vaccinate the Princess Louisa had proved completely successful. The noble example set by the royal family, in thus giving proof of their confidence in vaccination, was amply repaid by great future benefits to the community. It laid the foundation for an extensive and effective system of vaccinations, which were carried on under the immediate sanction of the King. His Majesty founded a royal Inoculation Institute in Berlin, the care of which was confided to Dr. Bremer, and his benevolent designs were seconded by this physician with the utmost zeal. He circulated tracts and exhortations to the public, and adopted a plan which had great effect both in increasing the number of vaccinations, and in enabling him to ascertain that they were properly performed. Assisted by generous individuals he collected funds sufficient to have a medal struck. This medal was given to those who caused their children to be vaccinated at the Royal Institute; not at the time of vaccination, but on the seventh day after, when they appeared to show the progress of the pustule.

The artists of Berlin cordially lent their aid. The drawing was made by Mr. Freitsch, the King's painter, and director of the Royal Academy. The die was sunk by the King's medallists, Loos and Son. The obverse exhibits a child pointing with the fore-finger of his right hand to the spot where vaccination is generally performed on the left arm. In the left hand the child holds a rose; and there is besides a garland of roses, and a cornucopia. The inscription commemorates Dr. Jenner's first vaccination in the following words — "Edward Jenner's beneficial discovery of the 14th of May, 1796."

The reverse states the object of the medal in

words to this effect. "In remembrance of protection afforded." Presented by Dr. Bremer, Berlin, 1803.

In Vienna, notwithstanding the unwearied efforts of Dr. De Carro, the Government for a time stood aloof, and actually confounding the vaccine inoculation with that of small-pox, issued an order forbidding vaccination within the walls of the city. As soon as the essential difference between the two affections was ascertained this order was revoked, and the practice of vaccination was carried on with the greatest energy and activity. Tracts were distributed among the people, and individuals, powerful from their rank and their talents, exerted their influence in diffusing the antidote. What Lord Egremont, Mr. Fermor, and others had done in England, was achieved upon a wider scale at Brunn, in Moravia, by Count Francis Hugh de Salm. This truly philanthropic nobleman, having obtained genuine Vaccine virus with appropriate directions for its employment, called in to his assistance two physicians, and sent a third to be fully instructed in the practice. He likewise held out rewards to the physician who should vaccinate the greatest number in that country; and himself drew up a popular treatise on the subject, which he distributed gratuitously to the clergy and schoolmasters of Bohemia and Moravia. Not contented with these generous exertions in the cause he came to England with a letter of introduction from Dr. De Carro* to visit in person the illustrious discoverer Jenner, and the other sedulous promoters of vaccination.

The good people of Brunn continue to commemorate the blessings conferred by the introduction of the Vaccine. They have erected a temple dedicated to Jenner. Here they annually hold a festival on his birthday. Some years ago they sent him a very interesting and warm-hearted account of their anniversary; which will be found in another page.

The practice was introduced into Hungary by Balthasar Nicolas de Bedecovitz, Lord of Kamor, living at Varasdin in Cröatia. He caused to be vaccinated on his lands at Stephaniez the children of one hundred and forty-nine families.

A sister of the late King of Poland, the Countess Zamoiska, who was living at Vienna while Dr. De Carro was carrying on his first vaccinations, desired him to furnish her with virus, which she transmitted to her daughter the Countess Mnieshek. One of this lady's own children was inoculated with this

* Extract of a letter from Dr. De Carro to Dr. Jenner, dated Vienna, 30th June, 1801.

Dear Sir,

Give me leave to recommend to your notice the bearer of this letter, Count de Salm, a friend of mine and a young man of very great merit. The propagation of your discovery in the Austrian Monarchy has been more forwarded by him than by all the imperial faculties of medicine. Open my work at page 91, and you will read part of what this nobleman has done for the two provinces of Moravia and Bohemia.

matter; and from this source was the practice diffused over Poland.

The ministers of religion were not slow in their exertions in this cause. The address to "Fathers and Mothers" drawn up by the faculty at Geneva, and which the officiating clergyman was in the habit of delivering to the parents and sponsors of every child presented to receive the rite of baptism, has been already mentioned. John Michel Körn, the pastor of Brunn Am Geburgh, struck with the mildness of the symptoms of vaccination, gave on the Sunday immediately following his observation of them an account of the discovery from his pulpit. The paternal and affecting simplicity of his exhortation so wrought upon the hearts of those who heard him that eighty persons were shortly after vaccinated. On each ensuing Sabbath he repeated his invitation; and a physician attended to vaccinate any one who came forward at the close of the service. I am determined, said this philanthropist in a letter to Dr. De Carro, to have no more smallpox in my parish.

In October 1801, the Vaccine was sent from Breslau to Moscow, by Dr. Friese. It was transmitted on some of Dr. De Carro's ivory points, and on threads. The Russian Court was at that time in the ancient capital, on account of the coronation of the late Emperor Alexander. The Empress Dowager zealously promoted the new practice. She desired that the name of Vaccinoff might be given to the first child who received the infection. The young Vaccinoff was then conveyed to St. Petersburgh in one of her Imperial Majesty's coaches, and placed in the Foundling Hospital, and a provision was settled on her for life.

Her Imperial Majesty continued to bestow every encouragement on the dissemination of the practice, and it soon became general throughout the Russian dominions. At a subsequent period the Emperor appointed a mission to carry the antidote to the more remote provinces of his empire, with the intention of ultimately reaching China and the whole of the northern parts of Asia.

On the 10th of August, 1802, the Empress, while residing at Pawlosk, sent a letter to Dr. Jenner signed with her own hand, which together with a valuable diamond ring she forwarded to him through Lord St. Helens, British Ambassador at the Court of Petersburgh. His Lordship on his arrival in England very politely wrote to Dr. Jenner, thinking that he was then in town. Finding that he was at Cheltenham his Lordship wrote a second note to this effect.

SIR,

Since writing my other letter to you of this date I have learned by the servant whom I sent with it that you are now at Cheltenham. I have to request, therefore, that if you do not purpose returning soon to London, you will have the goodness to commission some person to receive what I have brought you from the Dowager Empress of

Russia, consisting of a letter, and a small parcel containing (I believe) a ring set in diamonds. I shall probably remain in town till the 15th instant. I have the honour to be, with great regard,

Sir,
Your most obedient servant,
ST. HELENS.

DR. JENNER TO LORD ST. HELENS.

MY LORD,

I have been honoured with your Lordship's letters acquainting me with the distinguished mark of attention conferred upon me by the Empress Dowager of Russia. Sanctions like these, my Lord, from such exalted personages must necessarily be peculiarly pleasing to my feelings. They not only benefit me individually, but by blending with the general arguments for the universal adoption of vaccine inoculation, the annihilation of that dreadful disease the small-pox will be the more quickly accomplished.

May I request the favour of your Lordship to make known to the Empress the high value I set upon her Majesty s present, and to express my extreme gratitude and thankfulness for her goodness.

As I shall not be in London for some time to come your Lordship would oblige me in consigning the parcel to the care of Mr. Paytherus, Surgeon, No. 13, Norfolk-Street, in the Strand, who will faithfully transmit it to me.

I have the honour to be, my Lord,
Your Lordship's
Ever obliged, faithful
And obedient humble servant,
E. JENNER.

Cheltenham, 5th October, 1802.

Her imperial Majesty's letter was couched in the following terms.

MONSIEUR JENNER!

L'usage de la Vaccine en Angleterre ayant eu les succès les plus avantageux et les mieux attestés, je me suis empressée d'imiter cet exemple, en l'introduisant dans les establissemens pieux, qui sont sous ma direction. Mes soins remplissant parfaitement mon attente, je me plais à en rapporter le succès, et à en temoigner ma reconnaissance à celui, qui a rendue à l'humanité ce service signale. Ce motif m'engage, Monsieur, à vous offrir la bague ci-jointe comme une témoinage des sentimens d'estime et de bienveillance, avec les quels je suis votre affectionée,

MARIE.

Pawlosk ce 10 *Août*, 1802.

To this most gracious letter Dr. Jenner returned the following acknowledgement, in English as well as in French.

TO HER IMPERIAL MAJESTY, THE EMPRESS DOWAGER OF RUSSIA.

Berkeley, October 10*th*, 1802.

MADAM,

I know not how to express a just sense of the obligation I feel myself under to your Imperial Majesty for your great condescension in addressing to me a most gracious letter, and for your goodness in presenting me with a most valuable diamond ring; a gem, Madam, that I trust will descend, with your illustrious name annexed, to my posterity through ages!

That the discovery I have had the happiness to announce should, fortunately, have attracted your Imperial Majesty's attention; and that you should have been pleased to extend its benefits through the charitable establishments under your protection, are gratifications to me of the most pleasing kind. Not only on me has your Imperial Majesty conferred a favour by your invaluable present; it will be felt by the whole world: for sanctions like these will materially tend to extinguish prejudice, and hasten the universal adoption of vaccine inoculation; and thus will the annihilation of a disease the most destructive that ever preyed upon the human race be accelerated, and millions of victims rescued from an untimely grave.

Permit me, Madam, humbly to present to your Imperial Majesty those Essays which I have written upon this important subject.

Long may your Imperial Majesty enjoy those soothing reflections so incessantly poured into those bosoms which are formed, like yours, to soften the distresses of human nature!

I have the honour to be
Your Imperial Majesty's most grateful,
obliged and devoted
humble servant,
E. JENNER.

Madame,

Les paroles me manquent pour exprimer tout ce que je ressens de reconnoissance envers votre Majesté Imperiale pour avoir daigné m'addresser une très gracieuse lettre, accompagnée d'une bague de grand prix; mais infiniment plus précieuse pour être le don d'une Princesse encore plus distinguée par ses vertus, que par son haut rang dans les empires.

Ce present, pour moi inappreciable, sera confié après ma mort à mes descendans avec une legende qui marquera la source illustre d'où il m'est parvenu.

Il m'est assurément trés flatteur que la découverte, que j ai eu le bonheur d'annoncer au public, ait attiré l'attention de votre Majesté Imperiale ; et qu'elle ait bien voulu en étendre l'usage aux établissemens pieux sous sa protection.

Une telle exemple contribuera de la manière la plus efficace à éteindre les prejugés, et à hâter l'adoption universelle de la Vaccine.

De cette manière l'anéantissement d'une maladie la plus funeste peutêtre qui ait jamais affligée le genre humain, sera accéleré et des milliers de victimes sauvés de mort prématurée.

Permettez, Madame, que je saisisse cette occasion pour offrir à Votre Majesté, avec mon hommage respecteueux, quelques esquisses que j'ai écrites sur ce sujet important.

Puisse t'elle jouir pendant de longues années des douces reflections, qui sont le privilege precieux des âmes sensibles et bienfaisantes. Des telles reflections n'emanent que des cœurs formés, comme le sien, pour soulager les maux de la nature humaine.

J'ai l' honneur d'être
avec le plus profond respect
Madame,
De Votre Majesté Imperiale
le très humble devoué et reconnoissant
Serviteur,
EDWARD JENNER.

Berkeley, Compte de Gloucester.

This letter, together with a copy of his works on Vaccination, he forwarded to her Imperial Majesty, through Lord St. Helens. His Lordship, in reference to these affairs, writes thus:—

LORD ST. HELENS TO DR. JENNER.

Old Burlington Street, 14th December, 1802.

SIR,

I have received the honour of your letter and inclosure of the 9th instant; and yesterday I had the pleasure of a visit from Mr. Ring, who informed me that the volume of your publications on Vaccine Inoculation, which you intended as a present to the Dowager Empress of Russia, will be ready in the course of a few days.

I am sensibly obliged to you, Sir, for your attention in leaving open your letter to her Imperial Majesty, as I have perused it with very great satisfaction both in *the original and the translation:* which is, in my judgment, uncommonly well executed. I have therefore determined to send them both to her Imperial Majesty by the post of this evening, accompanied by a letter from myself explaining the circumstances that you have desired me to mention, and I shall take care to forward the volume of your Publications by the first messenger that may be dispatched for St. Petersburgh from Lord Hawkesbury's Office.

I hope I need not assure you, Sir, of the pleasure I feel in executing this commission: and that I shall be equally happy in availing myself of the first opportunity to assure you, in person, of the very sincere and high regard and esteem with which I have the honour to be, Sir,

Your most faithful and obedient servant,

ST. HELENS.

Among other distinguished individuals to whom Dr. Jenner sent his first work on vaccination was the celebrated Blumenbach. I believe it was delivered to him by Dr. Charles Parry, who was then pursuing his studies on the Continent. The ardent feelings of gratitude with which this eminent naturalist received the gift are thus expressed :—

DR. BLUMENBACH TO DR. JENNER.

Gottingen, Sept. 12, 1801.

SIR,

Accept my warmest and most cordial thanks for the very kind present of that immortal work by which you have become one of the greatest benefactors to mankind, as your great discovery is by far the most efficacious counterpoise to the ravages made by war and similar expedients for destruction of our fellow-brethren.

To show you in any way my grateful acknowledgment for your kindness to me I proposed you as a member to our R. S. and it needs not to tell you with what an unanimous applause they agreed to this proposition. I enclose here the certificate of your fellowship, and have the honour to be, full of the highest regard, Sir,

Your most obliged humble servant,

JNO. FREDK. BLUMENBACH.

P. S.—Give me leave, Sir, to tell you also that I, as a very warm friend and even teacher of Natural History, long very eagerly to see once your paper on the migration of birds, mentioned in your masterly observations on the cuckoo.

The Royal Society of Gottingen had the honour of setting the example to the other learned societies of Europe, by electing (unanimously) the illustrious author of Vaccination an honorary member of their body. Dr. Jenner received this mark of respect with great satisfaction.

Dr. Jenner to Professor Blumenbach.

Sir,

Among the favours conferred upon me, since I had first the happiness of calling the attention of mankind to the subject of vaccine inoculation, there is not one which I esteem more highly than that from the University of Gottingen.

Give me leave to request that through you, Sir, my sincere thanks be presented to the learned body who have so kindly made me the object of their attention; and to assure them that I shall ever retain a grateful sense of the obligation.

Permit me, particularly, to thank you for your very friendly letter. My observations on the migration of birds have not yet been published. I shall, if possible, present a paper on the subject to the Royal Society this winter: but latterly my attention (as you may suppose) has been so incessantly occupied by that of the cow-pox I have found it impracticable to wander into other paths of natural history; a branch of science which, I know, is the delight of your heart as well as mine, and which you have cultivated not only with ardour, but with effect.

A friend of mine, who has lately practised vaccine inoculation very extensively, has circulated the enclosed paper of directions. It contains some very useful and necessary rules, yet probably nothing with which you are unacquainted.

The state of feeling in France at this time, with respect to vaccination, may be gathered from the incident which I am about to mention. On the arrival of the British Ambassador, the Most Noble Marquess Cornwallis, at Amiens, the members of the Jury of Health and the Medical Committee of the Department of the Somme were so deeply impressed with the magnitude of the discovery of the "*immortal Jenner*" that they thought it a fit subject of congratulation to the British Ambassador. The address was duly presented, and afterwards forwarded to James Moore, Esq. by his gallant and most deeply lamented brother, the late General Sir John Moore, who was attached to the Embassy. This interesting document is printed in the Medical and Physical Journal, vol. 7, p. 201.

To elucidate the preceding narrative, I select from the communications of some of Dr. Jenner's foreign correspondents a few documents, which will convey much more accurate information of the state of feeling in the different countries from which they were written than I could effect in any other way. The list might have been swelled to a great amount; but I have inserted none except such as contain facts, of themselves, important; and which could not have failed to have been deeply interesting to Dr. Jenner.

DR. DAVIDS TO DR. JENNER.

Rotterdam, 24th March, 1801.

TO THE BENEFACTOR OF MANKIND, DR. JENNER.

SIR,

I was happy enough to introduce the cow-pox through the whole country with the greatest success, and the name of Dr. Jenner is adored. In a few days my translation of your essay on the cow-pox will be published.

The cox-pox inoculation was introduced just at the moment the small-pox made ravages through the whole country; but thank God not one is infected after the vaccine.

I will take the liberty of sending you one of my translations, and in the preface you will find an observation about an Arabian manuscript found in the Leyden Bibliothec:

In the spring I should be very happy to have a little fresh matter from the cow.

After assuring you of my sincere feelings and respect,

I remain,

Sir,

Your most obedient humble Servant,

L. DAVIDS, M. D.

DR. WATERHOUSE TO DR. JENNER.

Cambridge, 30th May, 1801.

MY DEAR SIR,

A few months ago I sent you a portion of infected thread, being the *first crop* of your own matter. I also sent you a portion of your own thread that you might see if it had lost any of its power by twice crossing the Atlantic. I now again send you a portion of your own virus, being a part of that *identical* thread which you, in conjunction with Mr. Ring, took from the arm of a patient on the *third day of last March* in London, that experiment may be made respecting its activity at a still greater length of time. I have about as much more left, which I will send in perhaps two months hence. It is now within four days of being *three months* old, and it had not lost its activity ten days ago.

A few days ago the Attorney General of this Commonwealth made me the following communication.

"On the 16th of the present month (May) 1801, when on the circuits, I passed through the town of Pelham, in the County of Hampshire, in company with *Chief Justice Dana*. We stopped at an inn kept by one Hyne. His wife, sitting in a chair, appeared to be indisposed. She informed me that she had the kine-pox; that she had taken it from a cow she had milked; that three others in the family had caught it in the same way from the same cow. I examined her arm; it was much inflamed, more especially above the elbow. There were sores very much like those in the human flesh made by small-pox taken in the natural way. Her husband said the cow's teats had been sore, but had recovered."

The Attorney General Mr. Sullivan is truly a man of intelligence. He is President of the Historical Society, author of the History of the Province of Maine (See Mass. Register, p. 170,) and has seen the kine-pox in his own family. I therefore am inclined to believe that we have the distemper among the *kine* of this region.

I have just received a packet from Mr. Dunning of Plymouth Dock. He quotes a passage from a letter of yours which has engrossed my mind almost entirely. It is this: "Are you inoculating your dogs yet in the West? That vaccine inoculation should give that disease which from its universality has been long termed *the distemper* is a new wonder in its history. But so it is. If you are not already acquainted with it, I will give it to you in my next."—What can this be?

You will doubtless rejoice with me that the vaccine inoculation is progressing here to my entire satisfaction. I began with the matter you, Drs. Lettsom, Woodville, Pearson, and Creaser sent me the 24th of March, and have inoculated not quite a hundred, and have not had one dubious case among them all. I have given the virus to most of the leading physicians in Boston and its vicinity. But, alas! poor human nature! thou art the same in *New*-England as in *Old*. Why should the sigh of sorrow arise in any one, when contemplating the progress of a practice so universally beneficial as that resulting from this new inoculation? But so it is—all the little contemptible *nonsenses* that have been uttered by imbecility and envy on your side the Atlantic have been retailed here. They have given a temporary check to the inoculation in the country villages, for which I am not sorry; but while the public see that all those characters throughout the country most distinguished for their talents and acquirements are firm believers in the Jennerian inoculation, it will and must prevail.

The inoculation finds no promoters in Philadelphia. I have had many letters lately on the subject from Virginia, where it is of vast importance to their plantations.

That the learned view it as a permanent blessing to this country may be inferred, among other things, from a passage in the history of Cambridge, just published by a worthy clergyman, viz: "The kine-pox was introduced at Cambridge this present year (1800) by Professor Waterhouse, who imported the *matter* from England. The first who was inoculated for this disorder in America was Daniel Oliver Waterhouse," a son of the professor.

Accept, dear Sir, the assurances of high consideration and esteem.

BENJAMIN WATERHOUSE.

DRS. BALLHORN AND STROMEYER TO DR. JENNER.

JENNERO
Summo Viro,
S.

Georgius Fredericus Ballhorn, Augustissimi Britanniarum Regis in Aula Electorali Medicus, et Christianus Fredericus Stromeyer, A. B. R. in Aula Electorali Chirurgus.

Transmittimus ad Te, vir illustris, et judicio tuo, quod tantum apud nos fidem habet, subjicimus observationes de insitione Vaccina Hannoveræ a nobis institutas. Libelli fronti insigne nomen tuum præposuimus. Ne graveris hoc, quod tenuis Musa nostra in Te conferre satagit, reverentiæ et amoris documento. Felices nos sanè puteremus, si res ipsa quam pertractavimus tuo assensu haud careat!

Materiam vaccinam, quam XVI mensibus circiter præterlapsis, benevole, ad nos transmisisti, Londinensi anteponimus. Londinensis enim sæpius effecit exanthemata satis molesta et pertinacia. Quæ quidem incommoda nunquam ab egregia illa materia, quam Tu in agris Anglicis collegisti, nobis subnata sunt. Maximopere itaque Te rogamus, ut iterum pauxillo hujus egregiæ materiæ, pro tua humanitate, nos donari haud graveris.

Vale, Vir Summe, et serva nobis tuam benevolentiam.

Scriptum Hannoveræ,
d. 10 *Junii,* 1801.

Extract of a letter from Dr. WATERHOUSE to DR. JENNER, dated Cambridge (America) Nov. 5th, 1801.

The characters in America most distinguished for wisdom and goodness are firm believers in *your* doctrine. They are not, however, over-forward in assisting me against this new irruption of the Goths. I do not wish them to do more than make cartridges, or at least hand them. At present they leave me too much alone, and it is probable will only come openly to my assistance when I do not *want* them. Had I not a kind of apostolic zeal I should at times feel a little discouraged. The natives of America are skilful in bush-fighting.

M. SAMARD TO DR. JENNER.

Avignon, December 12th, 1801.

SIR,

Your precious discovery is known to-day in the smallest villages of France. We have drawn our profit of its efficacy in this town to preserve our citizens from the dangerous effects of the natural small-pox; consequently we are indebted towards you of a portion of the common gratitude. We pay it with great satisfaction.

Receive our tribute with goodness. If we could do better nothing would be spared most certainly. Be flattered of being accounted amongst the Associates of our

Lyceum, because this title is the price of the sentiment and of the eminent service you have rendered to mankind.

I am, Sir,

Your most obedient and humble servant,

SAMARD.

Extract of a letter from DR. DE CARRO of Vienna to MR. RING, dated December 18th, 1801.

After nearly three years of success I need not tell you what I think of vaccination: I know of no encomium that can give an adequate idea of that blessing to mankind.

Remember me to Dr. Jenner. No medical man ever excited my admiration and veneration so much. He is not only great by the magnitude of his discovery, but he is also great by the manner in which he conducted his researches; by the perfection which he gave to them before he published his work, and by the extreme modesty with which he speaks of himself. His fame increases daily; but I blush for all sovereigns, and all governments, that have not hitherto bestowed any public mark of their gratitude on that immortal benefactor of mankind.

DR. MARCET TO DR. JENNER.

Dr. Marcet presents compliments to Dr. Jenner, and called upon him the other day to request he would have the goodness to let him have a little Vaccine virus, such as could be depended on as to its genuine origin.

It is to be sent to Copenhagen where Dr. Winslow, Professor of Medicine, wishes to promote the introduction of the cow-pox, and is very anxious to get some virus from the most respectable source. Some impregnated threads might be most easily and speedily sent in a letter;

but if Dr. Jenner thinks necessary to send it in any other shape Dr. Marcet will find some other conveyance.

Professor Pictet of Geneva, the editor of the Bibliothepue Britannique, and a natural philosopher with whose name Dr. Jenner is probably acquainted, will be here in about a fortnight on a visit to Count Rumford at whose house he is to reside. Professor Pictet in a letter which he wrote to Dr. M. a few days ago expressed a great wish to become acquainted with Dr. Jenner. Dr. M. hopes he will be allowed to introduce his friend to Dr. Jenner, and will be extremely happy in taking the first opportunity of bringing together two men who cannot fail to find much reciprocal pleasure in each other's conversation.

St. Mary Axe,
26th May, 1801.

Translated Communication from the College Advertiser of Copenhagen, dated 19th December, 1801.

(Abstract.)

The committee received his Majesty's commands to collect information concerning the vaccine inoculation—to form an opinion of the experiments made, and to propose such further arrangements as may be needful in his Majesty's dominions.——The committee observing a very great mortality amongst those who have the small-pox, we repeated, on the 24th of November, our former advertisements, and have received seventeen different communications.

The experiments and observations of the committee amount to 297 cases, and the greatest number thereof by me Winslow, Professor, partly tried in our presence and under our inspection ; those of other practitioners amount only to 408, and from the most exact observations the committee makes the following observations.

1st. That the vaccine inoculation, at least for some time, prevents the contagion of the natural small-pox, and notwithstanding the small-pox are at present of a very dangerous and epidemical nature, yet nothing creates a suspicion that those who were vaccinated have been infected with it. A few vaccinated have, eight or ten days after inoculation, caught the small-pox, but the committee attributes this only to a prior infection before the vaccine had taken effect.——From experiments of other nations, particularly the English, there are reasons to hope that the contagion of the natural small-pox throughout futurity can be entirely annihilated by the vaccine.

2nd. That the vaccination is not attended with danger, as the greatest number are hardly ill at all, and are not prevented following their usual occupations.

3d. That there is no ground for suspicion that the constitution of the vaccinated has been thereby impaired, or any other diseases produced.

4th. That the vaccination of persons of all ages, weak and robust, during dentition or not, appears to proceed in the usual way, without alteration in the constitution. And lastly,

5th. That the cow-pox does not infect but by immediate touch of the matter.

It is the opinion of the committee that those whose minds are unenlightened, whose circumstances are very limited, and whose prejudices call for wholesome advice and amongst whom the small-pox are particularly mortal, ought to be encouraged to the vaccine inoculation; also the soldiers with their wives and children, who have not had the small-pox, and the crews on board of ships bound on long voyages. That the vaccine inoculation should also be performed in the different institutions for the poor in this city, &c. all public schools, especially where the children are taught gratis, and every one of the lower class and of the populace, who have not their own physician, without expense: to perform which the committee proposes Mr. William Stebuss, Candidatus Chirurgiæ, Surgeon to the Royal Regiment of the Norway Foot Guards, and Amanuensis with me Winslow, Professor, as he has much practice, not only in performing vaccination under the immediate inspection of the committee, but also to mark the progress of the malady, and to collect and preserve the vaccine matter for further use and distribution.

To extend the benefits of vaccination in the country and the provinces, all governors, supreme judges, owners of landed property, bishops, clergymen, and schoolmasters, should be requested to encourage the same in their respective spheres. The committee has no doubt of the Rt. Rev. Bishop Balles' vigorous support in this case, as his own children have been vaccinated; nor of the physicians' ready co-operation. The committee foresee that if the vaccination should not be continued with every possible carefulness it will in a short time be extinct; and recourse must then be had to import from other places the cow-pox matter, which may prove very unsafe.

I Winslow, Professor, have therefore, agreeably to the request of the committee, not only every tenth day inoculated all those who are come to me; but I have also preserved the vaccine matter in proper glasses that safely can bear conveyance, and a quantity thereof has already been sent to several places in this country, as also to Sweden.

Dated Copenhagen, in the committee appointed by his Majesty for the Vaccination, the 5th December, 1801.

(Signed)

Aaskov; Gulbrand; Callisen; Winslow; Wiborg.

The foregoing having been laid before his Majesty by the college, the Monarch has been pleased to give his approbation to all the proposals of the committee, commanding the same to be communicated to the college for his German domains.

CHAPTER XII.

PRESENTATION OF PLATE BY HIS FRIENDS IN GLOUCESTERSHIRE.—FIRST PARLIAMENTARY GRANT.

THE preceding detail must have convinced the reader that Dr. Jenner considered every thing but his private interest in all that he did, from the time of the publication of his Inquiry. Indeed, he "never sought his own;" and would, at any time, have allowed whatever regarded his personal welfare or convenience to be put aside, provided he could thereby promote the benefit of others. To such an extent did this feeling prevail that he not only sacrificed his professional advantages, but offered likewise to disburse largely and liberally from his own private fortune in order to assist in conveying the salutary preventive to the most distant parts of the globe.

It became those who knew the generosity of his nature and the smallness of his own means to take care that he who had imparted a secret to his fellow-creatures, which in another age would have secured to him boundless wealth and honours all but divine, should not, whilst studying to preserve the lives of others, be himself compelled "to study to live."

He not only had relinquished all possibility of peculiar or private benefit by disclosing in the most unreserved manner the grand result of his labours, but he likewise spent a great deal of time, and incurred much expense in enabling others to profit by his discovery. When all this was done it is to be remembered that he was by no means in independent circumstances; he had a family and many relatives who looked to him for support and protection; and he was still compelled to pursue his practice as a physician.

That, of course, had been very much interrupted by the frequent change of his residence from Berkeley to Cheltenham and London; but, independent of this, his correspondence became so extensive as to occupy almost all his time and to make him a most laborious servant of the public, for their great and exclusive benefit; whilst there was nothing of advantage left to himself but the consciousness that he was so employed.

Under these circumstances it was thought that the magnitude of his discovery and the very disinterested manner in which he was sacrificing his time and his property in diffusing its blessings were fit subjects for the consideration of the British Parliament. After due deliberation, it was determined that his claims should be brought before the House of Commons by petition.

By this mode of proceeding the proofs of the utility of the discovery, and the right of Dr. Jenner to that discovery, would be placed in such an unquestionable shape as to put to silence all gainsayers; and it would demonstrate how much his own private fortune had suffered by his endeavours to serve others.

Previously, however, and in a degree preparatory to a direct application to the great national council, some of the chief personages in Gloucestershire began to take measures to give a testimony of the value in which he was held by those who knew him best and amongst whom his life had been spent.

An advertisement appeared in the Gloucester newspapers expressive of these sentiments. The concluding paragraph pointed also to a remuneration of a different kind, in some degree adequate to his deserts and to which he had the best-founded claim, a claim on the justice and gratitude of the British nation.

The late Earl of Berkeley took the lead in this becoming expression of public feeling. He wrote letters to those elevated by rank and fortune to co-operate with him in this measure. From some he received answers that could not well have been anticipated: such backwardness, however, did not retard the general sentiments of the county. Dr. Jenner's friends exerted themselves with an energy which more than counterbalanced any discouragements they met with. The Countess of Berkeley also applied herself with her usual earnestness to effect the object. Her ladyship, assisted by Mr. Henry Hicks, the Rev. Mr. Pruen, and other personal friends of Dr. Jenner, had the satisfaction, in a short time, of seeing such a list of subscribers as enabled them to give an order for a small service of plate, which was executed by Messrs. Rundells and Bridge, with appropriate devices and the following inscription:—

Presented
by the
Nobility and Gentry
of the
County of Gloucester
to their Countryman
Edward Jenner, M.D., F.R.S.,
as a Testimony of the high sense they entertain
of
those eminent abilities which discovered,
and
that disinterested philanthropy which promulgated,
the Vaccine Inoculation.

The sentiments entertained by the principal promoters of this design will not be uninteresting to those who venerate the character of Jenner. They are expressed with great propriety by the Earl of Berkeley, in the following letter:—

EARL OF BERKELEY TO THE DUKE OF BEAUFORT.

Berkeley Castle, Nov. 18*th*, 1801.

MY LORD,

Having received a proposal from a gentleman in this county for a subscription towards a piece of plate as a public mark of esteem to Dr. Jenner, the person to whom every father of a family owes the greatest obligation in preventing the dreadful effects of the small-pox, by finding out and introducing the vaccine inoculation.

Before the subscription paper is further offered to notice I take the liberty to convey it to your Grace for your approbation, supposing you to be convinced, as I am, of the general benefit it has been and will be to the human race.

Your name to it will sanction the proceeding, and tend to establish the general good that will arise to the world at large, and more particularly to that of the county of Gloucester.

I have the honour to be
Your Grace's most obedient
Humble servant,
BERKELEY.

His Grace the Duke of Beaufort,
&c. &c. &c.

Whilst these things were going on, Dr. Jenner was partly at Cheltenham and partly at Berkeley.

His affairs, however, did not permit him to remain long in this his favourite retreat. He was under the necessity of leaving it, with his family, on the 9th of December, for London.

No one, who has not seen the extent and variety of his correspondence, can form any idea of the labour and anxiety to which he was incessantly exposed at this and subsequent periods of his life. The subject of vaccination, though sufficiently simple, had been but imperfectly studied; and every individual thought himself at liberty to write to Dr. Jenner for that information which they might have obtained from his published writings. His patience was inexhaustible; and I believe he did all in his power to convey such information as was required of him. His own immediate interests now called for his presence in London; as it had been resolved that his claim should be brought before Parliament in its next session.

In a matter of so much moment it was highly desirable that nothing should be lost by overlooking or neglecting any of those measures which either custom or propriety had sanctioned, while attempting to arrest the attention of so distinguished an assembly as the British House of Commons. As the application ultimately referred to a money-grant it was necessary that the consent of his Majesty's ministers should be first obtained. Dr. Jenner had an interview with the premier on this subject, on the 12th of January, 1802. This conference was highly satisfactory. Unqualified approbation was expressed of the measure by the minister, and he fixed upon an early day for the presentation of the petition. Mr. Addington spoke likewise strongly in praise of the discovery, and of the benefit that all the world was likely to derive from it.

The next point regarded the selection of a proper individual to present the petition to the House. As the army and navy had both benefited largely by the practice of vaccination, it was thought advisable to entrust the petition to some of the official gentlemen connected with these departments of the state. Th arrangement however was not adopted. The eminent persons who had formerly evinced so warm an interest in the cause of vaccination did not neglect its author on this occasion. Among these are to be named the Honourable Admiral Berkeley, M. P.; W. Wilberforce, Esq. M. P.; Sir Henry Mildmay, Bart. M. P.; Lord Rous, Lord Berkeley, Lord Egremont, Lord Sherborne, &c. &c. &c. The sentiments of some of these distinguished persons may be collected from their letters written on this topic.

Lord Rous to his Sister, Lady Peyton.

Stenham Hall, January 8th, 1802.

Dear Sister,

I shall be extremely happy to be of every service to Dr. Jenner in my power, but with respect to writing myself to Mr. Pitt I am very doubtful of its being of any service; surely the Duke of York, as head of the army, and Lord Spencer as head of the navy, are the proper persons to take the lead with the Minister: it cannot be supposed that Mr. Pitt, who knows every thing, should not be acquainted with the introduction of cow-pox, though he may not have had time to inquire into the detail of it, or to inspect patients as you and I do. If I do write to Mr. Pitt I think it should be thus: That I have heard the subject is likely to be inquired into by Parliament, and that I shall be very happy either by letter or by a conference with him when I go to town, to state all that I have made myself master of, confident that if he personally entertains any doubts of the security of the practice, or the incalculable benefit likely to accrue to mankind, I shall be able to remove them.

In my own opinion there will be no occasion for very minute inquiries; the Minister will grant a reward, or he will not do it, and it will be decided whether he will do it long before it is brought before Parliament; at least it will not be worth trying for unless Mr. Pitt sanctions the application. Some of the medical men in England will certify that the cow-pox is completely established, and their testimony would outweigh a thousand certificates from those not of the profession.

* * * * *

LORD SHERBORNE TO DR. JENNER.

MY DEAR DOCTOR,

Many thanks for your circumstantial letter; I am sorry to say I do not know Mr. Addington, even by sight; they tell me the King is recovering very fast, and we may expect a drawing-room soon, which I will attend and I will then speak to Mr. Pitt. If patriot Grattan gets 50,000*l.* for his patriotism, the true patriot Jenner deserves much more: I am sure not less; and less would be perfectly shabby to think of. I perfectly recollect Grattan's business:—it was settled among his friends to propose 100,000*l.* for him; determining to ask enough, and fearing that sum should not be granted, one of his most particular friends was to get up afterwards and propose 50,000*l.* which was immediately granted, and he took 47,500*l.* for prompt payment.

I am, my dear Doctor,

Yours most truly,

SHERBORNE.

Sherborne, April 23d, 1801.

THE HON. ADMIRAL BERKELEY TO DR. JENNER.

Friday evening.

DEAR SIR,

I have arranged every thing with respect to the Committee, and as I find Mr. White was employed by you to draw up the petition, I consulted him upon the best means of conducting it. He wishes to see you with the *heads* of the allegations you mean to prove, and I have therefore desired him to write to you upon the subject, because he will put us in the way of calling evidence with the least inconvenience; as the respectable characters who are likely to appear will probably wish to be kept as little time as possible, and of course we ought to accommodate them as much as the nature of the case will admit. If you wish for any assistance which you may think me capable of affording before you see Mr. White, I hope you will believe you cannot afford me a greater satisfaction than by employing me: being with great truth

Sincerely yours,

G. BERKELEY.

I think if you was to write a letter to Mr. Wilberforce desiring him to second the motion, or to Mr. Lascelles, his colleague, the representatives of so large a county as Yorkshire might be thought most creditable in an affair of this sort, especially as the most proper person (the Marquis of Worcester) will not be in town.

I remain sincerely yours,

G. B.

FROM W. WILBERFORCE, ESQ. TO DR. JENNER.

Palace Yard, Feb. 24th, 1802.

MY DEAR SIR,

I have often thought of addressing you on the subject we conversed about formerly, that I mean of your valuable discovery becoming the topic of parliamentary discussion, with a view to your receiving some compensation for your eminent services to the community. I hoped long ere now to see the matter brought forward. I always intended whenever it should be so to give you my best assistance on a principle of duty. I really thought, as I told you, that there were reasons why I was by no means an eligible introducer of the subject, and I could not just now undertake it, on account of my being engaged to render a similar service (though contrary to my own judgment) to another gentleman. But are you aware that Friday next is the last day for presenting private petitions, and that a petition is the proper mode of bringing your discovery before Parliament?

If I can be of any use in advising you I shall be unaffectedly glad, and in rendering you any assistance I am able.

At all events, I am persuaded you will do justice to the motive which prompts me to address you thus frankly, and believe me with esteem and regard,

Dear Sir,

Your faithful servant,

W. Wilberforce.

Sir Henry Mildmay addressed him on the same subject: he said " I consider the value of your discovery is now so fully ascertained and confirmed by experience, that you would be unjust to your own interest if you were to postpone your application any longer. If I can be of any service to you in taking the part you may assign me in prosecuting your claims I shall be very happy to do so, be that part what it may; as, perhaps, there are few members of the House who can speak with greater authority on the extensive benefits the public have derived from your discovery than myself. I still think that your petition would come forward with greater effect from some person high in office, but should you be disappointed in all the promises you have received, I will very readily bring the subject before the House for you; or I will second it, whenever it does come forward: in whatever shape it may appear I shall consider myself bound in duty to give it every support in my power, and to give my reasons for doing so."

On the 17th of March, 1802, the following petition was presented.

To the Honourable the Commons of the United Kingdom of Great Britain and Ireland, in Parliament assembled.

The humble Petition of EDWARD JENNER, Doctor of Physic,

SHEWETH,

That your petitioner having discovered that a disease which occasionally exists in a particular form among cattle, known by the name of the cow-pox, admits of being inoculated on the human frame with the most perfect ease and safety, and is attended with the singularly beneficial effect of rendering through life the persons so inoculated perfectly secure from the infection of the small-pox.

That your petitioner after a most attentive and laborious investigation of the subject, setting aside considerations of private and personal advantage, and anxious to promote the safety and welfare of his countrymen and of mankind in general, did not wish to conceal the discovery he so made on the mode of conducting this new species of inoculation, but immediately disclosed the whole to the public; and by communication with medical men in all parts of this kingdom, and in foreign countries, sedulously endeavoured to spread the knowledge of his discovery and the benefit of his labours as widely as possible.

That in this latter respect the views and wishes of your petitioner have been completely fulfilled, for to his high gratification he has to say that this inoculation is in practice throughout a great proportion of the civilized world, and has in particular been productive of great advantage to these kingdoms, in consequence of its being introduced, under authority, into the army and navy.

That the said inoculation hath already checked the progress of the small-pox, and from its nature must finally annihilate that dreadful disorder.

That the series of experiments by which this discovery was developed and completed have not only occupied a considerable portion of your petitioner's life, and have not merely been a cause of great expense and anxiety to him, but have so interrupted him in the ordinary exercise of his profession as materially to abridge its pecuniary advantages, without their being counterbalanced by those derived from the new practice.

Your petitioner, therefore, with the full persuasion that he shall meet with that attention and indulgence of which this Honourable House may deem him worthy, humbly prays this Honourable House to take the premises into consideration, and to grant him such remuneration as to their wisdom shall seem meet.

The prime minister, Mr. Addington, (now Viscount Sidmouth) informed the House that he had taken the King's pleasure on the contents of the petition, and that his Majesty recommended it strongly to the consideration of Parliament. It was referred to a committee, of which Admiral Berkeley was appointed chairman. This honourable gentleman rendered most essential service on this occasion, and bestowed unceasing attention on all the details connected with this very interesting investigation. The points to which the committee chiefly directed their inquiries were

First, The utility of the discovery itself.

Secondly, The right of the petitioner to claim the discovery.

Thirdly, The advantage in point of medical practice and pecuniary emolument which he has derived from it.

Upon all these matters the committee reported, after examining a number of witnesses of the highest character and most extensive experience in the profession. The testimony, likewise, of persons not professional was admitted; among these we find his Royal Highness the Duke of Clarence, the Earl of Berkeley, Lord Rous, and Mr. Gardner. The result of this examination went to prove, in the opinion of the committee, that the discovery of Vaccine Inoculation was of the most general utility, and that it tends to eradicate, and, if its use become universal, must absolutely extinguish one of the most destructive disorders by which the human race has been visited.

A great mass of written evidence was also submitted to the consideration of the committee. It related to the extensive and successful practice of vaccination in every quarter of the globe; and showed the estimation in which the discovery was held in other countries. The committee likewise investigated the difference between the new practice and the inoculated small-pox, and that disease when caught in the natural way.

With respect to the new practice the oral depositions, as well as all the written documents from abroad, were uniform and decisive in favour of Dr.

Jenner's claim to originality in the discovery. Some pretensions were advanced adverse to these claims, as well as to the utility of the discovery itself, but after the closest examination they had no weight with the committee. As, however, they were urged with considerable pertinacity both before the committee, and in another form, it will be necessary to examine them.

Upon the last division of the subject the committee found that Dr. Jenner had not only reaped no advantage from his discovery, but that he was a considerable loser: that he had relinquished the prospects of emolument, by propagating and extending his important discovery, and rendering it rather of universal utility to the human race than of pecuniary or private advantage to himself.

As the evidence on which this report is founded has been already published at length by the Rev. George Jenner, I will only mention a few of the principal statements, and refer to that work for fuller information.

Sir Everard Home established the fact that Dr. Jenner had, in the year 1788, carried a drawing of the Variolæ Vaccinæ, as the disease appeared on the finger of a milker, to London. Mr. Hunter urged him to prosecute the inquiry, and mentioned the prophylactic powers of the cow-pox both privately and in his public lectures. Sir Everard proved his own confidence by inoculating one of his own children with it.

Dr. Ashe's evidence went to show that the vaccine had not been known abroad till after Dr. Jenner's publication of his "Inquiry" He considered the discovery of the utmost importance, and had vaccinated three of his own children.

Sir Walter Farquhar said that Dr. Jenner's was the greatest discovery that had been made for many years: that if Dr. Jenner had kept it a secret he might have made 10,000*l.* a-year.

Mr. Cline coincided in this opinion, adding further that he considered it the greatest discovery ever made in the practice of medicine.

Dr. Bradley never heard of any one as the discoverer except Dr. Jenner, and believed that he might have, at first, made 10,000*l.* per annum, and in five years double that sum had he kept the discovery a secret.

"*Two millions*" he said, "have been already vaccinated in the world; of these two millions not an individual is known to have died in consequence of the affection."

Dr. Sims considered it the greatest and most useful discovery ever made in medicine; and that Dr. Jenner, had he kept it a secret, might have died the richest man in these dominions.

Sir Gilbert Blane had at first been sceptical, and his prejudices were rendered stronger than ever by the events which had occurred at the Small-pox Hospital. Further inquiry removed his doubts, and led to a complete conviction of the truth of all that had been alleged by Dr. Jenner.

Dr. Baillie said "I think that the cow-pox forms

an extremely mild disease, and that when a patient has properly undergone it he is perfectly secure from the future infection of the small-pox: and further, if Dr. Jenner had not chosen openly and honourably to explain to the public all he knew upon the subject, he might have acquired a considerable fortune. In my opinion it is the most important discovery ever made in medicine."

Lord Berkeley's evidence was of a strong practical nature, drawn from the experience of his own family. He mentioned a very important fact, that an old servant of his "now seventy years of age, who had the cow-pox from milking cows when a *boy fifteen years old*, from that time has never been in the least cautious in guarding himself from the small-pox, but, on the contrary, has exposed himself repeatedly, without being sensible of its effects."

Lord Rous gave evidence to the same effect; his lordship's own child, three months old, was vaccinated the year before, and he was perfectly satisfied that it was protected from small-pox.

His Royal Highness the Duke of Clarence stated that he began Vaccine Inoculation in 1798; since which time he had caused it to be practised among his household and farm-servants. His Royal Highness brought forward some very strong facts drawn from his own observation, illustrative of the prophylactic virtues of cow-pox. His Royal Highness had offered to inoculate with cow-pox all his adult servants who had not had the small-pox. A postilion of the name of Johnston positively refused to be vaccinated. About eighteen months afterwards he caught the small-pox. It was of a virulent and confluent kind. Several children who had been previously vaccinated were constantly exposed in the room where this boy lay, but they escaped unhurt.

Mr. Gardner proved that Dr. Jenner in the month of March, 1780, had informed him of the nature of the cow-pox, and communicated at the same time his design of perpetuating that affection from one human being to another.

It is needless to specify particularly the evidence of the other gentlemen who gave their testimony in favour of Dr. Jenner's claims, but I cannot conclude this abridgment without alluding to what was said by Dr. Lettsom and Mr. Ring. The former gave some very valuable statistical information respecting the mortality of small-pox before and after the introduction of inoculation for that disease. He spoke from extensive experience of the virtues of vaccination, and of the accuracy and originality of Dr. Jenner's information. He thought, from the success of the Suttons in monopolizing small-pox inoculation, that Dr. Jenner might have kept his practice a secret, and gained incalculable riches.

Mr. Ring said that Dr. Jenner's discovery was beyond all comparison the most important and valuable ever made by man. This, he said, was the opinion of all his correspondents, and like all others who could form an accurate opinion, he believed that Dr. Jenner's annual emoluments might have been at least 10,000*l*.

The evidence of every other medical gentleman who was examined, with the exception of that of an adverse nature, which I am presently to notice, concurred alike in the utility of the discovery and in favour of Dr. Jenner's claim to originality.

To impugn this claim efforts of a very extraordinary kind were made. The individual who signalised himself on this occasion has already been made known to the reader both by the disasters which attended his early attempts at vaccination, and by his officious eagerness in propagating his own errors through the medium of a public institution.

Dr. Pearson learnt, from the correspondence which he opened with medical men in different parts of England, that cow-pox was much more extensively epizootic than had been imagined. It was known in many counties, but it was in Dorsetshire and Devonshire that it seems to have been particularly observed.

The Rev. Herman Drew had received information on the subject from Mr. Downe, surgeon, in Bridport; Mr. Bragge, of Axminster; and Mr. Barnes, of Collyton. The information received from those gentlemen was conveyed by Mr. Drew to Sir George Baker. It is impossible to ascertain the precise contents of these documents. It is, however, certain that they were not deemed of sufficient importance by that eminent and enlightened physician either to be given to the public or to render him very anxious for their preservation, as they remained unheeded, and have been allowed to perish. It is probable, therefore, that the knowledge obtained by Mr. Drew from his medical acquaintance in the country did not extend beyond the simple fact which had been observed in those districts where cow-pox prevailed; and, that in communicating it to Sir George Baker, he really did no more than Jenner himself had done when he told John Hunter what he had heard of cow-pox in Gloucestershire, nearly thirty years before he had completed his inquiry.

Of Dr. Pearson s statements before the Committee of the House it is by no means easy to give a lucid view. He sets out with asserting that he was *conversant* with the vaccine inoculation ever since January 1799, and as though he would bespeak credit for himself by exciting a partial sympathy for Dr. Jenner's claims, he thus proceeds: "I think it but justice to Dr. Jenner to state that I am acquainted with the practice of inoculation of persons for the small-pox, who on good evidence have been said to have gone through the cow-pox since June or July 1798—the result of which was that they could not receive the small-pox infection." He then goes on to say, in answer to questions put by the Committee, that his own knowledge of the vaccine inoculation was obtained by him "*in the first instance from Dr. Jenner;* afterwards I got *information* from other sources." These sources have been already mentioned—and, with respect to the information derived from them, Dr. Pearson declares that he *imagines* that *it* and Dr. Jenner's first publication "were independent of each other." He next states that Mr. John Hunter did not mention in his lectures the *inoculation* of the cow-pox, but simply that persons who had had that disease could not take the small-pox, and that it had not been known to prove fatal in any one instance; that the Rev. Herman Drew did not lay claim to the discovery of inoculating with vaccine matter from one human being to another; that *that* discovery was exclusively Dr. Jenner's; that the events mentioned in the documents handed in took place earlier than 1798, because immediately on the publication of Dr. Jenner's Inquiry in that year he, Dr. Pearson, wrote to the gentlemen who furnished the information, and they "*immediately communicated their cases of vaccine inoculation* without appearing to be acquainted with Dr. Jenner's work."

The reader, from this last sentence, would certainly believe that Dr. Pearson's correspondents had themselves practised vaccine inoculation, but the very reverse was the fact. Mr. Drew in his letter to Dr. Pearson merely states, on the authority of Mr. Dolling of Blandford, that he (Mr. D.) had inoculated many hundreds for the small-pox who said they had had the cow-pox, and that very few of them took the infection.

The same person mentioned the conduct of a mother who had caused five of her children to play with the teat of a cow, by which they were infected, and that they resisted small-pox afterwards. He added that a farmer of the name of Jesty, at Yetminster, had gone a step further than this good woman; that he inoculated his wife and children with matter taken from the teats of a cow that had the cow-pox. In about a week from the time of inoculation their arms were very much inflamed, the patients were so ill as to require medical assistance; yet they had since been inoculated for the small-pox by Mr. Trowbridge, but did not take it.

A letter from Dr. Pulteney to Dr. Pearson gave the same facts mentioned above, and on the same authority, namely, that of Mr. Dolling. In another letter from Mr. Drew he mentions that about twenty years before a woman inoculated her children with matter taken from the cow on the point of a large needle. This dame was not only a vaccinator herself, but her "*children have since inoculated their friends and neighbours whenever an opportunity has offered.*" *

Excepting the attempts of Farmer Jesty and Mrs. Rendall and her children, nothing like a case of vaccine inoculation is reported by any of Dr. Pearson's correspondents; and granting that these really were what they purport to be, they do not in the slightest degree affect Dr. Jenner's claims. They did not advance the knowledge or the practice of vacci-

* See Mr. Bragge's letter to Sir William Elford, Bart. Evidence, page 160.

nation beyond what casual observation and popular rumour had rendered common in many districts: if, indeed, they ever took place (which I think more than doubtful), they were quite unknown to Dr. Jenner, and had it not been for *his* publication they never would have been drawn forth from their obscurity.

The documents presented by Dr. Pearson were evidently intended to prove that vaccine inoculation had been practised by others before Dr. Jenner. His second examination before the Committee had a different object. It went to show that, though Dr. Jenner promulgated the practice of vaccination, he really knew very little about the matter; that his opinions as to its origin were erroneous; and that it required the experiments and labours of other observers to correct his mistakes: and that he and Dr. Woodville had the chief merit in the establishment of the Vaccine Inoculation. Thus, after making due allowance for the claims of farmer Jesty and the valuable and scientific investigations of others, nothing was left to Jenner save that of being the publisher of a provincial rumour, the nature of which he himself did not fully understand! What must have been the feelings of those who could utter statements capable of leading to such inferences?

The disasters which occurred at Petworth, the Small-pox Hospital, Hanover, and many other places, (for they all directly or indirectly may be traced to the same source) bear testimony alike to the accuracy of those who made them and to the great modesty of their pretensions!

Mr. Keate, Surgeon-general to the Army, was the next gentleman whose evidence tended to call in question the full merit of Dr. Jenner. Mr. Keate admitted that he was the person who published the "cases of inoculation which gave rise to further investigation and improvement, but to whom to attribute the discovery I am unable to say." He delivered in to the Committee certain manuscripts of a Mr. Nash or Naish, a surgeon of Shaftesbury, which were supposed to prove that *he* had practised vaccine inoculation. They only show that he could not infect with variolous virus those who had casually been affected with cow-pox. His mind had certainly been directed to the subject, but it has been ascertained, beyond the possibility of doubt, that his cases of alleged vaccination were really cases of small-pox inoculation. The documents will be found in a letter from Dr. Pew to Mr. Creaser which is printed at the conclusion of Mr. Jenner's "Report of the Evidence."

Among the testimonials in favour of Vaccine Inoculation the Committee selected a few of those which had been sent from foreign countries. One was an address of the members composing the Jury of Health, and the Medical Committee of the Department of the Somme, to his Excellency the minister plenipotentiary of England at Amiens. Another was from the Central Committee of Vaccination at Paris, signed by Portal, Halle, Sabatier, De Jussieu, Fourcroy, Parmentier, Huzard, Corvisart, &c. &c. &c.

The certificates, likewise, granted by Admiral Lord Keith and General Lord Hutchinson to Drs. Marshall and Walker were among the documents selected by the committee of the House of Commons.

Dr. Jenner, who was examined on the first day, corroborated the evidence which he delivered in the form of a printed paper, by vouchers from correspondents in various parts of the world referring to, at least, 100,000 cases of successful vaccination.

The Committee, while referring to the pretensions advanced in opposition to Dr. Jenner's claims, took, on the whole, a just view of these pretensions; but they were not equally correct in estimating the real nature of his merits. They truly observed that in various parts of England an opinion was current, among the common people employed in dairies, that the cow-pox was a preventive of the small-pox. "It appears, they add, not improbable that in some very rare instances this knowledge was carried one step further, and that the cow-pox was communicated either by handling the teat or by inoculation from the animal, for the purpose, and with the intention of securing against the danger of small-pox: but the practice of which Dr. Jenner asserts himself to be the original inventor is the inoculation from one human being to another, &c. &c." With all deference to the Committee, it must be remarked that this forms but a very small part of his claims to public gratitude and remuneration. His chief merits are, as has been already proved, of a very different description. They consist in the patient, laborious, and original investigations which enabled him to extract correct and scientific information from the most unpromising materials; to divest popular tradition of all its obscurity and uncertainty; and to elicit from vague and contradictory rumours the most accurate and valuable truths. It was not till all this was achieved that he ventured to perform his first inoculation: and, had it not been so, this most admirable and interesting experiment had been as valueless and inconsequential as the *alleged* vaccinations of Mrs. Rendall and Farmer Jesty.

When we duly consider the nature of the adverse testimony, it cannot fail to increase our admiration of Dr. Jenner's genius. The subject had been forced upon the attention of many professional gentlemen in different parts of England. It had been laid before some of the most eminent of the faculty in London, but no beneficial result followed.

Dr. Jenner's own endeavours to instigate his brethren were alike unsuccessful. His earnestness on this topic and the discouragements he met with prove at once his liberality and benevolence, and the constancy and perseverance of his nature. Let the reader recollect that the information communicated by him to Mr. Hunter was often mentioned by that enlightened physiologist both in public and private. From this source Dr. Pearson himself acknowledges that he derived that information which, for nine successive years anterior to the publication of Dr. Jenner's work, enabled him to detail to his pupils in his lectures the opinions of the latter as to the prophylactic virtues of cow-pox, and his design of propagating that affection by inoculation with a view to the extinction of small-pox.* Still further to prove how extensively Dr. Jenner's doctrines had been disseminated in London, I must repeat that they had been referred to by Dr. Adams in 1795, in his work on Morbid Poisons; and in the following year by Dr. Woodville in his History of Inoculation. It was, therefore, with a singularly bad grace that a physician, who was a public lecturer in London and who, for at least nine years, had been announcing to his pupils by far the most remarkable pathological fact that ever was observed, should, when the subject was divested of all its difficulties by Jenner and brought forward by him with all the modesty and simplicity that so peculiarly distinguished his character, have forgotten his former expressions of respect and admiration, and busied himself, with an activity and pertinacity that might have been useful in a better cause, to find occasion to rob Dr. Jenner of every title to distinction, whether as an original discoverer, or as an accurate observer.

* See Dr. Pearson's Inquiry, page 8.

The Report of the Committee of the House of Commons was brought up the 2nd of June, 1802. An abstract of the proceedings which took place on that occasion is here subjoined.

Admiral Berkeley, chairman of the Committee, said that in their investigations the Committee did not confine themselves (as is usually the case) to the examination of *petitioner's witnesses* exclusively—bnt their mode was "to sift out any case which could make against petitioner's evidence. This conduct, which certainly may appear to bear hard upon the petitioner, has proved a matter of fresh triumph to him, for although we descended to sift out information from every anonymous letter, though we raked the very kennels for information against this practice, all we were able to get is fully printed in the Report."—"Great as is that discovery (of the Longitude) I really cannot look upon it in any view to be compared with this of Dr. Jenner's, which is unquestionably the greatest discovery ever made for the preservation of the human species. *It is proved* that in these united kingdoms alone 45,000 persons die annually of the small-pox; but throughout the world what is it? Not a *second* is struck by the hand of time, but a victim is sacrificed at the altar of that most horrible of all disorders, the small pox, &c. &c. &c.

"I shall therefore move that a sum not less than 10,000*l.* be granted; but when I do this I declare I do not think it sufficient; if the House should think it right to adopt any *larger* sum, I shall hold myself free to vote for it."

Sir Henry Mildmay did not think the sum proposed at all adequate. By keeping the discovery a secret Dr. Jenner might as readily have realized 100,000*l.* as any smaller sum. He concluded by moving as an amendment 20,000*l.*

Mr. Bankes, Corfe Castle, was for acting under a sense of paramount duty as guardians of the national purse, and that public economy was to be consulted. Dr. Jenner could have fully remunerated himself had he kept his own secret; and he could even yet be remunerated by his practice, without a parliamentary grant.

Mr. Windham urged as one great merit of Dr. Jenner that he did not keep his discovery a secret; had Dr. Jenner done so, Mr. Windham scarcely knew what sum the House might not be called on to grant for the purchase of such an invaluable discovery as that which went to the complete eradication of such a dreadful disorder as the small-pox. Dr. Jenner could *now* enjoy no benefit from monopoly, as he had enabled, most generously and humanely enabled, every medical man to apply his discovery in practice, for the public safety and for the preservation of lives incalculable. He wished a reward to Dr. Jenner if it were only to encourage others to *divulge* their *discoveries* for public advantage. The sum proposed for Dr. Jenner was the *least possible.*

Sir James Erskine Sinclair followed on the same side. He noticed the *actual expense* which Dr. Jenner had incurred in prosecuting his inquiries, at the least 6000*l.*; if then 20,000*l.* were objected as too large, he would propose 15,000*l.* as in some measure remunerating.

Mr. M. A. Taylor thought, as Dr. Jenner's expenses had not been stated in the Report of the Committee as a ground for the vote, that the Committee or its Chairman should be instructed to *report progress.*

Mr. Hobhouse said that Dr. Jenner s expenses might very fairly be adduced in argument; as the Committee had considered them in framing the report and resolutions thereon.

Mr. Fuller thought the larger sum (20,000*l.*) due to Dr. Jenner—more especially as he could look to *no remuneration* by *patent.*

The Chancellor of the Exchequer said that whatever sum of money the House might vote as a *future reward for his merit* he had already received the highest reward in the approbation, unanimous approbation, of the House of Commons—an approbation most richly deserved, since it was the result of "the greatest, or one of the most important discoveries to human society that was made since the creation of man." That the value of the discovery was without example, and *beyond all calculation,* were *points not to be contested,* for they were made out by convincing evidence; and that he (Dr. Jenner) had precluded himself from *great emoluments*, by the generosity of his own conduct, was also most manifest: but he (the Chancellor) had also a duty to discharge towards the public in voting away the public money, and when he reflected on *the other advantages* that the Doctor must derive from this vote, *he* was for the smaller sum. In saying this he was rather pursuing the sense he had of his public duty, than his own feelings. He had, however, the satisfaction to reflect that this discussion had given to Dr. Jenner a *reward that would last for ever*, and also that the comfort of his family would be *amply provided* for in his *extended practice*, by means of the sanction of that House.

Mr. Grey said that, from the tenor of the Right Honourable Gentleman's speech, and from his owning the importance of the discovery, he hoped that he (the Chancellor) would have concluded by concurring with the amendment.

He had heard *no good reason* for limiting the sum to 10,000*l*. It ill became the House to diminish *their* reward, because Dr. Jenner's merit was of such a nature as to yield gratification to his own benevolent feelings. He therefore hoped that the House would vote for 20,000*l*.

Mr. Wilberforce stated that Dr. Jenner had been engaged in completing this discovery upwards of twenty years. He was not to be considered as an adventurer who might hope by it to push himself into practice. He had already attained to great celebrity in his profession, and an extensive practice, which he had sacrificed to completing this discovery. In every view he thought the larger sum ought to be granted.

Mr. Courtenay said, it appeared in evidence that 40,000 men were annually preserved to the state by Dr. Jenner's discovery—by this number 200,000*l.* were annually brought into the Exchequer; and certainly Dr. Jenner, the efficient cause, was well entitled to 20,000*l.*

The question was then put that the words ten thousand pounds do stand part of the resolution; when the Committee divided,—

Ayes, 59. Noes, 56.—Majority, 3.

Thus we have seen in what manner the assembled representatives of our country, supported by the declaration of the king's minister in his place in parliament, rewarded the discovery of Jenner!

Before we pass altogether from this important incident in his life it may not be an uninteresting part of our general subject to advert to the annals of former ages, and learn from history what was their estimate of the services rendered to society by the great masters of the healing art.

Among the civilized nations of antiquity Greece surpassed all others in medical research and knowledge: and many of her schools, as those of Rhodes, Crotona, Cnidos, Cos, and Smyrna had obtained high reputation at an early period. Democedes rendered Crotona illustrious by his humanity, not less than by his skill. He, whilst a captive to the Persians, restored to health Darius their king, who repaid the benefit with great wealth and honours. But the reward of Democedes stopped not here. He asked and obtained from the haughty and enraged monarch pardon and freedom for some Egyptian physicians who had unskilfully treated the disease of Darius in its commencement, for which error they had been condemned to death. See Herodotus L 3. c. 129.

But the brilliant fortune of the Coan school outshone that of all the rest, by producing the Father of Medicine, the great Hippocrates—not only did his fellow-citizens of Cos impress their money with his likeness, but all *Greece* joined in this tribute to his merits—Greece, then the most enlightened region in the world, " quæ communi consilio, quod venientem ab Illyriis pestilentiam prædixerat, discipulosque ad auxiliandum circa urbes dimiserat, honores illi, quos Herculi, decrevit." * The Athenians also enrolled him as a citizen, presented him with a golden crown, and granted to him and his posterity a public maintenance † τὴν ἐν πρυτανείῳ σίτησιν ἔδοσαν. This last was considered by far the highest honour that could be bestowed on any human being by the Athenians. He was received by the Thessalians and Argives with similar marks of profound respect. Honours the most splendid were accumu-

* Plin. Nat. Hist. L. 7. c. 37. † Soran. in Vit. Hippocr.

lated on him by the Coans, more especially because by his influence, and by his procuring an alliance with the people of Thessaly, he warded off from his own country a war which the Athenians were preparing to wage against Cos. The fame of Hippocrates and the prodigal offers of riches and of honours were not confined to Greece:—his high character for skill and science had reached "the king of kings" the mighty Artaxerxes, in whose immense army a pestilential disease had broken out. The Persian monarch was informed by Pætus that Hippocrates was the first of physicians, the father of health, the preserver, the reliever from pains and diseases, the leader of the heaven-born science.

Forthwith Artaxerxes orders Hystanes, his prefect of the Hellespont, to offer to Hippocrates as much *gold* as he might desire; and that he should rank with the first nobility of Persia. Hystanes sends this message of the king to Hippocrates, who immediately replies: "In answer to the letter you sent me as coming from the king, write back to him immediately, that whatever *we* require for food, for vesture, for dwelling, and all the necessaries of life, we have in abundance;—but to accept of the wealth of the Persians would not be just in me,—nor yet to set free from their diseases barbarians who seek to lord it over the Grecians." Hippocrates himself thus wrote to Demetrius: "The king of the Persians has sent for *us* to him, not knowing the 'word of wisdom' has more power with me than that of gold."

To the physicians of the school of Smyrna, especially to Erasistratus, honours scarcely inferior were paid than to the most dignified magistrates. Their effigies were stamped on the public money, and their names were inscribed on these coins and medals, in common with the images of such of the divinities as presided over health. Thus, on one side of the coin or medal was impressed the figure of the deity, as of Hygeia, on the other the effigies of Æsculapius, with the insignia of the medical art.——

No sooner was the amount of the parliamentary grant announced than every unprejudiced mind that had attended to the claims of Dr. Jenner felt its extreme inadequacy. In the early part of the preceding century the House of Commons had voted a much larger sum for importing a silk-throwsting machine from Italy: and the liberality of their grants on other occasions proved that they did not estimate Dr. Jenner's services by a money-price.

It gives me pleasure in regard to this point to mention the conduct of a physician of distinguished genius and learning, whose early opinions were unfriendly to vaccination. The late Dr. Beddoes, on the 4th of June, 1802, (only two days after the debate) wrote a letter to the editors of the Medical and Physical Journal, which contained the following expression: "With as high a sense of what mankind owe to Dr. Jenner as has been expressed by any of your correspondents, I cannot but be deeply mortified at the smallness of the parliamentary reward which he is likely to receive. The largest sum now proposed must, I think, be felt as very inadequate, and without a national subscription the communication of discoveries of immediate and general utility will be checked." "I hope to see by the newspapers, or your next number, that many others feel as I do. If professional men exert themselves they will find abundance of families who have received, or hope to receive, benefit from Dr. Jenner s labours, ready to help towards advancing his fortune; and he surely will not blush to receive such proofs of their sense of obligation. Probably those very members of Parliament who, from a sense of duty, showed themselves most sparing of the public purse, will be among the most forward to open their own."

The same subject was taken up in the subsequent number of the same journal, by Dr. Langslow of Suffolk. He asks "what is the amount of such a sum (10,000*l.*) when divided amongst the individuals of the populous Islands of Great Britain and Ireland, at this time, perhaps, amounting to fourteen millions? A very little calculation will inform us that this sum is no greater than amounts to the demand of one penny each. And let it be observed and remembered that every one of these individuals, without any exception, is nearly and dearly interested in this discovery. I cannot, therefore, but suppose that the motion for the vote of 10,000*l.* was carried without properly adverting to the value of the discovery, or the merits of the discoverer, whose candour, ingenuousness, and humanity must ever reflect the greatest honour on our nation."

Sir Gilbert Blane felt so strongly on the subject that he drew up an address, intended for circulation, of which the following is an abstract.

"Public demonstrations of gratitude towards the benefactors of mankind have ever been deemed just and politic. Our legislature has frequently acted on this view by rewarding individuals for eminent services to the state. National bounty was never extended on fairer grounds to any individual than in the case of the author of Vaccine Inoculation. It is, however, the universal voice of this as well as other nations that the remuneration given to Dr. Jenner is greatly inadequate to his deserts, and to the magnitude of the benefit his discovery has conferred on mankind: yet no blame can, on this score, be fairly imputed to our country or its representatives. The decision for the smaller sum was carried by a very small majority (of three), and even that majority was casual, and depended on circumstances always incidental to a popular assembly, such as the absence of members at the moment of division of the house, &c. &c.

"When, however, public opinion on this topic, both at home and abroad, is considered; as well as the great saving of human life throughout the world, which may, in a degree, be estimated by the former loss of life in this metropolis alone as arising from small-pox; 2500 persons, at least, falling victims to it annually; and throughout Great Britain and Ireland not less than 45,000; with a like proportion in most other countries in the world, even under the mitigation of that disease by means of inoculation: this practice, though a benefit to the limited number of the community who submit to it, or can avail themselves of it, has been injurious to society at large, by spreading the contagion far and wide, and thereby increasing the absolute mortality of that most pestilential and fatal disease. When in contrast with this dreadful yet true view of the existing state of small-pox it is considered that *Vaccine* Inoculation produces a very mild affection, unattended with any danger, and a preventive morally certain for each individual undergoing its salutary influence; but above all, that by means of it, when rendered universal, small-pox will be altogether annihilated: when, I say, all these points are fairly and fully considered and taken into account, the general voice respecting the inadequacy of the remuneration as yet bestowed on Dr. Jenner will be completely borne out: and but one universal wish prevail throughout the world that an additional and permanent testimony of public gratitude and justice should be given by the great community of mankind; a testimony which, whilst it shall convey a becoming and well-merited reward to the author, will at the same time tend to increase the utility of this great discovery by diffusing the knowledge of its high value, and thereby recommending it to universal practice; and thus hasten the consummation of its benefits to the human race, namely, the extinction of small-pox. It is, therefore, proposed that a general subscription should be opened in this and other countries, with a view to confer a substantial and perpetual boon on Dr. Jenner and his family.

"It is proposed that the sum so raised shall be placed in a fund, to be vested in trustees; and that the interest thereof should belong, as a perpetual annuity, to Dr. Jenner and his successive representatives bearing his name."

It is gratifying to observe the coincidence of feeling on this subject, among the liberal and eminent of his own profession. Dr. Lettsom, of London, (soon after the vote of the House of Commons for 10,000*l.* as a national remuneration for the discovery and promulgation of the vaccine inoculation), wrote to Dr. Jenner in this animated strain, "I was truly chagrined on seeing the niggardly reward voted by the House; and had double that sum been asked, it would have been granted: however, as an individual, I am not disposed to stop here; but immediately to set on foot a subscription that should invite every potentate and person in Europe, America, and Asia, because every avenue of the globe has received, or may receive, your life-preserving discovery. This subscription should not be for you, but it should be a fund the interest of which should be for ever devoted to the name of Jenner.'

These liberal sentiments were generally shared by almost every respectable professional man in the kingdom; and at a future time they were adopted by the nation at large through the medium of their representatives. Dr. Pearson alone, I believe, opposed the general feeling, and by placing himself on this "bad eminence" he has compelled me to advert again to his proceedings. On the fourth of July he sent a letter to the editor of the Medical and Physical Journal complaining of the conduct of the Committee of the House of Commons; and at a subsequent period he enlarged the statement of his grievances into a work purporting to be an Examination of the Report of the Committee, &c.

When he wrote this work he seems to have relied more on the forbearance of the House than on the justice of his cause. It is now quite unnecessary to examine this publication at length. The universal voice of the civilized world has alike vindicated the conduct of the Committee and the reputation of Jenner. Some of his friends thought that he might reply to Dr. Pearson's "invidious attack;" but he declined, observing that "as he had never thought it worth his while to notice any of Dr. Pearson's former remarks, he should now pursue the same line of conduct and treat his present publication with silent contempt." Two gentlemen, however, felt themselves called upon to expose the fallacies of Dr. Pearson. Mr. Hicks of Eastington, and Mr. Creaser of Bath, each published observations on his "*Examination*" of the Report of the Committee. The friends of Jenner, and of truth, must indulge the hope that the refutation then given to statements evidently written under feelings which, I trust, have passed away, has carried the same conviction to Dr. Pearson's mind as it has done to others.

From the time of Dr. Jenner's arrival in Bond-Street, on the ninth of December 1801, to the conclusion of the Parliamentary Inquiry his time was much occupied in collecting information and arranging facts for that interesting investigation. When he himself was examined before the Committee of the House his simple and unadorned narrative, and the evidence by which it was supported, gave so much satisfaction that they wished not to call any more of the faculty. "I could have given," he says in communicating this incident to a friend, "fifty times as much as they have gotten. One gentleman gave in upwards of 10,000 cases of successful inoculation.'

As the petition to the House of Commons did not specifically mention the expenditure and losses that he had incurred in establishing the practice of vaccination, he was therefore called upon to make them known to the Committee. He stated that the prosecution of the inquiry had led him into a train of inevitable expense, the greater part of which had arisen from having been under the necessity of living much in London. "Since I first made my discovery public I was compelled to adopt this measure from observing the confusion that was arising among practitioners in the metropolis, in some measure from a misrepresentation of my facts by some, and a too careless observance of them by others. Foreign nations, too, were sending deputies to inquire into the new practice, and as my aim was to diffuse the knowledge of it as widely as possible, and as expeditiously, this work I was confident could not go on so well by correspondence only, as by a constant personal intercourse. My receipts arising from the practice have gone but a little way in reimbursing me. My private affairs, as my time was so incessantly occupied in establishing the new practice, have of course experienced that derangement which neglect always brings on. This exposed me to a serious evil; and I never could have persevered, to the obvious injury of my family, had I not been buoyed up by a confidence in the generosity of my country." He then goes on to state that by leaving his place of residence in the country where he had been established many years in a pleasant and lucrative profession, which after so long an absence it was not probable he could ever regain, he sustained another serious evil. The minor expenses, such as postage, &c. &c. he forbore to mention.

After the Committee had finished its labours he expressed his sentiments to the following purport. He returned thanks for the candour and patience with which the inquiry had been conducted. He remarked that gentlemen of the highest rank and talent in the medical profession had proved that the most loathsome pestilence which ever afflicted human-nature might not only be stayed, but finally extirpated; and that the feeble efforts which ignorance or evil passions may have made to counteract the force of such evidence never could produce any permanent effect on the minds of those capable of appreciating it.

The result of the Parliamentary inquiry, and the rapid diffusion of the practice of vaccination, excited a strong desire in the public mind to know more of the character of the author. A short account of his life was given in the "*Public Characters*" of this year.

A confirmation of his opinion respecting the origin of the cow-pox was about this period made known by Dr. Loy, of Whitby, in Yorkshire. The experiments of this gentleman were quite conclusive. Dr. Jenner sent a copy of his pamphlet to his Royal Highness the Duke of Clarence, on which occasion he expressed himself as follows:—

"In obedience to the wish your Royal Highness expressed to me at Lord Grantley's, I have done myself the honour of sending you Dr. Loy's pamphlet on the Origin of the Cow-pox, which decisively proves my early assertions upon that subject. This discovery is the more curious and interesting as it places in a new point of view the traditionary account handed down to us by the Arabian physicians that the small-pox was originally derived from the camel. The whole opens to the physiologist a new field of inquiry, and I sincerely hope it may be so cultivated that human nature may reap from it the most essential benefit."

"The ardour your Royal Highness has shown in the Vaccine cause, and the personal favour with which you have honoured me, will ever be most gratefully remembered by

"Your Royal Highness's
"most humble and devoted servant,
E. JENNER."

A confirmation of Dr. Loy's printed work will be found in one of his letters to Dr. Jenner.

DR. LOY TO DR. JENNER.

SIR,

I have not yet had an opportunity of making any further experiments respecting the origin of the cow-pox, on account of the disease of grease having been of late remarkably rare in this country. From the evidence, however, I have had of the truth of your opinion, and from some observations which have been made on my experiments by my worthy preceptor Dr. Duncan, of Edinburgh, I consider myself in some degree called upon to pay more attention to this curious subject, and you may, Sir, be assured that you shall be informed of my success.

I have the satisfaction to mention that the subject inoculated with the grease matter on Experiment VI. has withstood the action of the small-pox, by way of repeated exposure to the natural disease. Several of those also who were inoculated with Vaccine virus, generated by inoculation with the equine, have been exposed more than once to the natural infection of the small-pox, but without the least effect. Dr. Duncan seems to conjecture that the persons on whom the experiments were performed might have previously had the small-pox; but any foundation for such a supposition is perfectly groundless. Most of the persons who were subjected to the experiments had never been within several miles of the small-pox till inoculated. And that the small-pox matter I made use of was good is proved by the same virus giving readily the disease to others.

There is not the least doubt but the experiments will remain successful; and that they were fairly performed many respectable gentlemen in this neighbourhood can testify. One gentleman at my request saw me inoculate one of his cows from the greased heels of his horse, with a lancet with which he himself supplied me at the time of experiment. This trial was successful.

I apprehend that Dr. Pearson's inveteracy will have little effect upon a man who, from the good he has rendered his fellow-creatures, has established his fame over all Europe.

I cannot, Sir, sufficiently express the obligations I am under for the notice you have bestowed on me; accept, however, through this medium the blessings of hundreds in this country for the favours you have conferred on them and their latest posterity.

Give me, Sir, the honour to subscribe myself
Your faithful friend and servant,
JOHN GLOVER LOY.

Whitby, December 26th, 1802.

After so many letters of a professional nature, it may be a matter of curiosity to peruse one written by a worthy Norwegian merchant, who had heard of the virtues of the Variolæ Vaccinæ.

MR. ISAACH ISAACHSON TO MESSRS. WOLFFS AND DORVILLE, LONDON.

Christiansand, 11*th May*, 1802.

(DUPLICATE.)

The purport of the present is to desire you would be so obliging as to procure from the famous Dr. Jenner some matter of vaccination for the inoculation of the cow-pox, so much in fame all over Europe. I could wish it to be sent by the post as soon as possible, and in the best preserved state. For the sake of a speedy conveyance, a small portion might also be sent by the post to Stockton, where a vessel with timber is going, and might take it home. If there exists any printed direction for the using of it, it might also be sent to Stockton, apprehending its weight too bulky to be sent by the post of the continent. You might address the parcel to Mr. Hutchinson of said place, desiring him to forward it with all possible haste and care. Please debtor my account for the amount, and excuse the trouble given by

Your obedient humble servant,

ISAACH ISAACHSON.

Almost every post brought the gratifying intelligence of a wider and wider range to the Vaccine practice. It had already been established in every country in Europe; but though at the period we are now treating of it had passed into Asia, and reached our possessions in India, the knowledge of this fact had not yet been communicated to Dr. Jenner; however, the tidings from Germany and the entire of the north of Europe, as well as from France, from Spain, and Italy, were of the most cheering aud animating nature. From Dr. Struve, of Gorlitz in Saxony, he received a letter dated April 16th, 1802, mentioning that the practice had been introduced into that country in January 1801. The writer then proceeds thus: "Great and honoured Sir—The practice of Vaccine Inoculation, your discovery, is highly esteemed and cultivated by the Germans, insomuch that not a single town or village can be found in which one or more persons have not been shielded by the Jennerian Ægis from the contagion of small-pox."

In a postscript Dr. Struve mentions an instance of a woman eighty-two years of age, then living, who caught the cow-pox by milking cows, in Moravia, seventy-two years before. She had since that time lived in the same apartment with her brothers when labouring under the small-pox; and nursed her son when dreadfully afflicted with that disease; yet totally escaped infection.

By an ordinance of the Austrian Government dated at Vienna, March 1802, a public and authoritative recommendation was given to vaccination. The prejudices which had at first opposed it were thus effectually overthrown, and a series of regulations were established, which soon rendered it general in Vienna; and in no long time small-pox was almost banished from that capital.

Before these regulations were enacted, a number of inoculations with vaccine matter had been performed publicly under the direction of M. De Franck, ancient counsellor, in presence of Count De Küfstein and a great number of physicians.

A copy of these regulations, together with the continuation of his Latin translation of Dr. Jenner's work, was transmitted by Dr. Careno, of Vienna, to Dr. Jenner, through the hands of the Baron Landolina, an accomplished Sicilian nobleman.

The intelligence from Spain was of a much more encouraging description than could have been anticipated from the state of that country. Dr. Vivas of Valencia, and Dr. Nadal of Merida, particularly interested themselves in the diffusion of the practice, and in obviating the prejudices existing against it. The former wrote a short dialogue in Spanish with that intention, a copy of which, with a long letter in Latin, dated on the 27th of May, 1802, he sent to Dr. Jenner.

In Paris the greatest enthusiasm continued to prevail. Mr. George Jenner was in that capital in June. He dined with the Vaccine Committee in company with Dr. Marshall. Jenner's portrait, a print by Smith, was hung up in the room crowned with a chaplet of flowers, and this motto attached to it, "*Viro de matribus, de pueris, de populis bene merito.*"

A feast of the same kind was celebrated at Breslau on the anniversary of the first vaccination in Silesia. It was attended by several persons of rank, and by the physicians and surgeons who were associated for the propagation of the discovery. The same engraved portrait was hung up in the room, and decorated with flowers and wreaths and emblematical figures, with the motto "*Huic Vota.*" After dinner a chorus was sung, composed by a celebrated German poet, Professor Fülliborn. A copy of this print by Smith was afterwards engraved by a German artist at Vienna.

Shortly afterwards the Central Committee of Paris sent to Dr. Jenner a most gratifying account of their proceedings.

Paris, ce 29 *Juillet*, 1802.

LE COMITÉ CENTRAL DE VACCINE, AU DOCTEUR EDW. JENNER.

MONSIEUR ET TRÈS HONORÉ CONFRÈRE,

Les C[yens] Huzard et Parmentier, membres de l'Institut National de France, veulent bien se charger de vous porter les temoinages d'estime et de consideration, dont tous les membres du Comité sont pénétrés pour votre zêle et votre génie observateur. Ils doivent vous instruire des efforts constants qui ont soutenu le Comité depuis l'époque à laquelle la découverte, dont vous avez enrichi le monde, a été connue en France. Ils vous diront que des succès constants ont couronné nos travaux, et que le comité a fait pénétrer dans toute la France les bienfaits de la Vaccine.

Aujourdhui, Monsieur, les rapports de tous les médecins Français sont unanimes; partout les épidémies varioleuses respectent les vaccinés, et il n'est presque plus de village en France qui ne bénisse l'ingenieux inventeur de la nouvelle inoculation.

Dejà votre gouvernement a acquitté en partie la dette de l'humanité : si les felicitations sincères des membres du Comité Central de Vaccine peuvent ajouter à la jouissance qu'a du vous faire éprouver la justice du parlement Britannique; croiez, Monsieur, qu'il ne vous reste à cet égard aucun vœu à former.

Le C[yen]. Huzard, notre plus célèbre artiste veterinaire, se propose vous entretenir sur l' origine du Cow-Pox. Le Comité qui a tenté avec lui l'inoculation des eaux aux jambes sur des vaches, n'a pas obtenu les mêmes resultats que vous. Nous esperons que vous voudrez bien donner à notre compatriote tous les renseignemens possibles. Il nous a promis de nous en rendre un compte fidèle. Nous ajoutons beaucoup d importance à recevoir sur cet objet les connoissances qui nous manquent très probablement: l' humanité vous doit la première observation de l'effet preservatif; la science vous devra des notions précises sur l'origine encore incertaine du bien que vous avez fait à tous les peuples.

Recevez, Monsieur, l'assurance de notre parfaite estime, et de notre consideration distinguée.

Au Nom du Comité,

THOURET,

Directeur de l'Ecole de Medecine, President.

HUSSON,
Secretaire.

M. Husson, the secretary, also wrote a private letter to Dr. Jenner in the following terms:—

"I hope to have the honour of a personal interview with you before the expiration of this present year, 1802. My object in going to reside some weeks in London is to see the celebrated man to whom the world will be indebted for the extirpation of a disease which has long dispeopled it. On my return to France, I shall be infinitely delighted in having to congratulate myself that I have had some acquaintance with you.

Sir,

Be assured of the admiration with which

I have the honour to be, &c. &c.,

HUSSON."

Dr. Sacco continued to labour with unwearied activity. In a letter written to Dr. Jenner he mentions that more than sixty thousand vaccinations had taken place in the Italian Republic; and one third of these had been performed by himself. He states that he was much impeded in his proceedings by sinister rumours of various kinds: among others it was affirmed that vaccination had been abandoned in England, and that it had been forbidden by the Government! With this letter were sent some medals struck to commemorate the introduction of the vaccine practice into the Italian Republic.

Nor was Dr. De Carro less active; he sent vaccine matter to Rome at the request of Cardinal Gonsalvi. That prelate was very anxious to introduce the practice into the Roman states, but it was much opposed by a "*faction*" of physicians.

Having taken a rapid view of Dr. Jenner's intercourse with the continent of Europe on the subject of vaccination, we must now turn our attention to other parts of the world. The intelligence from North America, as well as from Newfoundland and our West-Indian possessions, was of the most gratifying nature. The exertions of the President of the United States, Mr. Jefferson, have already been mentioned. To show how minutely and carefully he studied the progress and varieties of the cow-pox I insert an extract from a letter of his addressed to Dr. Waterhouse.

Washington, Jan. 14th, 1802.

"I have waited till I could inform you that some variolous, after vaccine, inoculations have proved that I had preserved the matter of the kine-pox in its genuine form. Dr. Coxe, of Philadelphia, has ascertained this, having received his matter from hence. To this is added your information that the matter I sent you produced the genuine disease, and consequently those in Virginia who received the matter from me are in security.

"Knowing how little capable the people in general are of judging between genuine and spurious matter from their appearance, or that of the sore, I endeavoured in the course of my inoculations at Monticello to find some other criterion for their guide. With this view, I was very attentive to discover whether there be not *a point of time*, counting from vaccination, when the matter is genuine in all cases: I thought the *eight times twenty-four hours* furnished such a point; I governed myself by it, and it has been followed here successfully by Dr. Gaut; but your experience, so much greater, can inform us whether this rule is a sure one, or whether any other point of time would be still more certain. To the eye of experience this is not necessary; but for popular use it would be all-important: for otherwise the disease degenerates as soon as it gets into their hands, and may produce a fatal security. I think some popular criterion necessary to crown this valuable discovery."

In allusion to the President's remarks Dr. Waterhouse observes, "In answer I quoted your opinion, and finally fixed the point of time with the president at "EIGHT TIMES TWENTY FOUR HOURS;" so it has gone forth as by an edict. I wish you had been one of the deliberative council. Adieu! B.W."

Not many months after the foregoing remarks were written Dr. Jenner had the satisfaction of receiving a letter from the gentleman who now fills the distinguished situation then held by Mr. Jefferson, announcing his election as a member of the American Academy of Arts and Sciences. *

* This diploma was signed by *John Adams,* who was at that time President of the United States, as well of this society.

While writing these remarks, I cannot avoid noticing here that intelligence has just arrived from America announcing the death of that venerable Statesman and Philosopher. This event has been rendered the more remarkable by the coincidence of the decease on the same day, and almost at the same hour, of another distinguished individual *Thomas Jefferson,* who succeeded Mr. Adams in the direction of the Government of the United States of America.

JOHN QUINCY ADAMS TO DR. JENNER.

Boston, 13*th July*, 1802.

SIR,

I have the honour of enclosing herewith a certificate of your election, by an unanimous vote, as a member of the American Academy of Arts and Sciences. And, in transmitting this testimonial of respect from my countrymen, I am sure of expressing their sentiments when I add that never since the institution of this society have its members enjoyed a more genuine and universal satisfaction, by the accession of a new associate, than when they acquired the privilege of reckoning among their numbers the name of Dr. Jenner.

I am very respectfully, Sir,

Your very humble and obedient Servant,

JOHN QUINCY ADAMS,

Corresponding Secretary to the

American Academy of Arts and Sciences.

The success of the vaccine practice in Newfoundland will be learned from the following letter:—

THE REV. J. CLINCH TO DR. JENNER.

MY DEAR FRIEND,

I will hasten to tell you the general result of my practice in the vaccine disease in the Island of Newfoundland. I informed you in a former letter that the matter sent me by your nephew produced the effect completely, although from the date it was kept four months.

I began by inoculating my own children, and went on with this salutary work till I had inoculated seven hundred persons of all ages and descriptions. Many opportunities soon offered at St. John's (where the small-pox was making great ravages) which afforded convincing proofs of the safety of the practice, to the inhabitants and servants in Trinity Bay. They saw (at first with astonishment) that those who had gone through the Jennerian inoculation were inoculated with the small-pox, and exposed to the infection without the least inconvenience, and I hope it will every day become more and more extensive; as nothing can be more certain than that it will annihilate the worst and most dreadful of all disorders, the small-pox.

Poole, 25th Jan. 1802.

Dr. Jenner sent out vaccine matter to Barbadoes in 1801 by Mr. Holder. This gentleman inoculated several sailors with it on the voyage, so that he was enabled to make use of recent virus on his arrival. The success was complete; and the inhabitants received the preservative with much thankfulness.

This year, as well as the preceding, the resources of chemistry were applied to ascertain the qualities of the vaccine lymph. Messrs. Dupuytren and Husson published their researches in Paris; and M. Hunold of Hesse-Cassel gave an account of a similar investigation, in his Annals.

About this time Dr. Jenner received from Mr. Bryce of Edinburgh, a copy of that gentleman's treatise on cow-pox, accompanied with a very polite letter, in which he very modestly says, that he had added only " some new hints" on the subject.

MR. BRYCE TO DR. JENNER.

SIR,

I presume to address you on the present occasion because I think it may be pleasant for you to learn that your favourite subject the inoculation of cow-pox, for which society must ever be indebted to you, has not been altogether neglected in this corner of the world.

Convinced as I am of the power of the cow-pox in shielding the human constitution from the attacks of small-pox, and being placed some time ago in a situation favourable for making observations on that interesting subject, I did not fail in marking down whatever occurrences appeared to me worthy of notice, and these I have now presumed to publish.

I have taken the liberty of desiring Messrs. Cadell and Davies, booksellers, to forward you a copy of this my publication.

You will no doubt, Sir, find a great deal of your own labours scattered throughout this essay; for your investigations on the subject have been so full and so satisfactory that it is impossible to say much on the subject without interfering with you. I trust, however, you will also find some new hints, which your ingenuity may improve farther than I have yet been able to do.

It would afford me much pleasure, Sir, if you will favour me with your sentiments on the two most prominent features of my publication; namely, the new mode proposed for obtaining and preserving the virus of cow-pox, and the proposed test of a constitutional affection. Should any objections occur to you on these points, which I have not foreseen, I trust that you will have the goodness to mention them freely; for this appears to me to be the only way by which we can arrive at truth in our investigations.

It affords me much pleasure, Sir, to observe that you are likely to reap some fruits from your labours. And while a grateful nation bestows a worldly remuneration to her benefactor, it shall be my care to——*(The remainder of this letter is wanting.)*

Of the " two most prominent features," as Mr. Bryce in his Essay terms them, the opinion of Dr. Jenner will be best learned from a perusal of extracts from his replies to Mr. Bryce's communications.

DR. JENNER TO MR. BRYCE, EDINBURGH.

DEAR SIR,

Before I enter on the subject of your present letter allow me to return you my most sincere thanks for your former one, and for your valuable publication on the cow-pox. Although you have found me so tardy in my acknowledgements, yet be assured the very handsome manner in which you mention my name in that work has not passed away without exciting in me a grateful remembrance of your own. You have anticipated some few observations that I had noted down for my next number.

You may easily conceive how incessantly I am toiling if you consider the immense extent of the correspondence that must necessarily fall upon me; and let this consideration make some allowance for my apparent neglect of your letter. I can with truth assure you that I have made efforts to answer it, fifty times since it has been in my possession.

* * * * *

Again, he writes thus to Mr. Bryce—April 5th, 1803.

"It afforded me great pleasure to find that the Vaccine virus I sent you put an end to all your solicitudes respecting the perfection of that you had been previously using, and that you have diffused it so widely through the towns and villages of Scotland.

"I doubt not that you will with much ease establish an institution for inoculation, on a plan equally useful with ours in London. I should send our plan, but it is not yet in print.

"I much admire your precaution in using a *test* of the certainty of infection; and your ingenuity in the manner in which you employ it. To all young vaccinators it cannot be too strongly enjoined. The *experienced* will determine from the character of the pustule. The evidence before the House of Commons evinces the propriety of your observations.

"I put your crust into the hands of my friend Ring, and he informed me yesterday that it had produced a good pustule. Experience now tells us that this is a good mode of sending the Vaccine virus to distant parts."

Among the many marks of public approbation and respect which were presented to Dr. Jenner at this period of his philanthropic career, none afforded him higher gratification than those offered to his fame by his professional brethren; inasmuch as these were the best testimonials of his merit as a discoverer, and of the greatness and universal advantage of the discovery itself.

He was in the early part of this year (1802) addressed on that subject in the warmest terms of congratulation and approval by the Medical Society of London, of which he had long been a member. At a full meeting held on the 29th of March it was unanimously resolved "That taking into consideration the important discovery of Dr. Jenner, the members of this society are of opinion that great benefit will accrue to the inhabitants of these islands, and to mankind in general, from the introduction of Vaccine Inoculation; and from their own experience, as well as from the extensive trials made in various parts of the world, that it will, in all probability, ultimately eradicate the small-pox, one of the most fatal diseases to which the human species is liable.

"Resolved, That a copy of this Resolution, signed by the President, be presented to Dr. Jenner."

The Suffolk Society of Surgeons on the 19th of April, after congratulating him on his great and happy discovery, and returning him their "grateful thanks for the invaluable advantages that the community at large had derived from his labours," entered into similar resolutions: as did also the Benevolent Medical Society of Essex and Herts, at their annual district meeting held at Hatfield, May the 3d, 1802.

But the proceedings of this kind which were calculated to convey to the heart and mind of Jenner, at this time, the greatest satisfaction, were those of the Physical Society of London, which holds its meetings at Guy's Hospital, Southwark.

At one of these meetings, unusually full, a memoir on the subject of Vaccine Inoculation underwent long, serious, and ample discussion. Dr. Jenner himself attended on four successive nights, having received an express invitation and request from the Society for that purpose.

On entering the theatre he was constantly received with universal and rapturous applause; and as no discussion was ever of greater importance or of deeper interest, none, probably, ever attracted in a higher degree the attention of professional and scientific men. At the conclusion of this animated debate, the Society expressed to Dr. Jenner their deep sense of the value of his discovery, "their opinion of its efficacy, and their profound veneration for its author." After professing their conviction of the prophylactic powers of vaccination, and that it is quite free from any such effects as that of "calling forth those latent seeds of disease, which occasionally arise after the small-pox, even under the best management," the Society thus expresses the sentiments of its members,—"We feel the warmest sensations of gratitude and respect for the liberal manner in which the author has communicated his discovery, uninfluenced by any motive of self-consideration; and we contemplate the discovery itself as a memorable epoch in the annals of medicine. The Society presumes to hope that the author of this happy discovery will meet with that reward from his grateful country which he justly deserves for having thus rendered himself the benefactor of mankind."

A new order of merit was instituted on this occasion, the members of which are called honorary associates. This distinction is to be conferred solely on the authors of some remarkable discovery in medicine.

The President, after an eloquent oration, presented to Doctor Jenner the subjoined Diploma.*

In addition to the honours already mentioned as having been conferred on Dr. Jenner about this period, he was elected a corresponding associate of the Medical Society of Tours; and also of the Medical Society of Paris.

* "Societas Physica, anno 1771 constituta, et in Nosocomio Thomæ Guy habita, omnibus ad quos hæ pervenerint literæ salutem. Cum meritissimus ornatissimusque vir Edvardus Jenner, M.D. R.S.S. &c. non modo ingenii acumine, felici artis medicæ cultura, animoque ad optimum quodque parato, jamdudum inclaruerit, sed etiam Variolæ Vaccinæ Institutione in lucem prolata stragem hominum variolarum morbo antehac illatam compescuerit, imo ferè penitus in futurum represserit, sicque dirum illud mortis telum obtundendo de genere humano optimè meruerit; notum facimus nos præfatum virum primum inter socios maximè honorandos ascivisse et retulisse. In cujus rei fidem has literas, meritissimis tantum concessas, manibus nostris signatas, expediri lubentissimè jussimus."

Although many of the clergy on the Continent had availed themselves of the opportunity which their sacred function affords to promote the practice of vaccination by addressing their congregations on the subject, the example was not followed in England. In the commencement of this year, however, one worthy clergyman, the Rev. Dr. Booker of Dudley, broke through the silence, and made an honest and energetic appeal to the good sense and good feeling of his hearers. He published his discourse (addressed chiefly to parents) on the duties and advantages of inoculating children with the cow-pox. Dr. Jenner was very much gratified with this publication. In writing to the author he observed, "You are not only entitled to my thanks and praises, but to those of the whole Island, and I heartily wish so good an example may be followed. As it has pleased Providence in his mercy to impart to us the means of annihilating a most afflictive malady, every effort should be made to counteract those prejudices which stick so firmly to the poor and uninformed. As far as I know, you are the first clergyman who has addressed his congregation on this subject. On the Continent, particularly in Germany and Switzerland, it has long occupied the attention of the pastors of the church; and their discourses have been attended with consequences the most beneficial. Many have done much good here by taking up the practice, and inoculating the poor around them. I shall lay before you a passage in a letter just received from the Rev. Mr. Finch, St. Helen's, Lancashire." 'A few years ago I was in the habit of burying two or three children every evening in the spring and autumnal seasons, who had died in small-pox; but now this disease has entirely ceased to call a single victim to the grave. Why? I have inoculated for the cow-pox upwards of 3,000 persons; and the small-pox is no longer in existence here.'

The idea of connecting religious services with the practice of vaccination had occurred to several individuals in this country as well as on the continent. The late eminent Dr. Darwin, of Derby, wrote to Dr. Jenner on the 24th Feb., 1802. "Your discovery of preventing the dreadful havoc made among mankind by the small-pox, by introducing into the system so mild a disease as the vaccine inoculation produces, may in time eradicate the small-pox from all civilized countries, and this especially: as by the testimony of innumerable instances the vaccine disease is so favourable to young children, that in a little time it may occur that the christening and vaccination of children may always be performed on the same day."

It is pleasing to find that Dr. Jenner, amid the momentous and engrossing occupations in which we have seen him engaged, should preserve both the playfulness of his character, and that delightful simplicity by which he was so much distinguished. The following dialogue written about the time that he was most exposed to the vexatious opposition directed against him before the Committee of the House of Commons, indicates very strongly these qualities.

DIALOGUE BETWEEN E. J. AND HIS SERVANT RICHARD.

R. A servant, Sir, has call'd just now,
And left a very handsome cow.
'Twas brought, he told me—let me see—
From some such place as *Hitaly*.
Well have I look'd her o'er and o'er,
And never saw her like before.
In no one point, Sir, does she fail,
Head, horn, neck, carcase, limbs or tail;
And here she is to take her station
In compliment to Vaccination.
J. Well, Richard, this is very good,
But how shall we contrive her food?
Instead of town, were we at home,
Among our meadows she might roam;
But here I've not one inch of pasture—
R. Oh, Sir, don't vex; that's no disaster,
The cow is only Paris Plaster!

Bond-Street, 1802.

CHAPTER XIII.

ADVERSE CLAIMS—FRENCH AND HINDOO.

As it has been necessary in the preceding chapter to treat of some of the claims which have been urged in opposition to those of Dr. Jenner, I consider this a fit opportunity to bring together such other attempts of the same description as have more recently appeared before the public. In this part of my work I have been called upon to speak more of the conduct of individuals than I could have wished, and, in doing so, I have been brought nearer to the confines of disputation than accords with my feelings. It has been, nevertheless, my solemn purpose to avoid giving just cause of offence to any man. I will continue to state such facts as are necessary to elucidate the character of Dr. Jenner; to delineate his efforts both as a medical philosopher and as a philanthropist; and this with the single object of presenting him to his fellow-men as he actually was. Could this be accomplished without seeming to attach blame to any individual it would be to me most gratifying. But as I cannot hope to be able to do this altogether, I trust I shall succeed in my endeavours to narrate the evidence without any admixture of prejudice, and to declare what truth demands "without wrath or doubting." I need scarcely repeat that these remarks do not apply to the vaccine controversy strictly so called, with which it is not my intention to intermeddle; they refer to questions of a personal nature, and it is impossible to uphold Dr. Jenner's reputation or to render the barest justice to him as a man, without meeting them in a manner the most open and unreserved.

The pretensions already mentioned have, I would hope, been sufficiently considered and refuted. It is now necessary to examine those of another individual. The quarter from whence this attempt has proceeded is calculated to excite no small degree of surprise in the minds of those who are acquainted with the history of vaccination either in this country or on the continent. Our neighbours in France seem not disposed to allow the ancient rivalship, that has so long interfered with the best interests of both countries, to subside. It unfortunately still appears that a degree of jealousy respecting the character of England is permitted to hold place in the breasts of scientific men, who ought to be well aware that truth knows no limits of country.

It is painful, therefore, after national harmony has been so long restored, to be compelled to observe a spirit of envy tending to keep alive an intellectual warfare, at all times unworthy of the republic of letters.

It would appear, from the article which has called forth these remarks, that England has a heavy debt of injustice to atone for to France. With a rapacious zeal she has appropriated to herself many discoveries in art and science which, of right, belong to her neighbour. A bill of indictment, containing many counts, has been drawn up against her; from which it might be thought that many of her recent inventions and improvements have been purloined from France, and it is more than insinuated that Vaccination is among the number. This charge is much to be lamented: neither country is so destitute of solid claims to respect and consideration, in the culture of human knowledge, as to render it becoming that either should attempt to disparage the other in matters of this kind. Let us hope that the time is approaching when higher and better principles will guide the whole family of civilized man; that truth in its purest form will be the aim and end of all their researches; and that their chief efforts will be directed to diffuse its blessings to every kindred and to every people. This object certainly was ever uppermost in the mind of Jenner, as from the first instant that he became fully acquainted with the character and virtues of vaccination he coveted nothing so much as to be able to render them accessible to every individual of our race.

In promoting this design he devoted his time, his talents, his fortune; and all this, too, when his discovery was undervalued, his motives calumniated, and his character traduced. Amidst no small suffering from causes of this kind, and with painful feelings of responsibility for the result of a practice which so immediately concerned the lives of his fellow-creatures, and the success of which was so often marred by the ignorance or incompetence of those who took upon them its administration, he continued to repay all the contumely and insult offered to himself with increasing and unremitting endeavours to heap greater benefits on mankind. France was among the first of the nations that experienced his benevolent purpose. The horrors of war had, at the time of the first publication, dissevered all national and personal intercourse, but they could not interrupt the course of Jenner's philanthropy. He set an example worthy of universal imitation; and endeavoured, as far as in him lay, to mitigate the evils attendant on a warfare conducted with a bitterness unusual in modern times: and it will hereafter be seen that some of the events in which he himself was personally concerned are almost the only bright and cheering incidents that give relief to the dark and dreary scene which Europe at that time exhibited.

All this was for a long period generously felt and

acknowledged by the French nation. Another feeling has, in certain quarters, begun to prevail, and very ill-founded attempts have been inconsiderately and, I would hope, unwittingly made to advance claims on the part of one of their countrymen, hostile to those of Dr. Jenner.

This event is in itself well calculated to cause astonishment; but that astonishment is increased when we consider who the individual is that brings it forward. M. Husson was one of the earliest votaries of the Jennerian practice. He studied the subject like a philosopher; he made himself perfectly acquainted with all the facts relating to the origin and progress of vaccination; he took a very active part in the dissemination of the practice, and has ever since held an important and confidential station in the Central Committee for Vaccination in France.

He was one of the earliest of the French authors who devoted their talents to the elucidation of this subject; and his "Recherches Historiques et Medicales" contain perhaps as complete and faithful a delineation of the origin, varieties, and advantages of the new inoculation as has appeared in any language. This publication, and his distinguished efforts in the cause, brought him into correspondence with Dr. Jenner, and from my personal knowledge I am certain that M. Husson occupied a high place in his estimation: in fact, it could not be otherwise when we consider the terms of respect and veneration with which that gentleman spoke of the author of vaccination both in his published writings and in his private letters. As a convincing proof that this representation is correct, the reader is requested to observe the following, among other similar sentiments, which are to be found towards the conclusion of his work. "L'examen de toutes ces objections prouve bien évidemment la faiblesse des moyens employés pour s'opposer aux progrès de la *découverte de Jenner*." And again, "La postérité bénira la mémoire de Jenner, et les siècles à venir le proclameront l'un des plus grands biénfaiteurs de l'humanité." Such *was* M. Husson in 1803.

"Hei mihi, qualis erat, quantùm mutatus ab illo"

in 1821.—In that year he published in the *Dictionnaire des Sciences Medicales* his elaborate article on vaccination: in which, after enumerating many scientific and literary thefts on the part of the English he concludes the list in the following terms, "Se sont également appropriés tout le mérite d'une découverte dont la première pensée leur a été donnée par un Français, et dont l'étude et la juste appréciation ont été, même de leur aveu, plus rigoureusement suivies parmi nous que parmi eux."

The sentiment expressed in this extract appears very strange when contrasted with that which precedes it; but it is still more remarkable when viewed in reference to the claim which it is designed to enforce. The whole history of this claim is so singular, and is so much at variance with well-ascertained facts, as to have filled with amazement all in this country who have heard of it. Had it not been brought forward with all seriousness, and with such circumstantial details as might impose upon the uninformed, it would scarcely be necessary either to examine it minutely, or to give it consequence by a formal refutation.

I trust that the following words were written without sufficient consideration: "Il parait que c'est en France, en 1781, que l'idée première de la possibilité du transport d'une éruption de la vache sur l'homme a eu lieu; que cette idée, émise par un Francais devant un medecin Anglais, a été communiquée par ce dernier au Docteur Ed. Jenner, qui ensuite aurait appliqué tout son attention à ce projet," &c.

I would willingly suppose that M. Husson never had read Dr. Jenner's own account of the origin of the vaccine inoculation, published in 1801, in which it is stated that his inquiry into the nature of cow-pox commenced upwards of five years before the time assigned for "l'idée première" of M. Rabaut. Of course the details which I have given relative to the first impression made on the mind of Dr. Jenner concerning cow-pox could not have been generally known. I am sorry, however, that I cannot persuade myself that M. Husson was unacquainted with other sources of information, which might have demonstrated to him that the allegations contained in the preceding quotation are untrue.

I know that he had seen the Report of the Committee appointed by the House of Commons to inquire into the claims of Dr. Jenner, because he quotes largely from that very Report in his "Recherches sur la Vaccine," third edition, 1803. He could not, therefore, be ignorant of the testimony of Mr. E. Gardner, who proved that, in the year 1780, Dr. Jenner had communicated to him the result of his past inquiries with regard to the Variolæ Vaccinæ; and his hopes of propagating the disease by inoculation, to the eventual extinction of small-pox.

The manner in which it is asserted that the *first idea* of vaccination was communicated to Dr. Jenner is, to the full, as worthy of credit as is the pretension "sur l'origine *vraiment Française* de la Vaccine.'

It appears that M. Rabaut, the protestant minister at Montpelier, was struck with the affinity between some of the eruptive diseases of the inferior animals, and the small-pox in man. His own observations applied chiefly to *sheep*, but a farmer in his neighbourhood informed him that the teats of cows were liable to a similar eruption, adding that it was a rare yet a mild disease. At this period, (1781) a Bristol merchant, Mr. Ireland, with his physician Dr. Pew, was in the habit of spending his winters at Montpelier. M. Rabaut, who was intimately acquainted with these gentlemen, observed, one day that the conversation turned on inoculation, "qu'il serait probablement avantagueux d'inoculer à l'homme la picotte des vaches, parce qu'elle était constamment sans danger."

They conversed on this subject, it is added, when Dr. Pew said that immediately on his return to England he would propose this new species of inoculation to his friend Dr. Jenner. Many years afterwards (1799) M. Rabaut, having heard of the discovery of vaccination, supposed that his own proposition had been realized, and wrote to Mr. Ireland to recall their conversation on that subject. Mr. Ireland replied by two letters (the originals of which M. Chaptal has read) that he remembered quite well all that had been said at Montpelier, the promise which Dr. Pew had made of speaking to Dr. Jenner, but he says nothing of what Dr. Pew *did* on his return to England.

Such are *the facts*, as they are somewhat whimsically called, on which the title to the French origin of vaccination is made to rest.

The refutation which has been already given might be sufficient: but there is something so peculiar in the getting up of the story; so inaccurate in its affirmations; so inconsequential in its deductions; and so remote from truth in its most material positions that I cannot allow it to pass without further exposure.

In the first place, then, it is worthy of remark that it is not even alleged that M. Rabaut had seen, or knew, the disease on the cow; and that all the information which he possessed on the subject was obtained from an agriculturist in his neighbourhood. Admitting that the farmer, good man! had really observed the disease as it exists in the cow; and, besides, knew something of its antivariolous virtues, it is truly remarkable that the faculty, although they had been seeking for it during twelve years in many departments of France, found no certain traces of it till 1810, at which time it is said to have been detected in the department of the Meurthe. It also may be worth while to notice that the disease of the cows is so rare in France that a single case of it reported to have been seen in the neighbourhood of Clairveaux (in 1822) is deemed worthy of special record in the Report of the Central Committee of Vaccination, for 1821-2. M. Husson was fully aware of both these occurrences. The first (in 1810) is stated in his article Cow-pox, vol. vii. of the Dictionaire Medicale; the second is attributed to him as Secretary to the Central Committee. Surely "L'agriculteur des environs de Montpelier" who first observed "la picotte sur le trayon des vaches" nearly fifty years ago, must have been mistaken in his observations, since it appears that nothing like an authentic proof that the disease existed in France was obtained till 1810; and that, too, after a diligent search on the part of all the medical and veterinary professors in that kingdom.

It is, therefore, certain that the cow-pox must be an extremely rare affection in France; and it will not imply a great degree of incredulity if, for the reasons already stated, any one should doubt the knowledge of M. Rabaut's informant, either as respects its existence or its antivariolous powers!

It now remains to make a few remarks on another part of this marvellous statement. It is distinctly affirmed by M. Husson that the *first idea* of the possibility of inoculating from the cow "émisé par un Français devant un medecin Anglais, a été communiquée par ce dernier au Docteur Ed. Jenner, qui ensuite aurait appliqué tout son attention à ce projet, &c."*

Disposed as I am to make every allowance for such inaccuracies as are apt to arise when an author is intent in supporting a favourite assumption, I feel myself constrained to treat the declaration in the preceding extract with the seriousness that truth demands. I have already said enough to vindicate the claims of Dr. Jenner, which, indeed, could not have been affected even though the strange pretensions of M. Rabaut had been admitted. My present object is to expose more fully the unjustifiable manner in which these pretensions have been urged, and to enter a solemn protest against the confident and unhesitating mode in which unfounded assertions have been promulgated. It ought not to have been said on slight grounds that Dr. Jenner had derived his *first idea* of vaccine inoculation from M. Rabaut, through the medium of an English physician; nor

* Vide Dictionaire Medicale, tom. 56, p. 394.

ought it to have been made matter of charge against Dr. Jenner that he had appropriated this idea without acknowledging in the most remote degree the source from whence he derived it; still less ought it to have been asserted that his countrymen had aided in this act of injustice, unless strong and sufficient evidence had been produced. Every one interested in the progress of science must feel the necessity of dealing with integrity and fairness in questions of this nature. Without qualities of this kind there is no certainty in the diffusion of knowledge, no security for the maintenance of truth, and the upright labourers in that cause must be deprived of their just distinction and reward.

Now, it has been demonstrated that Dr. Jenner's mind had been directed to the subject of cow-pox during his apprenticeship at Sodbury; that he carried the tradition of the country to London in 1770; and that he actually commenced his inquiry into its nature about five years thereafter. I need not, therefore, add that he could not have derived the first idea either from the Bristol merchant or Dr. Pew. On the subject of vaccination he certainly never had any communication with either of those gentlemen; and it is my conviction that he had no personal acquaintance with them whatever. I may further add that, if the Dr. Pew above-mentioned be the same person who lived near Shaftesbury in Dorsetshire, he never alluded to any circumstance connected with the pretensions of M. Rabaut, though he was particularly engaged in a correspondence with Mr. Creaser of Bath, in 1804, respecting the origin of vaccination in his own county. Excepting *this Dr. Pew* I know of none other who has been in any way concerned in this matter. M. Husson has affirmed that Dr. Pew, to whom he has assigned so conspicuous a part, was the friend of Dr. Jenner: for this assertion there is as little foundation, as for his alleged communication of M. Rabaut's pretended discovery to Dr. Jenner.

The same work which sets forth the claims of M. Rabaut recounts likewise the pretensions of the Hindoos to the knowledge of vaccine inoculation. An ancient Sanscrit work has been appealed to as an authority on this subject.

The subject was mentioned many years ago in the Bibliotheque Britannique. It has, more recently, been revived in the Dictionnaire Medicale, and in the Madras Courier of the twelfth of January, 1819. The writer of the last-named article refers to Sancteya Grantham, a medical work attributed to Dhawantari, which is said to be "undoubtedly an ancient composition."

From this work extracts are given. The first extract describes the method of performing the inoculation with fluid taken from a pock on the udder of a cow, or from the arm of a human subject, &c. The next more particularly describes the small-pox produced by the fluid from the udder of a cow, and appears, in short, to be an imperfect abstract of the opinions and descriptions of Dr. Jenner. It does not at all discriminate between the different sorts of pustules to which cows are liable; it is destitute of all the characters of fidelity and accuracy which give value to information of a scientific nature, and must, therefore, have been quite insufficient to have guided any one in the management of the very practice it professes to teach. This practice, if it did exist at all in India, must have been extremely rare, but the description given of it is not like that which would occur to any inquirer who had himself investigated the very singular properties of cow-pox inoculation. On the contrary, it wears the appearance of a delineation which had been made, not from original observation, but from materials obviously acquired from other sources and put together with studied ambiguity, the writer having been more anxious to maintain the semblance of antiquity than to convey precise information on a point of infinite importance. Had it been otherwise it is inconceivable that any precise knowledge on the subject of vaccination, had it ever been obtained by the Brahmins, could have been overlooked, and allowed to remain in obscurity till it was called into notice by the industry of British residents in India. To no people on the earth was the secret of vaccine inoculation of greater moment than to the inhabitants of the East, and it exceeds all powers of belief, to suppose that such a secret could have been possessed by the most influential and most respected caste, without being diffused universally, and the practice adopted with corresponding avidity.

The suspicions excited by the internal evidence are not a little strengthened by some circumstances which I am about to mention. I made it my business to inquire from eminent oriental scholars whether such a Sanscrit work existed, and whether from their experience of the habits and customs of the Hindoos there was reason to believe that *they* possessed any knowledge of vaccine inoculation.

For valuable information on these points I have to express a deep feeling of gratitude and obligation to Sir John Malcolm, G.C.B. a gentleman not less distinguished by his genius and skill as a commander than by the wisdom, learning, energy, and benevolence evinced in all his civil relations. Through his kindness I am enabled to state the following facts:—

On the introduction of vaccine inoculation into India it was found that the practice was much opposed by the natives. In order to overcome their prejudices the late Mr. Ellis, of Madras, who was well versed in Sanscrit literature actually composed a short poem in that language on the subject of vaccination. This poem was inscribed on old paper, and said to have been *found*, that the impression of its antiquity might assist the effect intended to be produced on the minds of the Brahmins while tracing the preventive to their sacred cow.

The late Dr. Anderson, of Madras, adopted the very same expedient in order to deceive the Hindoos into a belief that vaccination was an ancient practice of their own. It is scarcely necessary to observe that had any authentic record of such a practice existed, these gentlemen never would have resorted to such a contrivance to gain their object. It is further to be observed that small-pox inoculation was frequently practised by the Hindoos, but there is no proof whatever that they employed vaccination.

Shortly after the introduction of vaccination into Bengal, similar attempts were made to prove that the practice was previously known there also. As the account of this transaction is somewhat different from that which occurred at Madras, it is proper to mention it. A native physician of Barelly put into the hands of Mr. Gillman, who was surgeon at that station, some leaves purporting to contain an extract of a Sanscrit work on medicine.

This work is said to be entitled Sud'ha Sangreha, written by a physician named Mahadeva, under the patronage of Râjá Râjasin'ha. It contained a chapter on Masúrica or Chickenpock.

Towards the close, the author appears to have introduced other topics; and immediately after directing leeches to be applied to relieve bad sores he proceeds thus: "Taking the matter of pustules, which are naturally produced on the teats of cows, carefully preserve it, and before the breaking out of small-pox make with a fine instrument a small puncture (like that made by a gnat) in a child's limb, and introduce into the blood as much of that matter as is measured by a quarter of a ratti. Thus the wise physician renders the child secure from the eruption of the small-pox."

This communication was shown to Mr. Colebrooke and Mr. Blaquiere, both eminent Sanscrit scholars, and they both suspected that it was an interpolation. The first-named gentleman further adds that the original work, from which the extract purports to have been taken, was not exhibited to any one well versed in Sanscrit. I believe I may further add that Mr. Colebrooke made inquiries whilst in India, which fully satisfied him that no original work of the kind ever had existence. Sir John Malcolm has also been kind enough to ascertain that no such book is to be found in the library of the East India Company. From these statements it must be apparent, that the well-meant devices of those who attempted to propagate Vaccination in India, have led to the belief that the practice was known to the Hindoos in earlier times.

It is a providential arrangement that the satisfaction which well-regulated minds derive from the investigation of truth does not depend upon external things. The labours of those who cultivate this field are seldom duly estimated. The bad passions of their fellow-creatures too often interfere with their just claims to consideration, and cavillers and disputers either deny the accuracy of their statements, or attempt to disprove their originality.

This has been especially the fate of almost every inventor in medicine, and it was not therefore to be expected that Jenner should escape the common lot of his predecessors.

When Harvey published his great work on the circulation of the blood, he had not only to encounter a most violent opposition to his doctrines, but actually lost a great deal of public confidence, and experienced a sensible diminution of his practice as a physician. The strength of his demonstrations, however, at length vanquished all obstacles, and knowledge prevailed. Though his detractors were thus foiled, they were not defeated; they assailed him in another quarter, and contended (though his doctrines were true) that they were not his own, that traces of them were to be found in the works of Hippocrates, Aristotle, Plato, Galen, and Michael Servetus.

If men would but consider that "all knowledge (as has been happily expressed) is but the double of that which is," they would cease from these unjust attempts to rob those, who have brought that knowledge to light, of the praise which is their due. All the most splendid discoveries were in the first instance suggested by some apparently unimportant event; and who can tell how many more have been lost to the world, because those who are placed in situations to examine the changes and transmutations of different bodies, the habits or functions of plants and animals, or the processes in mechanic arts, are incapable, either through inattention or ignorance, of taking advantage of unusual occurrences, of following out the hints and admonitions which every day's business and observation supply; and remain blind, or unexcited spectators of events which, had they been duly investigated, might have immortalized their own names, and conferred unspeakable benefits on their species.

Without referring to the histories of scientific discoveries, which are already well known, I may mention one fact that, I believe, has not been formerly noticed, bearing strongly upon the subject now under consideration. Every one is aware of the occurrence that first led Galvani to investigate that branch of knowledge which now bears his name, and which, in the hands of able inquirers, has led to some of the most splendid results of modern science. His lady, who was in a declining state of health, was about to be fed by a restorative soup prepared from frogs. The animals, skinned and prepared for the kitchen, were lying in the philosopher's laboratory, near his electrical machine. The machine being in action, an attendant happened to touch the crural nerve of one of the frogs with the point of a scalpel, which was not far from the prime conductor. The muscles of the limb were instantly thrown into strong convulsive action. This phenomenon, which was by no means anticipated, was presented to the observation of the lady, in the absence of the professor; she was very much struck with the fact, and communicated it to her husband. He immediately set himself seriously to investigate the subject, and was soon rewarded by a rich harvest of discovery. I introduce this anecdote in order to give due effect to the following incident. A considerable number of years before Galvani's attention had been drawn to the subject of animal electricity, a fact which was not ascertained till he had advanced considerably in his inquiries had been casually brought under the notice of another individual, but it led to no results whatever. The occurrence to which I allude arose in the following manner:—A fish which was about to be prepared for dinner happened to be placed on a table in connection with two metals. A galvanic circle was formed, and the animal was thrown into convulsions: an event so unusual naturally attracted attention. It was reported to the gentleman of the house who was a medical man, and although he did not make one single attempt to prosecute so remarkable a branch of knowledge, he thought it worth while to record the circumstance, and an account of it was actually published in one of our scientific works.

I am exceedingly sorry that I am unable to specify the name of the individual, or the work where the fact is mentioned. I met with it many years ago, and made a particular reference, but I have mislaid my memorandum. I am, nevertheless, confident that implicit reliance may be placed on the accuracy of that part of the statement which it was my main object to bring forward; I mean the convulsion of the fish excited in the manner already described. Had the individual to whom that event occurred been possessed of the mind of Galvani, that philosopher would have been anticipated in his discoveries, but unquestionably his merit is not in the slightest degree diminished by the facts just mentioned, nor would it have been, though hundreds of such unproductive observations had been recorded before his time.

By a strange inversion, a dull man's blunders and incapacity are sometimes made to cast a shade over the brilliancy of genius. To such causes are we to ascribe the hard fate of many distinguished inventors. A fact which has been lying common and at waste, floating on the very surface of daily experience, is seized upon by some penetrating and inquisitive mind. Its relations to the different branches of human knowledge are examined and defined: it throws a light all around, and is a lamp to the feet of the inquirer, while he surveys other regions. Having thus explored a *terra incognita* up starts one, and says,—Sir, you have not the whole merit of this discovery; I knew that such a land, as you have visited and explored, existed, for I saw it, but did not approach it. Another says,—I was actually cast away upon the coast; I noticed some of the things which you have described. I did not examine them minutely, but I remember from your description that such things did exist, and I therefore am entitled to the merit and reward which you claim.

A process similar to this marked the discussions regarding the origin of vaccination. The subject had been forced upon the attention of many individuals; but as far as they were concerned all the information relating to it might have remained in its original and unsatisfactory state. All the pretensions, therefore, of the men that became wise by the labours of Jenner, who achieved what they were unable to accomplish, instead of detracting from his fame ought to raise it still higher.

CHAPTER XIV.

FORMATION OF THE ROYAL JENNERIAN SOCIETY.—DEPARTURE OF THE EXPEDITION UNDER DON FRANCISCO XAVIER BALMIS FROM SPAIN.

I HAVE not been able to ascertain exactly at what time Dr. Jenner left London after the conclusion of the Parliamentary inquiry, but I suspect it must have been towards the end of July. In the beginning of August he was at Berkeley, where he received a visit from Dr. Franck of Vienna. This distinguished physician carried with him a letter of introduction from Dr. Babington of London.

DR. BABINGTON TO DR. JENNER.

DEAR SIR,

Nine years absence in the service of his country has obliged my friend Dr. Franck, the bearer of this letter, to visit your part of the world on account of his health; and the interest which he takes in the welfare of society at large makes him desirous of being made known to one of its greatest benefactors.

The object of my writing is to request that you will allow me to introduce him to your kind attention; such as you may have occasional opportunities of showing him, without putting yourself to any inconvenience. Dr. Franck has too much good sense and politeness to wish to trespass upon your valuable time. He desires merely the honour of being acknowledged as an acquaintance, which, I flatter myself when you come to know him, you will not repent of having granted, and which will add to the many kindnesses already shown to, dear Sir,

Your very grateful and sincere friend,

WM. BABINGTON.

Aldermanbury, *Aug. 5th*, 1802.

Dr. Franck's inquisitive and vigorous mind did not rest satisfied with superficial views of any subject. He acquired an accurate knowledge of vaccination as well as of its early history. He published an account of his travels in England, and the manner in which he mentions some of the events that took place soon after the appearance of Dr. Jenner's "*Inquiry*" proves that his information respecting transactions, which were not generally known, was surprisingly accurate. I allude particularly to the attempts made by certain individuals in London to elevate themselves by depreciating the merits of the author of vaccination.

I find from his memorandum-book that Dr. Jenner was at Cheltenham on the first of September, and there, I believe, he remained till the beginning of December, when he returned with his family to

Berkeley. He hoped in this seclusion to enjoy a few weeks repose after his arduous and harassing labours. This most reasonable expectation was not permitted to be fully gratified.

Several respectable gentlemen, friends of Dr. Jenner and of their fellow-men, anxious to see the practice of vaccination fixed on a firm basis in the metropolis, resolved to endeavour to form an institution for that purpose. On the 3d of December, 1802, they held a preliminary meeting in Queen Street; Benjamin Travers, Esq. in the chair. There were only three other persons present, namely, Dr. Hawes, Mr. Addington, surgeon, and Joseph Leaper. The proceedings of this meeting were communicated to Dr. Jenner by Mr. Addington.*

MR. J. ADDINGTON TO DR. JENNER.

Dec. 3d, 1802.

DEAR SIR,

I persuade myself you will hear with pleasure of every thing designed to promote the extension and beneficial

* Another friend of Dr. Jenner's, the late respectable Joseph Fox, of Lombard Street, writing to him on this subject on the 4th December, observes "The plan which is in agitation is of the most extensive and liberal kind. It is even expected that the Royal countenance will be gained, but much depends upon you; all persons are looking towards you as the only proper person to lay the foundation-stone. It would be well if this could be done in the course of the present year, particularly as that is the memorable time, in which it received parliamentary sanction."

effects of your inestimable discovery. The resolutions which I have the honour to enclose are the beginnings of a scheme which contemplates a magnitude of operation, best characterized, perhaps, by saying that it aims to be in some degree commensurate with the claims of its object. I may add that for its support it has already the prospect of a very handsome contribution and, further, that it is likely to engage the attention and assiduities of some very respectable persons. The third resolution makes it unnecessary to inform you that it looks to your direction and assistance in its establishment and progress; and designs, whilst conveying to mankind the benefits of vaccination, to pay to the author of those benefits its first and best tribute of respect and gratitude. In fact it is wished to make it a *Jennerian* institution. The gentlemen therefore feel very desirous of knowing when it is probable that they may have the pleasure of seeing you in town, as they are anxious to proceed in the business without delay.

In their name, therefore, I beg the favour of you to give me this information by an early post; and I take this opportunity of saying that, for myself, I shall also be glad to consult you respecting another edition of my pamphlet which I am intending to publish. With the sincerest wishes for your happiness in proportion to the value of your services to mankind,

I remain,

Dear Sir,

Your obliged friend and servant,

J. ADDINGTON.

To this communication Dr. Jenner replied.

Berkeley, Dec. 10th, 1802.

My Dear Sir,

The only apology I can make for not giving a more speedy answer to your very obliging letter is this; it found me just returned with my wife and children to our pleasant home, where I promised myself a few weeks of domestic comfort after some years spent, I may almost say, in constant anxieties.

This was the pull on one side, and on the other was the delightful prospect held up to my view of an establishment about to be formed for the promotion of universal vaccine inoculation: an establishment to which I have been, for years, looking forward with a longing eye. I need not go farther into explanation, and shall only say that, if it be incompatible with the generous design to suffer me to remain here the time I had allowed myself, I will certainly comply with the wishes of my friends, and go to town.

Yet it must be observed that I humbly conceive and, I may add, ardently hope that my presence will not be absolutely necessary. By this post I have written to my friend Dr. Lettsom and requested him to have the kindness to be (as far as such a thing is admissible) my representative. In his judgment on the present occasion I can place every confidence. There is no one, I believe, whose mind has been more zealously employed on the advantages that society will reap from the adoption of the benevolent scheme now in agitation, or will be better able to point out judicious plans. I do not think the business will be very complex. The society would perhaps so far indulge me as to permit of my inspecting the outlines of the scheme which may probably be brought forward at the next meeting; and in the best manner in my power I shall contribute to its final arrangement.

Pray be good enough to present my best respects, and acknowledgments of civilities to Mr. Travers and Dr. Hawes and the other gentlemen of the society you are so kind as to call Jennerian; and accept the best wishes of, dear Sir,

Your faithful humble Servant,

EDW. JENNER.

Mr. Addington,
Spital Square, London.

Mr. Travers, too, had written to Dr. Jenner on the same subject, and received an answer of like import with the preceding: it contained also the following sentences: "Government, I have no doubt, will give due support to so just and laudable an undertaking." "I am warranted in my suggestion respecting Government from a long conversation I had in the summer with Mr. Abbott (Speaker of the House of Commons) who expressed a wish that on my return to town I would renew it; at the same time added that after the investigation in the House of Commons he thought it became a public duty to form institutions for gratuitous inoculation."

On the 19th of January, 1803, a numerous and respectable meeting was held at the London tavern, at which the Lord Mayor presided. The committee, which had been appointed at a former meeting to prepare an address and a plan for the regulation of the society, on this occasion presented the result of their labours.

The Lord Mayor read the address to the public from the châir, and it was, on the motion of Dr. Lettsom, unanimously adopted by the meeting. The next motion having been made and seconded, the Hon. Admiral Berkeley rose and said that "he had it in command from His Royal Highness the Duke of Clarence to apologize to the meeting for his non-attendance, he having been unavoidably prevented from doing himself the pleasure and the honour of attending on the present interesting occasion; but that his Grace the Duke of Bedford had a motion in his hand which had his Royal Highness been present he himself would have made."

On this the Right Hon. Chairman observed that a motion had already been made and seconded, and that consequently it must be first disposed of. The Hon. Admiral acknowledged the propriety of the observation, but said that the motion was proposed as a tribute of esteem to a benefactor of the world; and that if the previous motion could be waived, and the wish of his Royal Highness could be acceded to, the honour intended would be greater coming from a popular assembly than if it were conferred by any organized society.

On this the previous motion, namely, "that this meeting do form itself into a Society for the extermination of the Small-pox" was postponed, and on a motion of his Grace the Duke of Bedford, at the special request of his Royal Highness the Duke of Clarence, seconded by the Hon. Admiral Berkeley, it was resolved unanimously that the thanks of this meeting be transmitted to Dr. Jenner expressive of the high sense it entertains of his merit, and the great importance of his discovery, and particularly for the liberal offer of his assistance to accomplish the great object it has in view.

It was farther resolved that the Hon. Admiral Berkeley be requested to apply to His Royal Highness the Duke of Clarence to entreat his Majesty that he would be graciously pleased to become the patron of this institution, which the society considers would greatly promote its important object, that of rescuing a very large proportion of his Majesty's subjects from an untimely grave; and that he would permit it to be designated the Royal Jennerian Society for the extermination of the Small-pox.

The other resolutions referred to the opening of subscriptions, the appointment of treasurers and a committee for carrying into effect the objects of the society, &c. &c.

On the 17th of February another general meeting was held at the London Tavern. The report of the Committee, signed by John Julius Angerstein, Esq. the chairman, was presented to this meeting. It communicated the very gratifying intelligence that his Majesty had graciously condescended to become the patron: that her Majesty had with great benignity acquiesced in the request to become patroness: that his Royal Highness the Prince of

Wales and their Royal Highnesses the Duke of York, the Duke of Clarence, and the Duke of Cumberland, had evinced, in a most flattering manner, their willingness to accept the office of vice-patrons: that his Grace the Duke of Bedford had complied with their solicitation to fill the office of president; and that many prelates, noblemen, and gentlemen of the highest rank and respectability had consented to be vice-presidents of the society.*

The Committee likewise interested many ladies of rank in support of the Jennerian practice: and they also appointed a board of directors and a medical council.

His Majesty's assent was announced to the chairman of the committee by an official letter from Lord Pelham: that of the Queen also officially, by the Earl of Morton. His Royal Highness the Prince of Wales was pleased to express his sentiments on this occasion in a letter to the Earl of Egremont. This letter evinced an ardent wish for the success of the new institution, and the most cordial approbation of the indefatigable perseverance of Dr. Jenner in perfecting the discovery of vaccination. This his Royal Highness stated from a full conviction of the efficacy of the practice, and from a distinct perception

* I am indebted to Charles Murray, Esq. for the means of giving an account of the formation of this institution. He likewise most kindly forwarded to me some very important documents relating to the transactions I have recounted as occurring in the end of the year 1799.

of the incalculable advantages it promised to the world.

The gracious and beneficent mind of the Illustrious writer is displayed in every line; and the whole is truly characteristic of those great qualities which continue to add lustre to his still more EXALTED STATION, and shed so much of real glory on his REIGN.

His Royal Highness was pleased, on other occasions, to declare his admiration of this invaluable discovery; to patronize and to cherish it in every way; and at the same time, to announce the highest esteem for the worth and character of its author.

His Royal Highness the Duke of York, who, for reasons already specified, was constrained to withdraw from another Vaccine Institution, wrote to the Earl of Egremont that he would have great pleasure in giving every support in his power to the new Vaccine Institution; and most readily consented that his name should be added to the list of vice-patrons.

His Royal Highness the Duke of Clarence likewise kindly accepted the office of a vice-patron of the society in terms highly encouraging to the future prospects of the infant institution, as will be seen in the subsequent letter.

"TO THE HON. ADMIRAL BERKELEY.

Bushy House.

DEAR SIR,

I shall be truly happy to be of any use in assisting Dr. Jenner to disseminate his invaluable discovery throughout the British empire; and cannot but feel proud that my name should stand among those of the patronisers of your society.

I remain ever yours,
WILLIAM."

His Royal Highness the Duke of Cumberland also expressed his gratification in accepting the office of vice-patron of the society, through his secretary Colonel Stephenson; and he commanded him to add that "he should be at all times happy in having it in his power to contribute towards the promoting of so truly a benevolent undertaking."

Dr. Jenner remained at Berkeley till the first of February, 1803, when he left it for London, which he reached the next day. His short retirement in the country, though not free from interruptions, was very refreshing to him; he took long walks in his neighbourhood, and for a time was enabled to throw off the many cares which had, for years, oppressed him. The celebration of the birth-day of his eldest son, who on the 24th of January had completed his fourteenth year; his walk to Tortworth, the beautiful seat of Lord Ducie; his superintendence of the produce of his cider-mill; and similar incidents, which I find casually recorded among his memoranda, show how eagerly he returned to the habits and occupations in which he delighted. Such were his *horæ subsecivæ;* yet, whilst thus unbending his mind, he was neither idle nor indolent. Letters from various quarters poured in on him every post;

and the affairs of the new institution occupied much of his time, and of his thoughts.

On the 3d of February he took his seat, for the first time, as President of the Royal Jennerian Institution. This meeting was held at the London Coffee-house. At subsequent meetings a central house for the institution, in Salisbury Square, was secured; a resident inoculator and medical secretary was appointed, and Dr. John Walker was elected to fill this situation.

In order to render just honour to Dr. Jenner it was determined that the friends of the institution should annually celebrate his birth-day by a public dinner. The first festival took place on the 17th of May 1803. The chair was filled by the Earl of Egremont.

On the 2nd of March Dr. Jenner, accompanied by Lord Berkeley, Lord Egremont, Lord Grantley, Mr. J. J. Angerstein, the Lord Mayor, Mr. Travers, and Dr. Lettsom, went to the levee as a deputation from the Royal Jennerian Society to return thanks to his Majesty for his goodness in becoming the patron of that institution.

The early proceedings of this society were vigorous and prosperous. Thirteen *stations* were opened in different parts of the metropolis. In eighteen months they were enabled to announce that 12,288 inoculations had taken place, and during the same space of time 19,352 charges of Vaccine virus were supplied from the central-house to most parts of the

British empire, and to foreign countries. The effect of these exertions was immediately perceived by a striking diminution of the number of deaths from small-pox within the bills of mortality. In 1803 they amounted to 1173; in 1804 they were only 622. The contrast will appear still greater when it is considered that the deaths amounted to 2409 in the year 1800: and that the annual average of deaths, for fifty years previously, was 2018.

This society was also in correspondence with other institutions, and its medical council investigated with care and fidelity such cases of small-pox as were alleged to have occurred after vaccination.

It would be foreign to the object of this work to trace minutely the proceedings of this institution. Founded for a benevolent purpose, fostered by the most exalted patronage, and adorned by all the learning and talent of the medical profession in the metropolis, it promised to run a long career of usefulness and honour; but the best devised human schemes too often carry within themselves the elements of their own dissolution. This unhappily was the case in the present instance. Irregularities soon crept into the proceedings of some of the officers of the society, which in a short time led to its virtual dissolution. Though matters did not arrive at an extremity till the year 1806, I am desirous, for various reasons, to finish now what is to be said on the subject. It is an ungracious and unpleasant one, and caused Dr. Jenner much disquiet; and had not the character of the vaccine practice and his own personal feelings and conduct been deeply involved in the proceedings I am about to recount, I should feel great satisfaction in passing them by altogether.

"After the election of Dr. Walker, on the recommendation of Dr. Jenner, to the offices of resident inoculator and medical secretary, he published opinions on vaccination, and recommended a practice contrary to the positive instructions and printed regulations of the Society. Dr. Jenner, president of the Medical Council, considered it his duty to admonish him upon this conduct, and repeatedly represented to him, in the most friendly manner, the mischievous tendency of these innovations. These remonstrances, which were likewise made by others, were unavailing: the deviations from the practice enjoined by the authority of the Society, as avowed by Dr. Walker in his publications, became more alarming; and at length compelled Dr. Jenner, who thought them of the most dangerous nature, to prefer a complaint against him before the Medical Council, the only organ of the Society for determining medical points, and by whom he had been nominated to his situation."

In bringing forward his charges Dr. Jenner was moved strictly by a sense of public duty. Finding all his private efforts unavailing, he ceased to have any further communication with Dr. Walker, after the summer of 1805; submitting rather to lament in silence the fate of the Society than come before it as a public accuser.

Dr. Walker's instructions to the Vaccine Institution at Nottingham, in March 1806, at length roused Dr. Jenner's fears for the safety of the practice, to such a degree, that he felt the absolute necessity of making his sentiments on the conduct of Dr. Walker publicly known. The Medical Council assembled to hear the charges was unusually full, consisting of twenty-four physicians and surgeons of high respectability, who investigated them and examined Dr. Walker's defence with most serious and patient attention; and after long deliberation they came to the following resolutions:—

Resolved—That, in the opinion of this council, Dr. Walker has very materially deviated from the practice of vaccination enjoined in the printed instructions of this Society; and that such conduct is entirely irregular and unauthorised, as well as inconsistent with his duty as resident inoculator and medical secretary.

That this council are of opinion that Dr. Walker is highly reprehensible for publishing under various forms and at different periods, whilst holding the offices of resident inoculator and medical secretary to this Society, doctrines and opinions directly contrary to the instructions published under the sanction of this Society.

That all the resolutions of this meeting shall be submitted to a special general court; and that the fitness of Dr. Walker to continue in the situation of resident inoculator and medical secretary be there determined.

It is unnecessary here to mention the specific instances of misconduct which were established against the resident inoculator. They regarded even the very name of the affection ; the method of managing the pustule ; the characters of correct vaccination ; the precautions to be observed in conducting the practice, &c. &c. In all these respects Dr. Walker inculcated doctrines directly opposite to those which he himself was officially bound to distribute for the guidance of medical men throughout the world.

It is distressing to know that Dr. Jenner, and those who thought with him, could not act as became them on this emergency without having motives ascribed to them of a very unworthy nature. It can not be required of me to attempt to vindicate Dr. Jenner from any insinuations of this kind. His every action was at variance with whatever was, in the most remote degree, selfish or unjust. Indeed, if in this transaction the slightest blame can attach to him, it must be on the score of his forbearance, and not of a disposition to find fault.

A special general Court of the Royal Jennerian Society was held on the 25th of July, when a motion founded on the resolutions of the medical council was made by Dr. Denman, and seconded by James Moore, Esq., that Dr. Walker should be dismissed from his office. This motion was negatived by a majority of three.

The proceedings of this day afford a melancholy proof of the mode in which many public institutions are managed. On the morning of the meeting a supporter of the resident-inoculator, by a payment of twenty guineas, brought, at once, twenty persons to vote in his favour, in direct defiance of the spirit of the seventeenth law of the Society. A ballot was therefore demanded by Dr. Lettsom, Dr. Jenner, J. J. Angerstein, Esq., Mr. Blair, &c. &c. &c., in order to collect the sense of the whole society respecting the conduct of Dr. Walker. The ballot was fixed for the eighth of August, but the proceedings were rendered unnecessary by the resignation of the resident-inoculator on that morning.

Had not these unpleasant transactions terminated in the removal of the offending party, Dr. Jenner certainly would have withdrawn from the Society. No man can read the reasons which he himself assigned for such a step without honouring him for the dignified and becoming principle that guided him in this determination. He looked upon every thing that tended to endanger the character of Vaccination, as a serious evil: and he would gladly have relinquished every personal gratification, or distinction, rather than have compromised in any way the safety of the practice. In this spirit he wrote thus to a friend—" With respect to the Society, if I had seen it go to wreck from a failure of the system to which it gave sanction, that would have been heart-breaking indeed: but while I see that system flourish the world over, whether supported by societies or by individuals unassociated, my mind will be more at ease respecting the fall or continuance of the fabric in Salisbury Square than you could readily have conceived." To the same gentleman he had previously expressed his feelings on this topic thus:—"As president of the Medical Council, and in many other points of view, I am considered by the world as responsible, in a great measure, for the medical conduct of the Resident-inoculator at the Central-House. This renders my situation so truly unpleasant (for I will frankly tell you that I have lost my confidence in Dr. Walker) that I think it quite necessary that one of us should withdraw."

I would here most willingly close my remarks on this subject were it not that the individual, whose conduct was chiefly complained of, has continued to endeavour to associate the name of Dr. Jenner with measures which he disapproved of and disavowed; nay more, with which he absolutely refused to be at all connected. As far as he was concerned he looked upon the Society as nearly extinguished after the late schism. It lingered for some time, but on the establishment of the National Vaccine Institution in 1808, its finances being exhausted, its operations would appear to have ceased entirely. An attempt was made in 1813 to revive this Society and the chairman of a General Meeting, held for that purpose, Dr. Bradley applied to Dr. Jenner requesting his acceptance of the office of President. He, in an answer dated Cheltenham, September 3d, 1813, declined this proposal. It is desirable that this fact should be remembered, inasmuch as it proves that Dr. Jenner's name ought to be entirely disconnected from the proceedings of what is still unaccountably termed the Royal Jennerian Society.

The events in which Dr. Jenner was chiefly engaged while he remained in London related to the establishment of the infant Society. He occasionally took the chair as President of the Medical Council, and forwarded by every means in his power the objects of the institution.

Early in this year (1803) he was visited by Dr. Valentin, of Nancy. This learned and estimable physician had conceived the greatest attachment to the author of vaccination. His own mind had for many years been directed to the history of that dreadful disease which the new practice is calculated to subdue, and his elaborate history of small-pox was nearly printed when the tidings of Jenner's surprising discovery arrived in France. The value of this discovery was quickly appreciated by the enlightened mind of Dr. Valentin. The vaccine practice had commenced in Paris in the year 1800. In the month of October of the same year it reached Nancy, and soon spread to every department of the kingdom. Dr. Valentin not only assisted in diffusing the practice, but studied it with the spirit of a philanthropist. In 1802 he published the results of inoculations with the Vaccine in the departments of La Meurthe, La Meuse, des Vosges, and of Haut Rhin. This work contains an account of some experiments which he instituted at Nancy of a very interesting and novel nature. He proved that it was possible to vaccinate with success not only the cow but likewise the goat, the ass, the sheep, and the dog; and that matter taken from their pustules afforded protection from the small-pox. These very interesting facts I have had occasion to refer to in another chapter, and it is but justice to Dr. Valentin to observe that no experiments undertaken to elucidate the character of the Variolæ Vaccinæ were better devised or more successfully executed than those which he originated. They were afterwards repeated by Dr. Sacco, of Milan; an account of them was published in his quarto work in 1809.

Dr. Valentin arrived in London at the time in which the question touching the connexion between the cow-pox and the *grease* was much agitated. All the principal medical men in London were adverse to the opinion. Dr. Jenner's own evidence was treated with little respect, and the experiments of Dr. Loy did not produce the conviction they deserved. Under these circumstances the letter from Dr. Sacco, announcing his very interesting and conclusive observations on this subject, was received by Dr. Jenner. To gain information on all points connected with cow-pox Dr. Valentin visited the West of England. On his return to London he told Dr. Jenner that the opinion of many of the farmers did not coincide with his. Jenner replied "You have seen that Dr. Sacco, who was one of the most incredulous, has been converted;" and he mentioned at the same time the evidence just received from Suffolk, which had been collected by Lord St. Asaph. To this he added another circumstance which bore on Dr. Valentin's experiments. The latter had proved that the cow was not the only animal susceptible of the Variolæ Vaccinæ: Dr. Jenner, in like manner, had ascertained that the cow was not the only animal capable of receiving the infection of the *grease*. A sheep that had three lambs, of which two perished, being incommoded by the superabundance of milk was drawn by a servant who, at the same time, dressed the *greasy* heels of a horse. Pustules, similar to those of the Vaccine, appeared on the teats of the sheep: the same person who milked the sheep immediately afterwards milked two cows, and communicated the disease to them. From the cows thus infected a servant of the house received the cow-pox.

Dr. Valentin held intercourse with many of the other medical gentlemen who had distinguished themselves in this cause. Among those was the late Dr. Woodville. He, I believe, had at that time become sensible of the mistakes committed at the Small-pox Hospital, and of his harsh proceedings towards Dr. Jenner. This state of mind laid a secure foundation for the renewal of that intercourse which had been interrupted. Happily, peace-makers were found to assist in carrying on this good work; Dr. Lettsom, Dr. Saunders, and Dr. Valentin also aided: Jenner could not resist any overture of this kind. He never harboured resentments, and he found no difficulty in convincing Dr. Woodville that forgiveness and kindness were congenial to his mind. His magnanimity is the more to be commended because his sensibility was singularly acute, and he felt injuries done to the vaccine cause more severely even than wrong done to himself. He further remembered that Dr. Woodville had once watched over his eldest son, when in dangerous sickness, with the skill of a physician and the tenderness of a parent. This incident dwelt with him, whilst every occurrence of an unpleasant nature was gladly allowed to pass into oblivion.

When Dr. Valentin quitted London in June 1803 Jenner sent him a note which he requested him to publish in foreign journals. This note proves how much the pressure of his correspondence was felt by him at this time. He, therefore, begged Dr. Valentin to express in his name, through the medium of the public papers, his lively sense of gratitude for the many letters and printed works on Vaccine Inoculation which the friends of that practice, in France, had addressed to him. He entreated them to accept this general expression of his thanks; his numerous occupations, and the ill state of Mrs. Jenner's health entirely absorbing his time, and rendering it impossible for him to address them individually, as was his wish. He besought, likewise, foreign societies who had sent him diplomas to receive this apology, promising to make up for the delay on the first favourable occasion.

Dr. Valentin kept up a frequent epistolary correspondence with Dr. Jenner. The war which then raged between the two countries rendered this intercourse uncertain and difficult; the letters sometimes taking a circuit by Gibraltar, the Barbary coast, and Sicily: many of them afforded the most convincing proofs of the esteem and regard that were entertained by the writer for his friend in England.

The arrival of Dr. Sacco's letter, and the communication from Lord St. Asaph confirming the connexion between the grease and the Variolæ Vaccinæ, gave great satisfaction to Dr. Jenner. Those who were inimical to his doctrine as well as to the practice of vaccination itself were becoming very clamorous, and evinced their hostility in various ways. The parliamentary discussion; the Royal patronage; and the increasing fame of the author, with which all "Europe rang from side to side," excited a peculiar animosity hurtful to the public, and highly discreditable to those who exercised it. Every effort was made to undervalue his individual merit as a discoverer, first by upholding the pretensions of others, and next by denying the utility of the discovery itself.

He, being in the zenith of his fame and standing on the firm basis of truth, had no occasion to notice such assaults. He regretted them more on account of the check given to vaccination than of the injustice done to himself. At the time that he was made the object of every uncandid remark, one of his old associates in the meetings at Alveston hearing of these most unfounded aspersions, wrote to him and retraced the history of their discussions at the meetings of that society; thereby confirming the statement already given of his unquestionable claims to the discovery.

Mr. Shute to Dr. Jenner.

Bristol, Sept. 12th, 1803.

My dear Sir,

I have often mentioned in many companies that to my certain knowledge the idea of vaccination had been long maturing in your own mind; as an investigation into the phenomena of the cow-pox, more particularly with respect to its property of securing a person in future from the small-pox, had been often urged and introduced at a medical meeting of which I was a member, but that I did not think that that interest had been excited in the minds of the majority of the members, which subsequent experience proved it to have so well deserved. And at the very time that I professed to require further proofs of the decided efficacy of vaccination, I constantly asserted that, if any benefits were to result from the practice, the idea was your own, and that no person could for a moment pretend to deprive you of one particle of the honour of the discovery.

* * * * *

If the enemies of vaccination were ignorant and intemperate, it must be confessed that its friends were, at times, over zealous and injudicious. This was shown both by what they did individually, and the measures they wished to recommend for general adoption. Some practised vaccination without acquiring a suitable knowledge of the disease. They neglected to watch its progress, and consequently permitted many to believe that they were protected from small-pox who never had the cow-pox. There can be no doubt that a very large proportion of the failures which, of late years, have been reported is to be ascribed to this ignorance and inattention. Those who did not practise vaccination themselves wished to see its general adoption enforced by a Parliamentary enactment.

An authoritative enforcement of this kind has certainly been of the greatest service in foreign countries, but the habits and modes of thinking in England do not admit of such compulsory interference. Part of the following letter from Sir Henry Mildmay contains this sentiment, and also alludes to another point in which Jenner was personally interested.

SIR H. S. MILDMAY, Bart., M. P., to DR. JENNER.

MY DEAR SIR,

I beg the favour of you to send me some vaccine matter which is to be immediately forwarded to Italy. Should you be enabled to give me this assistance, I think it extremely probable that it will considerably contribute to the general object we have in view, as I have reason to believe that the persons for whose use it is designed are of the first consequence in that part of the world. Should you be able to oblige me I will thank you to direct it to Marquis Douglass, Duke of Hamilton's, Grosvenor Place.

I own that I am sorry to see that the zeal of some of our friends carries them to a point which, if persisted in, I fear will tend to injure the cause. I see a plan advertized (and indeed I heard of it last year) to prohibit by Act of Parliament the use of small-pox inoculation. I am confident it will not pass the House, and I fear will create an alarm in the country that what we cannot prove by reason and experience we are determined to carry by force. Confident as I am of the complete efficacy of the vaccine matter I have not a doubt that, if left to take its own course, its adoption in lieu of small-pox inoculation will very shortly become universal, but if attempted to be enforced by Act of Parliament, I fear that a prejudice against it will be created which will, for some time at least, retard its progress.

When you come to town be so good as to let me see you, as I shall be glad to converse with you on the subjects of last year, and to take any part you please in any future measures for obtaining a more substantial and adequate compensation for the great blessing you have been the means of disseminating over the world.

Lady Mildmay begs to be kindly remembered to you, and I remain,

My dear Sir,

Your truly faithful,

H. S. MILDMAY.

Stanhope Street, Jan. 31, 1803.

The request of the Medical Society of Plymouth to Dr. Jenner to sit for his picture has been already mentioned. The thanks of that body were conveyed to him by Dr. Woolcombe in terms highly complimentary.

SIR,

The members of the Medical Society of Plymouth have deputed me to convey to you their thanks for your compliance with their wishes in sitting to Mr. Northcote. They have directed a proof impression of the engraving made from the picture to be sent to you, of which they request your acceptance.

While with pleasure we observe the progressive influence of your brilliant discovery, with peculiar satisfaction we regard the just tributes of public applause which are paid to the beneficent discoverer.

I am, Sir,

With the highest Respect,

Your obedient Servant,

WILLIAM WOOLCOMBE.

Plymouth, Feb. 7, 1803.

DR. JENNER.

The Rev. Dr. Booker, of Dudley, who some time before had printed a sermon on the subject of vaccination, did not remain satisfied with that effort, but exerted himself in promoting the practice in a very efficient way, as will more fully appear from the following letter.

Dudley, 17*th March*, 1803.

Conceiving myself bound in strict propriety to communicate to you whatever comes to my knowledge that relates to your great medical object, I send you the inclosed, together with the substance of my reply, viz: that by means of the liberality of our surgeons in general, who inoculate the poor, gratis, with the vaccine, and the extensive encouragement given to that practice by Lord Dudley in paying for the inoculation of all the children of his numerous miners, &c. your salutary discovery is gaining a wide welcome throughout this populous neighbourhood.* To these means I assured Mr. Addington rather than to my humble recommendation of the measure, must be attributed its growing success.

You will see, however, in the annexed address to parents on the subject that I have done more than recommend it from the pulpit. One of these printed forms I give to every person who brings a child to be baptized either at church or at my own residence, or when sent for to baptize abroad. By this means I distribute about

* The higher and middle classes universally adopt it, and the prejudices of the lower are subsiding apace; the offer to them from the churchwardens and overseers tended, I think, rather to awaken suspicion. That from the surgeon and humane noblemen was the most effectual.

twenty a week, and have the satisfaction to learn that the expedient has produced the desired effect. It influences at a time when mankind are easily convinced of the precarious tenure of infantine existence; and when they are ready to embrace any offer of security from one of the most alarming diseases to which a child is exposed. This expedient occurred to me before I knew that a similar one was used in Switzerland. As the printer, by my desire, keeps the form unbroken in his types, I have any number of copies taken off at a small expense whenever I want them. Should any alteration suggest itself to your better judgment I shall have pleasure in receiving and adopting it."

* * * * *

This year the aboriginal Americans began to participate in the blessings of the vaccine discovery. A very interesting account of this event was sent to Jenner by his correspondent Dr. Waterhouse.

Cambridge, April, 8th, 1802.

My Dear Sir,

Four hours ago I received your polite and very interesting letter dated 24th February. The ship had but thirty-two days passage. I have just put your Lombardy virus into a fine female arm of about two-and-twenty years' standing, and shall give you the result by the return of the ship which brought it. I am highly gratified by your written, printed, and engraven communications. I shall send one of each to the President of the United States directly. You have executed the very plan I had in contemplation, viz. printed directions on a page of a sheet for a common letter; for my numerous correspondents in these States have rendered the repetition of the task respecting directions so tiresome that I had resolved on a printed letter like yours. Now I shall just reprint yours, with the additional weight of your name to it.

Dr. Rush has come out full and strong in praise of the new inoculation, and has sent me a copy of an eloquent lecture of his on the blessings of the Jennerian discovery. I believe I informed our friend Dr. Lettsom that the vaccine inoculation was carrying on its salutiferous powers into the wilderness of the new world. If I did not, I will repeat it here.

Last December a grand embassy of certain tribes of the Indians came to the city of Washington while the Congress was sitting, or as they phrase it, while the sixteen *fires* or *lights* were burning. Our Government continued to do every thing to ameliorate their condition. They had sent them seventy ploughs, ten looms, and fifty spinning-wheels, with every common utensil in husbandry, besides establishing blacksmiths, bricklayers, &c. They had taught them to plant orchards, to rear and manage horses, to use scales and weights and measures (for heretofore the white traders used to put in the scales their *foot* or right *hand* against their beaver and ermine skins.) In short, Washington, Adams, and Jefferson have done every thing to civilize that shrewd people. The chief of this embassy was named *Little Turtle.* The President one day sent for this warrior and his interpreter, and told him that he had a matter of great importance to communicate to him, for the benefit of the whole nation of his *Red Children*, for these savages always call him *Father.* He then told him that the GREAT SPIRIT had lately made a precious donation to the enlightened white men over the great water, first to a single person, and from him to another on this side the waters, and then explained to him the history of the cow or kine-pock as a gift from Heaven to preserve them from the small-pox, and even to banish it from the earth. The chief heard him with marked attention, and desired first to receive the benefits of it himself. This was performed soon after by the Rev. Dr. Gautt, chaplain of Congress, and also upon nine or ten more warriors in his train. On their departure the President caused them to be supplied with the virus; and the interpreter (a white man) took a copy of the directions for conducting the process I had transmitted to the President.

When the Minerva returns I may possibly write you a *letter* instead of this *receipt* for your valuable favours; as it is, you will receive the cordial wishes of an affectionate friend,

BENJAMIN WATERHOUSE.

It was not till some years after this that the Mohawks and other Indians connected with our Canadian provinces received the advantages of vaccination. It will hereafter be seen that they expressed their gratitude in *their* most emphatic manner to Dr. Jenner.

The benevolent founder of the Humane Society, Dr. Hawes, proposed Dr. Jenner as an honorary life governor at the half-yearly court held on the 24th of March, 1803. "Every mind, every hand, and every heart" he observes "were unanimous in Dr. Jenner's election."

To illustrate his private habits and pursuits, as well as to evince the sentiments entertained by his friends of the measures adopted both for his personal advantage and the diffusion of the vaccine practice,

I insert two documents which may be acceptable to the reader.

MY DEAR DOCTOR,

As you gave us a very satisfactory account of the cuckoo, I trouble you with this to inform you that a woman in this parish has one she has kept in a cage all the winter. If you wish I should make any inquiries as to food, &c. I will do it with pleasure.

I hear you are appointing apothecaries in every parish to inoculate for the Cow-pox. You would oblige me much if you would appoint Mr. Cole, Mount Street.

Your friends made a poor business of your application to Parliament, as I think you the greatest patriot that ever existed; and you ought to have had at least 50,000*l.* All here unite in most kind regards to you and Mrs. Jenner,

And believe me to be,

My dear Doctor,

Your's most faithfully,

Sherborne, SHERBORNE.

April 19*th*, 1803.

THE DOCTOR JENNER.

MY DEAR DOCTOR,

I see by the order of your Society that five guineas, or upwards, in one donation constitute a governor; or one guinea annually. As Lady Sherborne and I would wish to have our names in such good company as long as we can, I have desired a person to pay two guineas for us, as annual subscriptions. I only now wish that the first medical man who can be wicked enough to inoculate for the Small-pox, and the patient should die, that the coroner's inquest may bring in their verdict wilful murder, and that he may be hanged *in terrorem* of his brethren; not as a warning to himself, as I once heard an Irish judge tell a man he had condemned to be hanged the next day— "take that, my friend, as a warning."

As to the cuckoo, it departed this life last Friday.— A neighbour persuaded the woman who had it to carry it out at her door; the day being uncommonly cold, and the cage uncovered, it died at night. It was very tame and flew about the house, and when called would perch on the woman's shoulder and head. The food given was the yolk of an egg hard-boiled, and water and raw meat. At night she put it in some feathers in the crown of a hat in a cupboard, with a light bit of flannel thrown over it. The cage was a common wicker cage, such as you see magpies kept in. The air of a cottage could not be very warm, though this woman baked for the whole parish with the wood she stole from me.

I am, my dear Doctor,
Yours faithfully,
SHERBORNE.

Sherborne,
May 2nd, 1803.

His Royal Highness the Duke of York, whose generous mind early perceived the beneficial effects that would arise to the army from the practice of vaccine inoculation, and under whose immediate auspices the first mission that left our shores to diffuse that practice to distant countries was undertaken, found it necessary this year to issue an order enforcing increased attention to the examination of recruits, that each might be vaccinated who had not had either small-pox or cow-pox. This order is dated November 15th, 1803.

In October Dr. Jenner was gratified by a poetic effusion from the celebrated author of the "Bath Guide," Christopher Anstey, Esq.

Of this gentleman's *native* Muse the inspirations are known to all the admirers of playful and elegant satire; but as his Camœnæ Latinæ are, perhaps, less familiar to the English reader, a few stanzas from the *Carmen Alcaïcum* addressed to Dr. Jenner, are here given.

O! qui secundo natus Apolline
Incumbis arti pæoniæ, studens
Arcana Naturæ, gravemque
More novo prohibere morbum,

Jennere, laudes an sileam tuas?
Dum mente sanus, nec cythara carens,
Turpive succumbens senectæ
Rura vagor per amœna Cheltæ?

Furore quod non ante domabili
Tot dira Pestis quæ peperit mala,
In gentis humanæ levamen,
Te medico superata cessit;

Quippe arte mira quæ tibi contigit,
Puris benigni guttula, ab ubere
Inserta vaccino lacertis,
Corporeas penetrat meatus,

Brevique facta in vulnere pustula,
Propulsat Hostem, nec sinit amplius
  Inferre morborum cohortes
  Innumeras, comitemque mortem.

Te mater ambit filiolo cavens
Ut tuto ab atra corpore sit lue,
  Innupta te virgo decentes
  Sint memori sine labe malæ:

Utcunque nostris laudibus invidens
Gens quæque grates dat tibi debitas;
  Te Gallus extollit, tuamque
  Obsequiosus adorat artem.

Nec longiori carmine te morer,
Mentemque curis utilioribus
  Jennere seducam,—valeto.—
  Teque, tuosque, precor, labores

Deus benigno numine prosperet;
Et dum perennis gloria Laureæ
  Insignit Heroas Brittanos,
  Civica te decorat Corona.

Soon after the publication of Mr. Anstey's ode, a translation, or paraphrase, in English verse appeared from the pen of "honest John Ring," one of the Doctor's earliest and warmest *Vaccine* friends in the metropolis.

Though as a metrical version it possesses merit, and is animated with the true spirit of ardent friendship, it has not been introduced here, from an earnest desire not to swell these pages with matter at all extraneous to their great object, a just exhibition of Jenner's life and character. A brief quotation, however, may not be deemed altogether irrelevant.

> " Jenner, farewell !—nor shall the bard detain
> From nobler studies by too long a strain,
> Nor from its object alienate a mind
> Intent on labours useful to mankind.
>
> May Heaven, to whom my suppliant voice I raise,
> Prosper thy labours and prolong thy days !
> While deathless heroes, who maintain our fame
> And add new glories to the British name,
>
> Around their brows unfading laurels twine,
> The civic crown, O Jenner ! shall be thine.

It may here be worthy of remark that the profits of Mr. Ring's poem were given to the Royal Jennerian Society for the extermination of small-pox.

Dr. Jenner was not without tributary verses from other sources. The Rev. Thomas Alston Warren, of Kensworth vicarage, near Dunstable, had published an address to his parishioners on the subject of Vaccine Inoculation. It very successfully combated the errors and prejudices of his poor parishioners, and was very gratifying to Dr. Jenner. But the reverend gentleman did not stop here. He accompanied it with a complimentary letter, and poem.

The author of the Farmer's Boy too, who had been patronized by Dr. Jenner, sang "his wood-notes wild" in praise of his benefactor.

Every extension of the vaccine practice in foreign countries was followed by increasing respect and veneration for the author. This feeling was almost daily expressed by letters and addresses from individuals and public bodies. About this time he was elected foreign associate of the Society of the School of Medicine of Paris. A similar honour was conferred on him by the Medical Societies of Avignon and Nismes.

Among the other honours conferred on Jenner this year I have to add that of Doctor of Laws from the University of Cambridge, Massachusetts. The diploma was sent to him by Dr. Waterhouse in November.

His personal influence was felt to be so great that many strangers applied to him to accomplish that which all the interest of our Government could not effect. On the sudden breaking out of the war, after the peace of Amiens, many of our countrymen were most unjustly detained in France. During the preceding year Jenner had received civilities from General Andreossi, then French Ambassador at our Court. Dr. Jenner's first attempt to obtain the liberation of some of the *Detenus* was through the medium of that gentleman. His next application was addressed to the National Institute of France. An English nobleman of high rank had exhausted all his interest to obtain permission to leave Paris. Jenner having become acquainted with this incident offered to intercede for his enlargement. To that circumstance the two following letters refer.

TO THE MARQUESS OF HERTFORD.

November, 1803.

MY LORD,

Since I did myself the honour of addressing your Lordship on the subject of Lord Yarmouth's liberation, I received a solicitation to intercede in behalf of the Peploes, a family detained at Paris on their way to Spa.

Having made an acquaintance with General Andreossi during his embassy here, I wrote to him, but received no answer. However, I have since heard that this family has obtained permission to go into Germany; so that it is probable, although General Andreossi has not written, that he may have interested himself for them; and not in vain.

On reflection, my Lord, I think my chance of success would now be greater by addressing a body than any individual. My letter I consign to the care of your Lordship. Whether it may be necessary to seal it I cannot determine: that it may meet with success is my most ardent wish.

E. JENNER.

To the National Institute of France.

Gentlemen,

Pardon my obtruding myself on you at this juncture. The Sciences are never at war. Peace must always preside in those bosoms whose object is the augmentation of human happiness.

Permit me, then, as a public body with whom I am connected, to solicit the exertion of your interest in the liberation of Lord Yarmouth, a young nobleman at this time detained with his family in France.

Lord Yarmouth, the son, the only son of my valued friend and patron the Marquess of Hertford: He stands high in my estimation for being among the foremost who encouraged my scheme of Vaccination when in its infancy, and contending with the prejudices of the world.

There is another family of the name of Peploe in whose behalf some months ago I solicited the interference of General Andreossi, a gentleman with whom I have had the honour of becoming acquainted during his residence in London; but alas! I have received no answer to my letter, nor heard any thing of my friends.

Should I be so fortunate as through your kind interference to see my friends restored to those who are suffering on their account the most painful solicitude, I shall ever be ready most gratefully to acknowledge the obligation you will have conferred upon me.

I have the honour to be, Gentlemen,
with high consideration,
&c. &c. &c.
E. Jenner.

In October 1801 the Royal Economical Society of Madrid had elected Dr. Jenner an honorary member. The first notification of this event I believe never reached his hand. The second was conveyed to him by Lord Holland. His lordship's secretary, Mr. Allen, on transmitting the diploma observes " There is no country likely to receive more benefit from your labours than this (Spain), for on the one hand the mortality among children from the small-pox, and its consequences, has always been very great; and on the other hand the inoculation for the cow-pox has been received with the same enthusiasm here as in the rest of Europe; though I am sorry to add that the inoculation of the spurious sort has proved fatal to many children at Seville, who have fallen victims to the small-pox after they had been pronounced secure from that disease." Mr. Allen adds, " as one of the many proofs of the estimation in which the cow-pox and its discoverer are held in Spain, I have enclosed a small engraving, which has just been thrown off here in order to be prefixed to a dissertation on the cow-pox about to be published by the Royal Academy of Medicine." The print spoken of by Mr. Allen was an engraving of Jenner—under it was written " Edward Jenner, English physician, to whom the world is indebted for the Discovery of Vaccine, the wonderful and only preservative from the small-pox."

The duplicate of the diploma is dated May 31st, 1803. On the 15th of the following August another learned body in the Spanish capital conferred its academic honours on Jenner. This was the same royal medical foundation mentioned in Mr. Allen's letter. But the Spanish Government itself, towards the conclusion of this year, gave a proof of the value which it attached to vaccination more gratifying to the feelings of Dr. Jenner than almost any other incident that had occurred in its history. He had but a short time before learnt by advices from our settlements in India, and by letters from Dr. de Carro, of Vienna, that the efforts to transmit the preservative to Asia had completely succeeded. He now had the satisfaction of knowing that the Spanish monarch had resolved to fit out an expedition for the express purpose of carrying to all the possessions of the crown of Spain beyond the seas, and to those of several other nations, the inestimable gift of vaccine inoculation. The plan adopted on this occasion was precisely that which had been repeatedly, though unsuccessfully, recommended by Dr. Jenner to persons in authority here.

It is said that an individual, not otherwise favourably known to Europe but then high in influence at the Court of Spain, I mean Godoy, the Prince of Peace, had the merit of suggesting this truly noble enterprize to the king.

An expedition was prepared, which set sail from

Corunna, on November 30th, 1803, under the direction of Dr. Francis Xavier Balmis, surgeon-extraordinary to the king. There were on board twenty-two children, who had never undergone the small-pox, selected for the preservation of the vaccine fluid, by transmitting it successfully from one to another during the voyage.

After an absence of nearly three years the able conductor of this philanthropic expedition returned to Madrid, having circumnavigated the globe, and having more than realized the most sanguine expectations that were entertained at his departure.

I will not dwell, at present, on the many interesting events connected with this memorable voyage. They will be recorded at length in a future part of this work, as they unfold themselves in connexion with the history of Jenner.

It is impossible to attend to the facts which have rapidly passed in review before us, without experiencing feelings of gratitude and admiration for the unexampled rapidity with which individuals and nations were incited to simultaneous and successful efforts for the extension of this discovery.

In this respect mankind seem to have acted with a spirit of unanimity not usual in human affairs. In a time surprisingly short every moral obstacle, every geographical boundary gave way ; and nations, not less differing from each other in language, in habits, in religion, than in climate and every outward circumstance, speedily abandoned their prejudices and eagerly received from the hands of strangers the proffered blessing. The aboriginal American, the followers of Brachma and Confucius, the blind and obstinate votaries to Mohammedan fatalism, alike concurred to embrace and cherish this salutary gift of their bountiful Creator.

As there is something peculiarly affecting in the history of the origin and progress of small-pox, a scourge universal in its infliction and carrying terror and devastation in its course, so there is in the nature and character of its antidote that on which the heart can dwell with thankfulness and wonder. Who can know that a power has been put into the hands of man to control the greatest of all his physical ills, and not be devoutly grateful for the boon? Who can reflect that there are those who disregard it, and not lament such obstinacy and blindness?

Whatever sentiments such persons may entertain of vaccination itself, they cannot but confess that some phenomena have attended its march over the globe, such as in no former instance distinguished any of man's inventions. In little more than six years after its promulgation it was known in every clime. Proceeding eastward and westward from our own island, it traversed the circumference of the globe: and, had the practice been pushed in all countries with the same zeal and perseverance as it has been in some, small-pox by this time might probably have been known only by name, and by the melancholy records of its former ravages throughout the World.

INDEX.

A.

B.

C.

D.

G.

H.

I.

K.

L.

M.

N.

O.

P.

R.

S.

W.

Y

LONDON:
PRINTED BY S. AND R. BENTLEY, DORSET-STREET.